Evolution of Biological Diversity

DATE DUE

OCT 0 9 2002		
MAY 1 3 2003		
MAR 0 2 2005		

Demco

Evolution of Biological Diversity

Edited by

ANNE E. MAGURRAN

School of Environmental and Evolutionary Biology,
University of St Andrews, St Andrews,
Fife, KY16 9TS, Scotland, UK

and

ROBERT M. MAY

Department of Zoology,
University of Oxford,
South Parks Road,
Oxford, OX1 3PS, UK

Originating from contributions
to a Discussion Meeting of
the Royal Society of London

OXFORD
UNIVERSITY PRESS

OXFORD
UNIVERSITY PRESS

Great Clarendon Street, Oxford OX2 6DP

Oxford University Press is a department of the University of Oxford
and furthers the University's aim of excellence in research, scholarship,
and education by publishing worldwide in

Oxford New York

Athens Auckland Bangkok Bogotá Bombay Buenos Aires Calcutta
Cape Town Chennai Dar es Salaam Delhi Florence Hong Kong Istanbul
Karachi Kuala Lumpur Madras Melbourne Mexico City Mumbai
Nairobi Paris São Paolo Singapore Taipei Tokyo Toronto Warsaw

and associated companies in Berlin Ibadan

Oxford is a registered trade mark of Oxford University Press

Published in the United States
by Oxford University Press Inc., New York

British Library Cataloguing in Publication Data

Data available

Library of Congress Cataloging in Publication Data

Evolution of biological diversity / edited by Anne E. Magurran and
R. M. May
'Originating from contributions to a discussion meeting of the
Royal Society of London.'
1. Biological diversity—Congresses. 2. Evolution (Biology)—
Congresses. 3. Natural selection—Congresses. I. Magurran, Anne E.,
1955– . II. May, Robert M. (Robert McCredie), 1936– .
III. Royal Society (Great Britain)
QH541.15.B56E96   1999   578.7–dc21   98-47315
ISBN 0 19 850305 9 (Hbk)
ISBN 0 19 850304 0 (Pbk)

1 3 5 7 9 10 8 6 4 2

Typeset by Joshua Associates Ltd., Oxford
Printed in Great Britain by
Antony Rowe Ltd, Chippenham, Wiltshire.

Preface

The chapters in this book derive partly, although with significant additions and some rearrangements, from a Royal Society discussion designed to explore the evolutionary origins of biological diversity in relation to population differentiation and population structure. We have attempted to provide a coherent and unified narrative and are most grateful for the cheerful co-operation of the various authors in this task. Our essential intention has been to produce a book which summarises key findings, discusses unresolved questions, and points to important issues for future research.

As shown by a brief glance at recent issues of *Nature* and *Science*, not to mention more specialised journals, there are many new reports of marked genetic and morphological variation among and within populations of the same species. Yet the relation between such intraspecific variation and the processes of speciation remains poorly understood. This book begins by documenting examples of the pronounced levels of variation found within particular species today, and goes on to consider this variation in relation to speciation. The role of selection—both natural and sexual (to use a conventional distinction which ultimately may not be sensible)—in generating and maintaining reproductive isolation, and in speciation, will be a recurring theme. Later chapters place particular emphasis on the behavioural and ecological bases of differentiation. The concluding chapters survey a wider horizon in space and time, indicating how these issues of population structure and differentiation relate broadly to fundamental aspects of evolution and to practical applications in conservation biology.

The chapter by Coyne and Orr sets the scene, reviewing recent advances, both empirical and theoretical, which are shedding light on the underlying genetic basis of reproductive isolation. The chapter gives a synoptic review of these theoretical studies, and of all the relevant data, emphasising the persistent patterns that are seen. It also points to several important, and increasingly tractable, questions about speciation that deserve more work.

The following seven chapters document variation within and among populations from a range of viewpoints. Tregenza and Butlin survey data on the levels of polymorphism of marker genes at various levels, from individuals through demes and populations, to species. These data are primarily from allozyme studies, but also increasingly from DNA-based markers and DNA sequences. They ask whether such variability in marker genes is a reliable indicator of genetic variability in quantitative traits which affect local adaptation or mating behaviour, and, ultimately, speciation; alternatively, marker gene variability may be dominated by drift. Harwood and Amos draw together much data from repetitive DNA sequences (minisatellites and microsatellites), mainly from humans and marine mammals, to show how population structure, and especially population size, can affect rates of evolutionary change. Beginning with a review of work on birds, amphibians and fish, Skúlason, Snorrason and Jónsson examine species having distinct morphs which use different food resources and/or habitats. Concentrating on the salmonid arctic charr, they examine how genetic and evolutionary factors combine to produce polymorphisms in resource use, and relate all this to an evolutionary model.

A recurring theme throughout the book is the degree to which sexual selection can be a precursor to reproductive isolation, and thence to speciation. Sexual selection, in the form based on female choice of males, is reviewed by Price. Drawing examples from birds (especially Darwin's finches and the Old World leaf warblers), he confirms the importance of sexual selection as a mechanism which can cause rapid divergence in male traits, thus generating a propensity to mating isolation between populations. Turner takes the oft told tale of the cichlid fishes of the African Great Lakes, and gives it two new twists. First, he demonstrates that this species swarm is (sadly, 'was' may be increasingly accurate) both more diverse, and more recent, than was thought earlier. Second, he discounts the 'labile jaw' explanation for their explosive radiation, and instead suggests a mechanism based on reproductive isolation through female preference for male courtship traits. In the most diverse lineages of these fishes, sexual selection is accompanied by fine scale population structure, suggesting allopatric or parapatric diversification. But in other lineages, such geographical structure is unlikely, and Turner argues for sympatric speciation by sexual selection. Partridge and Parker ring changes on this theme, highlighting the different selective pressures on males and females, and the consequences that this 'sexual conflict' can have for both sympatric and parapatric speciation. Magurran looks at the extent of morphological and behavioural variation among populations of the same species—especially the classic studies of guppies in Trinidad—to ask how population differentiation can arise through selection, both natural and sexual, as well as stochastic effects. Noting the puzzling absence of speciation among the Trinidadian guppies, despite high levels of selection and very fast rates of evolution, Magurran suggests that sexual conflict may be the explanation: high incidence of 'sneaky mating' could produce sufficient gene flow to undermine female choice, and effectively prevent reproductive isolation.

The next two chapters provide a rather different perspective on the question of how and when population differentiation may translate into the formation of new species. Tautz and Schmid survey our emerging understanding of the connections between genes and their expression in developmental processes. Using experimental studies of *Drosophila*, they show that genes with especially high divergence rates can be a source of rich polymorphisms. They speculate that such genes may be involved in the evolution of quantitative traits, with consequences for adaptation to changing circumstances, but this work is still at an early stage. Another, and increasingly powerful, approach to evolutionary questions is to ask about the differences in diversity patterns among extant groups. Such patterns are, however, confounded to a degree by phylogenetic relations: to what extent is an observed similarity between two species the result of shared environmental circumstances as distinct from shared ancestry? Barraclough, Vogler and Harvey review the 'Comparative Method', which has been developed to deal with these difficulties, and apply it to several problems: identifying factors correlated with diversity among higher taxa; illuminating patterns in phylogenetic relations at the species level; and determining the relative importance of natural and sexual selection in generating population differentiation.

The final four chapters seek to further broaden the book's scope. Gould argues that, although the conventional discourse of evolutionary biology focuses at the level of constituent organisms within populations, selection can operate meaningfully at higher and lower levels. Gould sets his case against the changing social values of the past 150 years or so, and suggests (a bit mystically, some may say) that there are new insights to be gained by thinking of selection at the level of species, or higher. Gaston and Chown widen the discussion of spatial and geographical aspects with an overview of patterns in the distribution of species' geographical ranges. These patterns are richer than the commonly attributed lognormal species–range size distribution. Gaston and Chown analyse a model showing how such patterns are shaped by the interplay among processes of speciation, extinction, and transformation (the changes in species' sizes over their lifetimes in the fossil record). They suggest that geographical range sizes is an interesting new theme which unites these three processes.

If we are to understand how population differentiation and structure relates to the diversity of living things, we must do so not just for today but across the sweep of the fossil record. So what of rates of speciation, and patterns of biodiversity, through time? Sepkoski pulls together a range of palaeontological data, to suggest that rates of origination of new species have been extremely variable throughout the fossil record. Overall, however, he reads the record as one of secular decline in such rates throughout the Phanerozoic, albeit with variation in speciation and extinction rates among taxa of a factor 10 or so. He also notes a general decline in speciation rates within given taxa over their histories in the fossil record, with rates systematically remaining higher in tropical regions. On a note of high contemporary relevance, Sepkoski points to pulses of extinction and speciation associated with episodes of climate change. Even so, some moderate climate oscillations have had little effect of speciation and extinction, despite causing significant changes in the geographical ranges of species. Conway Morris' final chapter builds on this and looks at the long-term evolutionary history of diversity and ecosystems. Conway Morris draws attention to yet-unresolved problems in reconciling data based in Linnean taxa with fundamental questions increasingly posed in terms of phylogenetic systematics. These problems are amplified in current dissonances between conventional fossil record approaches and molecular data approaches to estimating when major groups, such as diatoms, angiosperms, or mammals, arose. Such uncertainties undercut our ability to tackle fundamental questions about how we interpret the record of life on earth. We think these final two chapters, read against the background of the earlier chapters and especially in conjunction with those by Gaston and Chown and by Barraclough *et al.*, raise issues which are likely to shape the agenda for understanding how biological diversity has evolved, and what—given humanity's impact—its probable future is.

Oxford R. M. M.
St Andrews A. E. M.
January 1999

Contents

Contributors

William Amos Department of Zoology, Downing Street, Cambridge CB2 3EJ, UK.

Timothy G. Barraclough Department of Biology and NERC Centre for Population Biology, Imperial College at Silwood Park, Ascot, Berkshire, SL5 7PY, UK.

Roger K. Butlin Ecology and Evolution Programme, School of Biology, University of Leeds, Leeds LS2 9JT, UK.

Steven L. Chown Department of Zoology and Entomology, University of Pretoria, Pretoria 0002, South Africa.

S. Conway Morris Department of Earth Sciences, University of Cambridge, Cambridge CB2 3EQ, UK.

Jerry A. Coyne Department of Ecology and Evolution, University of Chicago, 1101 E. 57 Street, Chicago, Illinois 60637, USA.

Kevin J. Gaston Department of Animal and Plant Sciences, University of Sheffield, Sheffield S10 2TN, UK.

Stephen Jay Gould Museum of Comparative Zoology, Havard University, Cambridge, MA02138, USA.

Paul H. Harvey Department of Zoology, University of Oxford, South Parks Road, Oxford OX1 3PS, UK.

John Harwood School of Environmental and Evolutionary Biology, University of St Andrews, Fife KY16 8LB, UK.

Bjarni Jónsson Hólar College, Hólar í Hjaltadal, 551 Sauðárkrókur, Iceland.

Anne E. Magurran School of Environmental and Evolutionary Biology, University of Saint Andrews, St Andrews, Fife KY16 9TS, UK.

Robert M. May Department of Zoology, University of Oxford, South Parks Road, Oxford OX1 3PS, UK.

H. Allen Orr Department of Biology, University of Rochester, Rochester, New York 11794, USA.

Geoffrey A. Parker Population Biology Research Group, School of Biological Sciences, University of Liverpool, Liverpool L69 3BX, UK.

Linda Partridge   Galton Laboratory, Department of Biology, University College London, Wolfson House, 4 Stephenson Way, London NW1 2HE, UK.

Trevor Price   Department of Biology 0116, University of California at San Diego, La Jolla, CA 92093, USA.

Karl Schmid   Zoologisches Insitut der Universität München, Luisenstrasse 14, 80333 München, Germany.

J. John Sepkoski, Jr   Department of the Geophysical Sciences, University of Chicago, 5734, South Ellis Avenue, Chicago IL 60637, USA.

Skúli Skúlason   Hólar College, Hólar í Hjaltadal, 551 Sauðárkrókur, Iceland and Institute of Biology, University of Iceland, Grensásvegur 12, 108 Reykjavík, Iceland.

Sigurður S. Snorrason   Institute of Biology, University of Iceland, Grensásvegur 12, 108 Reykjavík, Iceland.

Diethard Tautz   Zoologisches Insitut der Universität München, Luisenstrasse 14, 80333 München, Germany.

Tom Tregenza   Department of Zoology, University of Melbourne, Parkville, Victoria 3052, Australia.

George F. Turner   Division of Biodiversity and Ecology, School of Biological Sciences, University of Southampton, Southampton SO16 7PX, UK.

Alfried P. Vogler   Department of Biology and NERC Centre for Population Biology, Imperial College at Silwood Park, Ascot, Berkshire, SL5 7PY and Department of Entomology, The Natural History Museum, Cromwell Road, London SW7 5BD, UK.

1

The evolutionary genetics of speciation

Jerry A. Coyne and H. Allen Orr

1.1 What are species?

Any discussion of the genetics of speciation must begin with the observation that species are real entities in nature, not subjective human divisions of what is really a continuum among organisms. We have previously summarized the evidence for this view and counterarguments by dissenters (Coyne 1994). The strongest evidence for the reality of species is the existence of distinct groups living in sympatry (separated by genetic and phenotypic gaps) that are recognized consistently by independent observers. To a geneticist, these disjunct groups suggest a species concept based on gene flow. As Dobzhansky (1935, p. 281) noted:

> Any discussion of these problems should have as its logical starting point a considera-
> tion of the fact that no discrete groups of organisms differing in more than a single gene
> can maintain their identity unless they are prevented from interbreeding with other
> groups. . . Hence, the existence of discrete groups of any size constitutes evidence that
> some mechanisms prevent their interbreeding, and thus isolate them.'

This conclusion inspired Dobzhansky (1935) and Mayr's (1942) biological species concept (henceforth 'BSC'), which considered species to be groups of populations reproductively isolated from other such groups by 'isolating mechanisms'—genetically based traits that prevent gene exchange. The list of such mechanisms is familiar to all evolutionists, and includes those acting before fertilization ('prezygotic' mechanisms, such as mate discrimination and gametic incompatibility), and after fertilization ('postzygotic' mechanisms, including hybrid inviability and sterility).

Like most evolutionists, we adopt the BSC as the most useful species concept, and our discussion of the genetics of speciation will accordingly be limited to the genetics of reproductive isolation. We recognize that this view of speciation is not universal: systematists in particular often reject the BSC in favor of concepts involving diagnostic characters (Cracraft 1989; Baum and Shaw 1995; Zink and McKitrick 1995). We have argued against these concepts elsewhere (Coyne *et al.* 1988; Coyne 1992a, 1993a, 1994) and will not repeat our contentions here. We note only that the recent burst of work on speciation reflects almost entirely the efforts of those adhering to the BSC. In fact, *every* recent study on the 'genetics of speciation' is an analysis of reproductive isolation.

1.2 Why are there species?

One of the most important but neglected questions about speciation is why organisms fall into many discrete groups instead of constituting a few extremely variable 'types.' The answer to the question of why species exist may not be the same as the answer to 'how do species arise?' The only coherent discussion of this problem is that of Maynard Smith and Szathmáry (1995, pp. 163–167) who give three possible reasons for the existence of discrete species: (1) species might represent stable, discontinuous states of matter; (2) species might be adapted to discontinuous ecological niches; (3) reproductive isolation (which can arise only in sexual taxa) might create gaps between taxa by allowing them to evolve independently.

The second and third hypotheses seem most plausible. There are several ways to distinguish between them. One is to determine if asexually-reproducing groups form sympatric taxa just as distinct as do sexually-reproducing groups. Although such comparisons are hampered by the rarity of large asexual groups, bacteria seem reasonable candidates. While little work has been published, two studies (Roberts and Cohan 1995; Roberts *et al.* 1996) show that forms of *Bacillus subtilis* from the American desert fall into discrete clusters in sympatry. Moreover, coalescence models (Cohan 1998) show that a combination of new mutations conferring ecological difference, periodic selection, and limited gene flow can produce distinct taxa of bacteria living sympatrically.

The only empirical work on clustering in asexual eukaryotes is that of Holman (1987), who, studying successive revisions of taxonomic monographs, determined that the nomenclature of bdelloid rotifers (not known to have a sexual phase) was more stable than that of sexually-reproducing relatives. From this he concluded that because they are recognized more consistently, asexual rotifers are actually *more* distinct than their sexual counterparts. While intriguing, this result is hardly conclusive, and we badly need similar studies based not on nomenclature but on genetic and phenotypic cluster analyses. We hasten to add that while studies of clustering in asexuals are worthwhile, they must not be overinterpreted. While such studies might show that ecological specialization can produce discrete asexual forms, it does not follow that such specialization explains clustering in sexuals: the maintenance of discrete forms by natural selection in sexuals is far more difficult and could be far rarer.

We suggest two other approaches not discussed by Maynard Smith and Szathmáry. First, if distinct niches alone (and not reproductive isolation) can explain the existence of species, then sympatric speciation should be common. The whole premise that geographic isolation is essential for speciation rests on Dobzhansky and Mayr's idea that the swamping effect of gene flow prevents the evolution of reproductive isolation. But if reproductive isolation is not important in explaining the *existence* of species, adaptation to distinct niches could occur in sympatry, and phylogenetic analyses should often reveal that the most recently evolved pairs of species are sympatric. Although there is some evidence for sympatric speciation based on niche use in fish (e.g., Schliewen *et al.* 1994), we know of no other studies featuring similarly rigorous phylogenetic analyses. Perturbation experiments could also address this problem. For example, sexual isolation between sympatric species could be overcome by hybridization in the laboratory, and the resulting (fertile) hybrids re-introduced into their original habitat. If the 'ecological-niche' explanation is correct, the hybrids should revert to the parental types (or to similar but distinct types). But if reproductive isolation helps maintain species distinctness, the hybrids should either revert to a single parental type or remain a hybrid swarm. Such a study would, however, probably require an unrealistic amount of work.

Of course, the existence of species in sexual taxa could well depend on *both* distinct adaptive peaks and reproductive isolation. But it is hard to believe that ecological niches alone can explain distinct species, if for no other reason than such species, lacking prezygotic isolating mechanisms, would hybridize. If they were then to remain distinct, hybrids would have to suffer a fitness disadvantage due to their inability to find a suitable niche, and this disadvantage is a form of postzygotic isolation.

1.3 Studying reproductive isolation

What is novel about speciation?

Some of our colleagues have suggested that speciation is not a distinct field of study because—as a by-product of conventional evolutionary forces like selection and drift—the origin of species is simply an epiphenomenon of normal popula-tion-genetic processes. But even if speciation is an epiphenomenon, it does not follow that the mathematics or genetics of speciation can be inferred from traditional models of evolution in single lineages. Under the BSC, the origin of species involves reproductive isolation, a character that is unique because it requires the joint consideration of *two* species, and usually an interaction between the genomes of two species (Coyne 1994). The distinctive feature of the genetics of speciation is, therefore, epistasis. This is necessarily true for all forms of postzygotic isolation, in which an allele that yields a normal phenotype in its own species causes hybrid inviability or sterility on the genetic background of another (see below). Epistasis also occurs in many forms of prezygotic isolation. Sexual isolation, for example, usually requires the coevolution of male traits and female preferences, so that the fitness of a male trait depends on whether the choosing female is conspecific or heterospecific.

These complex interactions between the genomes of two species guarantee that the mathematics of speciation will differ from that describing evolutionary change within species, and suggest that speciation may show emergent properties not seen in traditional models. Indeed, such properties have already been identified for postzygotic isolation (e.g., see discussion of the 'snowball effect' below).

Two motives usually underlie genetic analyses of speciation. First, just as with quantitative-trait-locus (QTL) analyses of 'ordinary' characters, we would like to understand the genetic basis of cladogenesis. That is, we would like to know the number of genes involved in reproductive isolation, the distribution of their phenotypic effects, and their location in the genome. Second, we expect genetic analyses of reproductive isolation to shed light on the *process* of speciation, as different evolutionary processes should leave different genetic signatures. The observation of more genes causing hybrid male than hybrid female sterility (see below) has suggested, for example, that these critical substitutions were driven by sexual selection (Wu and Davis 1993; True *et al.* 1996).

What traits should we study?

Because speciation is complete when reproductive isolation stops gene flow in sympatry, the 'genetics of speciation' properly involves the study of only those isolating mechanisms evolving up to that moment. The further evolution of reproductive isolation, while interesting, is irrelevant to speciation. This point is widely recognized but understandably often ignored in practice. If speciation is allopatric, and several isolating mechanisms evolve simultaneously, it is hard to know which will be important in preventing gene flow when the taxa become sympatric. *Drosophila simulans* and *D. mauritiana,* for example, are allopatric, and in the laboratory show sexual isolation, sterility of F_1 hybrid males, and inviability of both male and female backcross hybrids. We have no idea which of these factors would be most important in preventing gene exchange in sympatry, or if other unstudied factors—like ecological differences—would play a role.

It seems likely, in fact, that *several* isolating mechanisms evolve simultaneously in allopatry and act together to prevent both gene flow in sympatry and allow coexistence. (Although reproductive isolation is sufficient for speciation, different species must coexist in sympatry in order to be seen.) There are two reasons why multiple isolating mechanisms seem likely. In theory, no single isolating mechanism except for distinct ecological niches or some types of temporal divergence can at the same time completely prevent gene flow *and* allow coexistence in sympatry. Two species solely isolated by hybrid sterility, for example, cannot coexist: one will become extinct through excessive hybridization or ecological competition. Species subject only to sexual isolation are ecologically unstable because they occupy identical niches. Second, direct observation often shows that complete reproductive isolation in nature often involves several isolating mechanisms. Schluter (1998), for example, describes several species pairs having incomplete prezygotic isolation. When hybrids are formed, however, they are ecologically unsuited for the parental habitats, and do not thrive.

We know little about the temporal order in which reproductive isolating mechanisms appear. The only study comparing the rates of evolution of different forms of reproductive isolation is Coyne and Orr's (1989b, 1997) analysis of pre- vs. postzygotic isolation in *Drosophila*. In these studies, prezygotic isolation (mate discrimination) between species is on average a stronger barrier to gene flow than is postzygotic isolation (Fig. 1.1). This disparity, however, is due entirely to much

a.

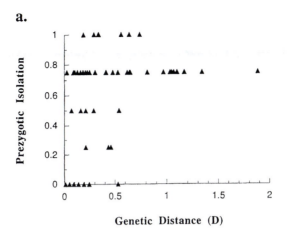

Genetic Distance (D)

b.

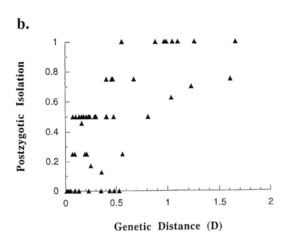

Genetic Distance (D)

Fig. 1.1 Strength of reproductive isolation in *Drosophila* plotted against Nei's (1972) electrophoretic genetic distance (an index of divergence time). Each point represents the average among pairs within a species group, so points are evolutionarily independent. (a) Prezygotic (sexual) isolation; (b) postzygotic isolation. See Coyne and Orr (1989b, 1997) for further details.

faster evolution of sexual isolation in sympatric than allopatric species pairs, suggesting—as we discuss below—the possibility of direct selection for sexual isolation in sympatry. Among allopatric taxa, pre- and postzygotic isolation arise at similar rates. When both pre- and postzygotic forms of reproductive isolation are considered simultaneously, the total strength of reproductive isolation increases quickly with time (Fig. 1.2). We are unaware of any analogous data on the rate at which total reproductive isolation increases in other taxa. It would certainly be worth obtaining such information, as it would allow one to see if the 'speciation clock' ticks at the same rate in different groups.

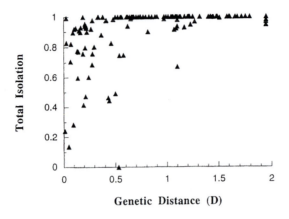

Fig. 1.2 Total reproductive isolation in *Drosophila* (combining both pre- and postzygotic isolation) plotted against Nei's (1972) genetic distance. Each point represents a single pair of species. See Coyne and Orr (1989b, 1997) for further details.

In other taxa, the data are far more impressionistic. In many groups of plants, such as orchids and the genus *Mimulus*, temporal or pollinator isolation sometimes evolve faster than postzygotic isolation, because related species produce fertile offspring when forcibly crossed in the greenhouse but fail to hybridize in nature (Grant 1981). Students of bird evolution have noted that prezygotic isolation often seems to evolve well before hybrid sterility and inviability (Prager and Wilson 1975; Grant and Grant 1996).

But these conclusions must be seen as preliminary. We require additional and more-systematic studies in which different forms of reproductive isolation are assessed among pairs of species that diverged at about the same time. Such work is especially practicable in plants, as ecological differences can be studied in the greenhouse, pollinator isolation can be studied *in situ*, and postzygotic isolation can be studied through forced crossing.

A summary of genetic studies

Because there are relatively few studies of the genetics of speciation, we have summarized them all in Table 1.1. We used two criteria for including a study in this table. First, the character studied must be known to cause reproductive isolation between species in either nature or the laboratory, or be *plausibly* involved in such isolation. Second, the genetic analysis must have been fairly rigorous, using one of three methods.

(1) classical-genetic analyses, in which species differing in molecular or visible mutant markers are crossed, and the segregation of reproductive isolation with the markers examined; here we included only those studies in which markers were distributed among all major chromosomes;

Table 1.1 Summary of existing genetic analyses of reproductive isolation between closely related species. The reference to each study includes a summary of the results, emphasizing any interesting features. The text gives our criteria for including studies in this table

Species pair	Trait	No. of genes	reference (see below)
Drosophila heteroneura/D. silvestris	head shape	9	1
Drosophila melanogaster/D. simulans	hybrid inviability	≥ 9	2
	female pheromones	≥ 5	3
Drosophila mauritiana/D. simulans	hybrid male sterility	≥ 15	4
	hybrid female sterility	≥ 4	5
	hybrid inviability	≥ 5	5
	male sexual isolation	≥ 2	6
	female sexual isolation	≥ 3	6
	genital morphology	≥ 9	7
	shortened copulation	≥ 3	8
Drosophila mauritiana/D. sechellia	female pheromones	≥ 6	9
Drosophila simulans/D. sechellia	hybrid male sterility	≥ 6	10
	hybrid inviability	≥ 2	11
	female sexual isolation	≥ 2	12
	ecological isolation	≥ 5	12a
Drosophila mojavensis/D. arizonae	hybrid male sterility	≥ 3	13
	male sexual isolation	≥ 2	14
	female sexual isolation	≥ 2	14
Drosophila pseudoobscura/ D. persimilis	hybrid male sterility	≥ 9	15
	hybrid female sterility	≥ 3	15
	sexual isolation	≥ 3	16
Drosophila pseudoobscura USA/Bogota	hybrid male sterility	≥ 5	17
Drosophila buzatti/D. koepferae	hybrid male inviability	≥ 4	18
	hybrid male sterility	≥ 7	19
Drosophila subobscura/D. madeirensis	hybrid male sterility	≥ 6	20
Drosophila virilis/D. littoralis	hybrid female viability	≥ 5	21
Drosophila virilis/D. lummei	male courtship song	≥ 4	22
	hybrid male sterility	≥ 6	23
Drosophila hydei/D. neohydei	hybrid male sterility	≥ 5	24
	hybrid female sterility	≥ 2	24
	hybrid inviability	≥ 4	25
Drosophila montana/D. texana	hybrid female inviability	≥ 2	26
Drosophila virilis/D. texana	hybrid male sterility	≥ 3	27
Drosophila auraria/D. biauraria	male courtship song	≥ 2	28
Ostrina nubialis, Z and E races	female pheromones	≥ 1	29
	male perception of pheromones	2	29
Laupala paranigra/L. kohalensis	song pulse rate	≥ 8	30
Spodoptera latifascia/S. descoinsi	pheromone blend	1	31
Xiphophorus helleri/X. maculatus	hybrid inviability	2	32
Mimulus lewisii/M. cardinalis	8 floral traits	(see note)	33
Mimulus guttatus/M. micranthus	bud growth rate	8	34
	duration of bud development	10	34
Mimulus, four taxa	(see note)	(see note)	35
Mimulus guttatus pops.	hybrid inviability	2 (system 1)	36
		≥ 2 (system 2)	36
Mimulus guttatus/M. cupriphilus	flower size	3–7	37
Helianthus annus/H. petiolarus	pollen viability	≥ 14	38

Table 1.1 (*cont.*)

1. (Templeton 1977; Val 1977). Sympatric species, biometric analysis. Head shape difference conjectured (but not known) to be involved in sexual isolation between the two species. Templeton (1977, p. 636) concludes that shape difference 'is a polygenic trait determined by alleles with predominantly additive effects.' No confidence limits are given for either of the two estimates of gene number (nine in one study and 10 in the other).
2. (Pontecorvo 1943). Sympatric species, genetic analysis. (Both species are cosmopolitan human commensals.) Gene number is a minimum estimate and includes factors from both species. Results based on analysis of 'pseudobackcross' hybrids between marked triploid *D. melanogaster* females and irradiated *D. simulans* males.
3. (Coyne 1996a). Sympatric species, genetic analysis. (Both species are cosmopolitan human commensals.) Genes affect ratio of two female cuticular hydrocarbons that appear to be involved in sexual isolation.
4. (True *et al.* 1996; Wu *et al.* 1996). Allopatric species, genetic analysis. True *et al.*'s analysis implicates at least 14 regions of the *D mauritiana* genome that cause male sterility when introgressed into a *D. simulans* background. Wu *et al.* (1996) report 15 genes on the *D. simulans* X chromosome causing hybrid sterility; by extrapolating to the entire genome, they estimate at least 120 loci causing hybrid male sterility. This latter figure may, however, be an overestimate, as some of the X-linked regions studied were known in advance to have large effects on sterility. See also Coyne (1989), Davis and Wu (1996), and Coyne and Charlesworth (1986, 1989).
5. (True *et al.* 1996). Allopatric species, genetic analysis. Estimate based on homozygous introgression of *D mauritiana* segments into *D. simulans*; we have given our minimum estimate of gene number based on the size of introgressions. Regions causing inviability invariably affect both males and females. No large effect of the X chromosome for either character. There could be many more factors affecting both female sterility and inviability, as 19 sites of 87 cytological positions produced hybrid inviability and 12 sites caused female sterility. Density of male steriles is probably higher than that of female steriles: 65 of the 185 regions caused male sterility. See also Davis *et al.* (1994) and Hollocher and Wu (1996).
6. (Coyne 1989a, 1992b, 1996b). Allopatric species, genetic analysis. Character studied was sexual isolation between *D mauritiana* females and *D. simulans* males. The major chromosomes had different effects in the two sexes, implying that different genes are involved in female discrimination vs. the male character that females discriminate against.
7. (True *et al.* 1997). Allopatric species, genetic analysis. Character measured was shape of the posterior lobe of the male genitalia. The interspecific difference in this character is thought, but not proven, to be involved in shortened copulation time between *D. mauritiana* males and *D. simulans* females, which itself causes reduced progeny number in hybrid matings (see note 8 below, and Coyne 1983).
8. (Coyne 1993b). Allopatric species, genetic analysis. Shortened copulation time between *D. simulans* females and *D. mauritiana* males contributes to the reduced number of progeny in interspecific crosses.
9. (Coyne and Charlesworth 1997). Allopatric species, genetic analysis. Ratio of two female cuticular hydro-

carbons that appears to affect sexual isolation between these species.
10. (Coyne and Kreitman 1986; Coyne and Charlesworth 1989; Cabot *et al.* 1994; Hollocher and Wu 1996). Allopatric species, genetic analysis. At least three genes on the X, two on the second, and one on the third chromosome cause sterility in hybrid males.
11. (Hollocher and Wu 1996). Allopatric species, genetic analysis. Two of three large non-overlapping regions on the *D. sechellia* second chromosome (the only chromosome analysed) cause inviability when introgressed as homozygotes into *D. simulans*.
12. (Coyne 1992). Allopatric species, genetic analysis. Only two of three major chromosomes are involved in female traits leading to reduced mating propensity of *D. sechellia* females with *D. simulans* males.
12a. (Jones 1998). Allopatric species, genetic analysis. Trait studied was adult ability to tolerate octanoic acid, a major component of the fruit *Morinda citrifolia*, which is toxic to all Diosophila species other than *D sechellia*. Tolerance to the chemical involves at least five semidominant alleles, with a major allele on chromosome three.
13. (Vigneault and Zouros 1986; Pantazidis *et al.*1993). Sympatric species, genetic analysis. Trait studied was sperm motility in hybrid males. The Y chromosome and two autosomes affect the character, while two other chromosomes do not. The X chromosome was not studied.
14. (Zouros 1973, 1981). Sympatric species, genetic analysis. Different pairs of chromosomes (and hence genes) are involved in sexual isolation among males vs. females (see note 6 above).
15. (Orr 1987, 1989b). Sympatric species, genetic analysis. Large X-effect was observed in both male and female hybrid sterility. Genes causing male and female sterility are probably different given the different locations and effects. See also Wu and Beckenbach (1983).
16. (Noor 1997). Sympatric species, genetic analysis. In each of the two backcrosses, one X-linked and one autosomal gene affect the male traits discriminated against by heterospecific females. Because the two X-linked genes map to different locations, the minimum estimate of genetic divergence between the species is three loci. See also Tan (1946).
17. (Orr 1989a,b; Orr unpublished). Genetic analysis between US populations and an allopatric isolate from Bogota, Colombia. Hybrid male sterility is due largely to X-autosomal incompatibilities. Recent work shows that at least three genes are involved on the Bogota X chromosome and two on the US autosomes (Orr, unpublished). Nonetheless, large regions of genome have no discernible effect on hybrid sterility (e.g., chromosome 4, half of chromosome 2, Y chromosome, etc.).
18. (Carvajal *et al.* 1996). Sympatric species, genetic analysis. Male inviability in backcrosses is caused by at least two factors on the X chromosome of *D. koepferae*. Two X-linked loci from *D koepferae* cause male lethality on a *D buzzatii* genetic background; this inviability can be rescued by co-introgression of two autosomal segments from *D koepferae*. Therefore, at least four loci are involved.
19. (Naveira and Fontdevila 1986; 1991). Sympatric species, genetic analysis. This is probably a gross underestimate of gene number, particularly on the X, as

Table 1.1 (*cont.*)

heterospecific X chromosomal (but not autosomal) segments of *any* size cause complete male sterility. Marin (1996), however, argues that Naveira and Fontdevila overestimated the number of autosomal factors causing hybrid male sterility.

20. (Khadem and Krimbas 1991). Sympatric species, genetic analysis. Male sterility quantified as testis size. Both X-linked and autosomal factors affect testis size in hybrids; the X chromosome has the largest effect.

21. (Mitrofanov and Sidorova 1981). Sympatric species (*D. virilis* is a cosmopolitan species associated with humans), genetic analysis. Character measured was reduced viability of offspring from backcross females.

22. (Hoikkala and Lumme 1984) Sympatric species, genetic analysis. Four factors (at least one on each autosome) affect the interspecific difference in number of pulses in pulse train of male courtship song. No apparent effect of X. Character is possibly but not yet definitely known to be involved in sexual isolation between these species.

23. (Heikkinen and Lumme 1991). Sympatric species, genetic analysis. F_1 males are weakly sterile. Backcross analysis uncovered X-2, X-4, X-5 and Y-3, Y-4, Y-5 incompatibilities. Slight female sterility was also detected in backcross.

24. (Schäfer 1978). Sympatric species (*D. hydei* is a cosmopolitan species associated with humans), genetic analysis. F_1 hybrids are fertile, but backcross hybrids show severe sterility. Female sterility involves a 34 incompatibility; some hint of X-A incompatibilities were also suggested, but not proven. Male sterility involves 34, X-3, X-4, Y-3, Y-4, and Y-5 incompatibilities.

25. (Schäfer 1979). Genetic analysis, see note 24 for geographic distribution. Backcross hybrid inviability involves chromosomes X, 2, 3, and 4. Chromosomes 5 and 6 play no role. Thus, while a modest number of genes are involved, lethality is not truly polygenic.

26. (Patterson and Stone, 1952). Sympatric species, genetic analysis. F_1 females having *D montana* mothers are inviable. Analysis showed that inviability results from an incompatibility between a dominant X-linked factor from *D. texana* and recessive maternally-acting factor(s) from *D. montana*. While the *D. texana* factor appears to be a single gene (mapped near the *echinus* locus), nothing is known about the number or location of the maternally-acting factor(s).

27. (Lamnissou *et al.* 1996). Sympatric species (*D virilis* is a cosmopolitan species associated with humans), genetic analysis. Male sterility appears only in backcrosses. Sterility is primarily caused by a $Y_{tex}5_{vir}$ incompatibility, although milder $Y_{tex}23_{vir}$ incompatibility also occurs. (23 is fused in *D. texana*; thus 23 behaves as a single linkage group in backcrosses.)

28. (Tomaru and Oguma 1994; Tomaru *et al* 1995). Sympatric species, genetic analysis. These species show strong sexual isolation in the laboratory and differ in male courtship 'song.' Genetic analysis showed that both major autosomes carried at least one gene affecting the interpulse interval (IPI) of the male mating songs, but the X chromosome had no effect (Tomaro and Oguma 1994). In a later study, Tomaru *et al* (1995) tested *D. biauraria* females with conspecific males whose wings and antennae had been removed, and played synthetic courtship song during these encounters. Con-

specific matings occurred more frequently and femal rejection behaviors less frequently when artificial songs contained the conspecific IPI than when they contained the longer IPI characteristic of *D. auraria*. This indicates that the song differences analysed genetically probably contribute to sexual isolation.

29. (Roelofs *et al.* 1987; Lofstedt *et al.* 1989). Genetic analysis, sympatric races of European corn borers in New York State, USA. Characters studied were female pheromone blend (ratios of two long-chain acetates, whose interspecific differerence is apparently due to a single autosomal locus). The electrophysiological response to these pheromones by male antennae sensilla is due to another, unlinked autosomal locus, and the behavioral response of males to female pheromones is due to a single X-linked factor.

30. (Shaw 1996). Allopatric species, biometric analysis. Trait studied was male song pulse rate of two species of Hawaiian crickets, which may be involved in sexual isolation. Reciprocal F_1 crosses indicate no disproportionate effect of the X chromosome.

31. (Monti *et al.* 1997). Sympatric species, genetic analysis. Trait studied was ratio of two pheromones, apparently due to a single factor. The *timing* of female emission of the pheromones, which differs between the species, appears to be polygenic, and the authors give no estimate of number of factors. Male perception of traits must, of course, also differ if there is to be reproductive isolation, so that, as in corn borers (see note 28), sexual isolation must be due to changes in at least two genes.

32. (Wittbrodt *et al.* 1989). Sympatric species, genetic analysis. Hybrid inviability between *X. helleri* (swordtail) and *X. maculatus* (platyfish) is caused by appearance of malignant melanomas, which are often fatal. These cancers are caused by the interaction between a dominant, X-linked oncogene encoding a receptor tyrosinane kinase, and an autosomal suppressor that is either missing in the swordtail or dominant in the platyfish. Some backcross hybrids inherit the oncogene without the suppressor, yielding melanomas.

33. (Bradshaw *et al* 1995). Sympatric species, genetic analysis. Traits studied by QTL (quantitative-trait locus) analysis include flower color, corolla and petal width, nectar volume and concentration, and stamen and pistil length. All of these traits affect whether a flower is pollinated by bumblebees (*M. lewisii*) or hummingbirds (*M cardinalis*). Species are at least partially reproductively isolated by pollinator difference. For most traits, the species difference involved at least one chromosome region of large effect. The difference in carotenoids in petal lobes was governed by a single gene.

34. (Fenster *et al.* 1995) Sympatric species, biometric analysis. Bud growth rate and duration reflect difference in flower size between these species: *M. micranthus* is largely selfing and apparently derived from the outcrossing *M guttatus*. It is not known whether this difference causes reproductive isolation in nature, but this seems likely given the decrease in gene flow caused by selfing. No evidence for factors of large effect. Lower bound of 95% confidence interval for gene number is 3.2 for bud growth rate and 4.6 for duration of bud development.

35. (Fenster and Ritland 1994). Biometric analysis. Four taxa of controversial status, named *M. guttatus, M. nasutus,* (both outcrossing), and *M. micranthus* and

Table 1.1 (*cont.*)

M. laciniatus (predominantly selfing). The latter three taxa are sometimes classified as subspecies of *M. guttatus*. Traits studied included differences in flowering time, corolla length, corolla width, stamen level, pistil length, stigmaanther separation. Selfing species have shorter and narrower corollas, shorter stamens and pistils, and less stigma anther separation than outcrossers. Flower characters are therefore associated with breeding systems. Minimum number of genes for character differences averaged across all taxa varied between five and 13 per trait. Standard errors are large. Our impression is that these phenotypic differences involve several to many genes, with no single locus causing most of the difference in any character between any two taxa. Within a cross, positive genetic correlations were often seen between many traits, so that genes causing these character differences are not necessarily independent.

36. (Macnair and Christie 1983; Christie and Macnair 1984, 1987). Allopatric populations, genetic analysis. Complementary lethal loci are polymorphic (one in each population) in two North American populations of *Mimulus guttatus*. There are two separate systems of inviability. The first involves only two genes, both polymorphic for complementary lethals. The other involves one locus (possibly the gene involved in copper tolerance) that interacts with an unknown number of genes in nontolerant population.

37. (Macnair and Cumbes 1989). Sympatric species, biometric analysis. Seven flower-size characters studied, some of which (e.g, height, width, pistil length) are probably related to difference in breeding systems between these species *M. cupriphilus* selfs much more often than does *M. guttatus*, and this difference in breeding systems may contribute to reproductive isolation. Stamen/pistil length ratio is also important in reproductive isolation, but it was not studied genetically. Each character difference was due to between three and seven genes. High genetic correlations were observed between many characters, so that traits are not genetically independent.

38. (Rieseberg 1997). Sympatric species, genetic analysis. Within the co-linear portions of the genome of these two sunflower species, Rieseberg estimates that at least 14 chromosomal segments are responsible for inviability of hybrid pollen. Thus, approximately 40 genes are involved if one assumes a similar density of factors in the rearranged portions of the genome.

(2) simple Mendelian analyses in which segregation ratios in backcrosses or F_2s indicated that an isolating mechanism was due to changes at a single locus; or

(3) biometric analyses, in which measurement of character means and variances in backcrosses or F_2s yielded a rough estimate of gene number.

Table 1.1 also gives the actual or minimum number of genes involved for each isolating mechanism. The footnotes give more detail about the type of genetic analysis, whether the species pair was sympatric of allopatric, and a brief summary and critique of the results.

We must add several caveats. First, despite our attempts to comb the literature comprehensively, we have surely missed some studies. Second, the quality of the analyses is uneven: some classical-genetic studies involved detailed mapping experiments with many markers, while others relied on only one marker per chromosome. Third, most studies have underestimated the number of genes causing a single reproductive isolating mechanism. This may reflect a limited number of markers, failure to test all possible interactions between chromosomes, or the use of biometric approaches, which nearly always underestimate true gene number. Finally, data are given for *single* isolating mechanisms, but several mechanisms may often operate together to impede gene flow in nature.

The most striking feature of Table 1.1 is the imbalance of both species and isolating mechanisms. Roughly 75% of all the studies involve *Drosophila*, with only seven other pairs of taxa, mostly plants from the genus *Mimulus*. Moreover, nearly two-thirds of the *Drosophila* work is on hybrid sterility and inviability. There is only one published genetic study of ecological isolation. We obviously must extend such studies to other groups and other forms of reproductive isolation.

For convenience, we discuss pre- and postzygotic isolation separately.

1.4 Prezygotic isolation

Although there are many forms of prezygotic isolation, including differences in ecology, behavior, time of reproduction, gametic incompatibility, and (in plants) differences in pollinators, only one form—sexual isolation—has been the subject of much theory and experiment.

Sexual isolation

Many evolutionists have noted that closely related animal species, particularly those involved in adaptive radiations, seem to differ most obviously in sexually dimorphic traits. This impression has been confirmed by two phylogenetic studies of birds. Barraclough *et al.* (1995) found a positive correlation between the speciosity of groups and their degree of sexual dimorphism, suggesting a link between sexual selection and speciation. (It is not likely that species are simply *recognized* more easily in strongly dimorphic groups, for the authors note that females of such species can also be distinguished easily.) Mitra *et al.* (1996) found that taxa with promiscuous mating systems contain more species than their nonpromiscuous sister taxa. Promiscuously-mating species are, of course, more likely to experience strong sexual selection.

The connection between sexual selection and sexual isolation may seem obvious. After all, it is a tenet of neo-Darwinism that reproductive isolation is a by-product of evolutionary change occurring within populations. Two populations undergoing sexual selection may readily diverge in both male traits and female preferences, and the natural outcome of this would be sexual isolation between the populations. This idea is much easier to grasp than, say, the notion that adaptation among isolated populations would cause sterility or inviability of their hybrids; under sexual selection, the pleiotropic effect of the diverging genes (sexual isolation) is obviously connected to their primary effect (exaggeration of male traits or female preferences).

Nonetheless, consideration of the role of sexual selection in speciation was remarkably late in coming. The earliest paper even mentioning such a possibility appears to be that of Haskins and Haskins (1949), who posited that sexual selection in guppies might lead to female recognition of only conspecific males as mates, which would serve as an isolating mechanism among sympatric species. This line of thought then vanished from the literature until Nei (1976) made a model of sexual selection in which one gene controlled the male trait and another the female preference. Exploring the conditions that could lead to the joint evolution of these traits, Nei noted that such a process could cause reproductive isolation. More recently, Ringo (1977) proposed that sexual selection might explain both the adaptive radiation and the strong sexual dimorphism of Hawaiian *Drosophila*.

The most extensive theoretical work on this problem is that of Lande (1981, 1982). In 1981, he showed that sexual isolation could be the by-product of sexual selection if random genetic drift in small populations triggered the 'runaway' process suggested by Fisher (1930). Later workers (e.g., Kirkpatrick 1996) have shown, however, that such instability is unlikely if natural selection acts on female preference. Lande's 1982 model, an explicit quantitative-genetic treatment of clinal speciation via sexual selection, is more realistic. Here he showed that *adaptive* geographic differentiation in male traits could be amplified by the evolution of female preferences, resulting in reproductively isolated populations along a cline. Data that may support this model come from the guppy *Poecilia reticulata*, in which local populations have differentiated so that females prefer to mate with local rather than foreign males (Endler and Houde 1995).

There is only one theoretical analysis of sympatric speciation resulting from sexual selection. Turner and Burrows (1995) constructed a genetic model of a male trait affected by four loci and a female preference for that trait affected by a single locus with two alleles. Under some conditions, this model produced sympatric taxa showing complete sexual isolation; but these results may be highly dependent on their assumptions. (One such assumption was that the preference locus showed complete dominance, and more realistic assumptions of intermediate dominance and additional loci would almost certainly reduce the probability of speciation.)

The only theoretical study of *allopatric* speciation resulting from sexual isolation is that of Iwasa and Pomiankowski (1995). Their quantitative-genetic model of sexual selection on a male trait and on female preference assumes a fitness cost to increased female preference. This cost results in a cyclical fluctuation of both trait and preference. Different populations undergoing the same selection regime could then fall out of synchrony, causing sexual isolation. Kirkpatrick and Barton (1995), however, note that this model neglects the possibility of stabilizing selection acting on the male trait around its optimum value for survival, and that such selection (even if very weak) could eliminate the cycling. Moreover, related species do not usually show different stages of elaboration or diminution of the same male-limited trait, but instead the exaggeration of completely different traits. [Some species of bowerbirds, for example, have different forms of elaborate male plumage, while others lack such plumage but have males who build elaborate bowers (Gilliard 1956).] A useful extension of Iwasa and Pomiankowski's model might incorporate either novel environments or genetic drift that could launch populations on different trajectories of sexual selection. Schluter and Price (1993) also suggest that, if the secondary sexual traits of males reflect their fitness or physiological condition, differences among habitats that change the female's ability to *detect* different traits could lead to sexual isolation among populations.

Sexual isolation is often asymmetric, i.e., species or populations show strong isolation in only one direction of the hybridization. This is a common phenomenon in *Drosophila* (Watanabe and Kawanishi 1979; Kaneshiro, 1980), is also seen in salamanders (Arnold *et al.* 1996), and may be ubiquitous in other animals. Kaneshiro (1980) offered an explanation of this pattern based on biogeography, but newer data contradict his theory (Cobb *et al.* 1989; David *et al.* 1974). Arnold *et al.* (1996) proposed that mating asymmetry is a transitory phenomenon that decays rapidly as populations diverge, but such asymmetry is often seen among even distantly related species of *Drosophila* (Coyne and Orr 1989b). (Of course asymmetry must eventually disappear, because sexual isolation will, with time, become complete in both directions.) There are likely to be other explanations for mating asymmetry, such as the combination of open-ended female preferences and sexual selection operating in only one of two populations. If asymmetry does prove ubiquitous, it may provide important clues about how sexual selection causes speciation.

Unfortunately, there are too few data about the *genetics* of sexual isolation to confirm or motivate any theory. The few studies listed in Table 1.1 show only that sexual isolation may sometimes have a simple basis, as in races of *Ostrina nubialis*, where complete sexual isolation is apparently based on changes at only three loci (one each for differences in female pheromones, male perception, and male attraction), or can be more complicated, as in *Drosophila mauritiana/D. sechellia*, where differences in female pheromones alone involve at least five loci. One conclusion that seems reasonable, based as it is on three independent studies (Table 1.1), is that for a given pair of species, the genes causing sexual isolation of males differ from those causing sexual isolation in females. This is not surprising, as there is no obvious reason why genes affecting male traits should be identical to those affecting female perception. (It is possible that sexual selection could cause male and female genes to be closely linked by selecting for genetic correlations between trait and preference, but there is no theory addressing this possibility.) This lack of correlation is worth verifying, however, as its absence is assumed in many models of sexual isolation (e.g., Spencer *et al.* 1986).

Given the likely importance of sexual selection in animal speciation, more genetic analyses are clearly needed. Fortunately, the advent of QTL analysis has made such studies feasible in any pair of species that produces fertile hybrids.

Reinforcement

One of the greatest controversies in speciation concerns reinforcement: the process whereby two allopatric populations that have evolved some postzygotic isolation in allopatry undergo selection for increased *sexual* isolation when they later become sympatric. Reinforcement was introduced and popularized by Dobzhansky (1937), who apparently considered it the necessary last step of speciation. Its wide popularity may have reflected its appealing assumption of a creative (and not an incidental) role for natural selection in speciation (Coyne 1994). While theoretical studies of reinforcement appeared only recently, two analyses of *Drosophila* (Ehrman 1965; Wasserman and Koepfer 1977) supported the idea: in species pairs with overlapping ranges, sexual isolation was stronger when populations derived from areas of sympatry than from allopatry.

In the 1980s, however, several critiques eroded the popularity of reinforcement. First, Templeton (1981) pointed out that the pattern of stronger sexual isolation among sympatric than allopatric populations could be caused not by reinforcement but by 'differential fusion,' in which species could *persist* in sympatry only if they had evolved sufficiently strong sexual isolation in allopatry. Thus, stronger isolation in sympatry might reflect not direct selection, but the hybridization and fusion of weakly isolated populations. Moreover, it became clear that some of the data offered in support of reinforcement were flawed (Butlin 1987). Finally, the first serious theoretical treatment of reinforcement (Spencer *et al.* 1986) showed that even under favorable conditions (e.g., complete sterility of hybrids), extinction of populations occurred more often than reinforcement.

Recently, however, a combination of empirical and theoretical work has resurrected the popularity of reinforcement. In an analysis of 171 pairs of *Drosophila* species, Coyne and Orr (1989b, 1997) found that recently diverged pairs show far more sexual isolation when sympatric than allopatric (Fig. 1.3). [An independent analysis of these data by Noor (1997a), making less restrictive assumptions, arrived at similar conclusions.] Although in 1989 there was no theoretical work showing that reinforcement was feasible, two such studies have appeared recently. Liou and Price (1994) and Kelly and Noor (1996) showed that reinforcement can occur frequently even if hybrids have only moderate postzygotic isolation. The important difference between these models and that of Spencer *et al.* (1986) is that the former explicitly allow for sexual selection, which greatly enhances the evolution of sexual isolation.

Newer data also support reinforcement. These include a re-analysis of the earlier literature, finding many more possible examples (Howard 1993), and two new studies of species pairs with partially overlapping ranges (Noor 1995; Saetre *et al.* 1997). The work of Saetre *et al.* on two species of European flycatchers is especially interesting, as the reduced hybridization in sympatry is caused by the divergence of male plumage occurring only in that area, a difference presumably caused by sexual selection.

Although the pattern of stronger isolation in sympatry shown in Fig. 1.3 might in principle be explained by several processes, including reinforcement and differential fusion (Coyne and Orr 1989b), reinforcement seems most likely for several reasons. First, differential fusion posits that cases of strong sexual isolation in sympatry form a *subset* of the levels of isolation seen in allopatry: while allopatric taxa might show either strong or weak isolation, the latter cases disappear by fusion upon geographic contact, leaving us with a pre-existing set of strongly isolated taxa. This is not, however, the pattern seen in *Drosophila*: Figure 1.3 shows that *no* recently diverged allopatric taxa have sexual isolation as strong as that seen among sympatric taxa of the same age. Furthermore, differential fusion predicts that *postzygotic* as well as prezygotic isolation will be stronger in sympatry, as the probability of fusion should decrease with *any* form of reproductive isolation. The *Drosophila* data also fail to support this prediction: while prezygotic isolation is much stronger in sympatry than allopatry, postzygotic isolation is virtually identical in the two groups (Coyne and Orr 1997).

Finally, it should be noted that reinforcement does not necessarily require the pre-existence of hybrid sterility or inviability, but might also result when allopatric populations evolve some behavioral or ecological difference that leaves hybrids behaviorally or ecologically maladapted. Stratton and Uetz (1986), for example,

a.

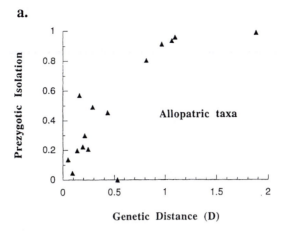

b.

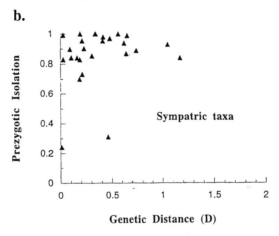

Fig. 1.3 Prezygotic isolation in *Drosophila* plotted against Nei's (1972) electrophoretic genetic distance. Each point represents the average among pairs within a species group. (a) Allopatric taxa; (b) sympatric taxa. See Coyne and Orr (1989b, 1997) for further details.

observed that hybrids between two species of wolf spiders were 'behaviorally sterile,' so that hybrid males were rejected by females of both species, while hybrid females refused to mate with any male. Similarly, Davies *et al.* (1997) observed that hybrid females between two species of butterflies suffer a reduced tendency to mate.

The data and theory reviewed above strongly suggest that reinforcement of sexual isolation can occur. This conclusion represents one of the most radical changes of views about speciation over the last decade. Future work must determine whether reinforcement is rare or ubiquitous across animal taxa, and whether it occurs in plants.

Ecological isolation

Ecological isolation must play a role in maintaining biological diversity since—even if it is not a primary isolating mechanism—ecological differences are required for the sympatric coexistence of taxa. Consider, for instance, the fate of a newly-arisen polyploid plant species. Although all auto- or allopolyploids are automatically postzygotically isolated from their parental species (hybridization produces mostly sterile triploid hybrids), any polyploid that does not differ ecologically from its ancestral species will be quickly driven extinct through either competition or the production of sterile hybrids with its ancestors, rendering it unavailable for study.

'Ecological isolation' actually subsumes three phenomena:

1. Individuals of different species live in the same region and may encounter each other, but confine mating and/or reproduction to different habitats so that hybrids are not formed [e.g., the sympatric, host-specific *Drosophila* that breed on different cacti (Ruiz and Heed 1988)]. This is a form of prezygotic reproductive isolation, and the *only* type of reproductive isolation that can by itself both cause complete speciation and allow persistence of species in sympatry. This form of isolation may be the most common result of sympatric speciation (Rice and Hostert, 1993).

2. Species live in different subniches of the same area and rarely, if ever, come into contact (e.g., the spadefoot toads *Scaphiopus holbrooki hurteri* and *S. couchi*). Although their ranges overlap extensively, heterospecific toads almost never meet because they are restricted to different soil types [Wasserman 1957]). Although this situation corresponds to Dobzhansky's (1937, p. 231) definition of ecological isolation, such species are effectively allopatric.

3. Species live in different subhabitats of the same area and sometimes come into contact with one another and mate, forming hybrids that are not well adapted to available habitat. Such cases may be common, and have been studied extensively in stickleback fish (Schluter 1996). They constitute examples of *postzygotic* isolation that, while technically a form of hybrid inviability, depend on ecological details of the environment and not on inherent problems of development. Like any other form of postzygotic isolation, 'ecological inviability' could trigger reinforcement between sympatric species (Coyne and Orr 1989b). Unlike other forms of postzygotic isolation, however, ecological inviability may—depending on how population size is regulated in the different habitats—allow two species to coexist without prezygotic isolation. In addition, postzygotic isolation based on ecological divergence need not involve complementary gene interactions (epistasis) between alleles of two species. Instead, the genetics of this type of ecological isolation would presumably resemble the genetics of ordinary adaptation (about which we unfortunately know little [Orr and Coyne 1992]).

Schluter (1996, 1997) makes a strong case for the importance of ecological isolation (especially type 3) in the origin and persistence of species. Unfortunately, there has been only one study of the genetics of ecological isolation. The island endemic *Drosophila sechellia* breeds exclusively on the normally-toxic fruit of *Morinda citrifolia*. All of *D. sechellia*'s relatives, including its presumed ancestor *D. simulans*, succumb to *Morinda*'s primary toxin, octanoic acid. Although *D. sechellia* and *D. simulans* were originally allopatric, they have recently become sympatric and breed on separate hosts. Recent work (Jones, 1998) shows that *D. sechellia*'s resistance to octanoic acid involves at least five mostly dominant alleles, distributed over all three of the major chromosomes, with the largest effect mapping to chromosome three.

Pollinator isolation

Pollinator isolation is probably a common form of reproductive isolation in plants (Grant 1981). A variant of this is the isolation of insect-pollinated plants from self-compatible species, a mechanism described in *Mimulus*. The few data at hand (Table 1.1) show that the difference between outcrossing and inbreeding is due to several genes, but that differences in flower shape, color, or nectar reward that attract different pollinators may be due to one or a few major genes (Prazmo 1965; Bradshaw *et al.* 1995). This latter observation, if common, might suggest a rapid form of speciation.

Postmating, prezygotic isolation

Biologists have begun to appreciate that sexual selection is not limited to obvious behavioral and morphological traits that act before copulation, but can include 'cryptic' characters acting between copulation and fertilization. Such selection (as well as sexually antagonistic selection [Rice 1996]) can lead to female control of sperm usage, male–male sperm competition within multiply inseminated females, and the mediation of such competition by the female. Selection acting between copulation and fertilization has also been invoked to explain the striking diversity of male genitalia among animal species, bizarre conformations of female reproductive tracts, and postcopulation 'courtship' behavior by males (Eberhard 1985, 1996). Moreover, the relatively rapid evolution of proteins involved in reproduction (Coulthart and Singh 1988; Metz and Palumbi 1996; Tsaur and Wu 1997) is also consistent with sexual selection.

Just as sexual selection acting on male plumage or courtship behavior can pleiotropically produce sexual isolation, so post-copulation, prezygotic sexual selection can produce 'cryptic' sexual isolation detectable only after fertilization. Such isolation may take the form of either blocked heterospecific fertilization, such as the 'insemination reaction' of *Drosophila* (Patterson 1946), or the preferential use of conspecific sperm when a female is sequentially inseminated by heterospecific and conspecific males. This latter phenomenon has recently been described in grasshoppers (Bella *et al.* 1992), crickets (Gregory and Howard 1994), flour beetles (Wade *et al.* 1994) and *Drosophila* (Price 1997). In three species of *Drosophila*, *single* heterospecific inseminations produce large numbers of hybrid offspring, but females doubly inseminated by both conspecific and heterospecific males produce very few hybrids. Such reproductive isolation thus depends on *competition* between conspecific and heterospecific sperm, and resembles inter-specific pollen competition in some plants, in which heterospecific pollen tubes grow more slowly than conspecific tubes, yielding reproductive isolation detectable only after double pollination (Rieseberg *et al.* 1995). The only genetic analysis among these studies is that of Price (1997), who showed that, among females, 'conspecific sperm precedence' was dominant, so that F_1 female hybrids between *D. simulans* and *D. mauritiana* induce the same sperm preference as do *D. simulans* females. Gametic incompatibilities may be among the earliest-evolving forms of reproductive isolation, and so are worthy of more attention.

1.5 Postzygotic isolation

Postzygotic isolation occurs when hybrids are unfit. Evolutionists have, historically, pointed to three types of genetic differences as causes of these fitness problems: species may have different chromosome arrangements, different ploidy levels, or different alleles that do not function properly when brought together in hybrids. Although each of these modes of speciation has enjoyed its advocates, it is now clear that the latter two are by far the most important.

Speciation by auto- and allopolyploidy is clearly common in plants, as about 60% of angiosperms are of polyploid origin (Masterson 1994). We will not, however, discuss polyploid speciation here as it has been thoroughly reviewed elsewhere (e.g., Grant 1981; Ramsey and Schemske 1998). Instead, we concentrate on the question of whether postzygotic isolation in animals is based on chromosomal or genic differences. We will conclude that genes play a far more important role in hybrid sterility and inviability than do structural differences in chromosomes.

Chromosomal speciation

The notion that chromosome rearrangements are a major cause of hybrid sterility was once very popular (White 1969, 1978), and still enjoys a few adherents (e.g., King 1993). The idea of chromosomal speciation, however, suffers from both theoretical and empirical difficulties. The theoretical problems have been discussed elsewhere (Lande 1979; Walsh 1982; Barton and Charlesworth 1984), but the empirical problems have not been widely recognized.

The first such difficulty is that many species producing sterile hybrids are *homosequential*, that is, they do not differ in chromosome arrangement. M. J. D. White (1969, p. 77), the greatest champion of chromosomal speciation, argued that such cases 'represent only a small fraction of the total number of species complexes that have been extensively studied.' But this argument is potentially misleading: many of the taxa showing fixed chromosomal differences are old, and may have accumulated many—if not all—of their chromosomal differences *after* the actual speciation event. Resolving this issue requires systematic comparisons of the extent of chromosomal divergence with both the age of taxa (as determined by fossils or molecular data) and the strength of postzygotic isolation. In addition, although sterile hybrids often suffer meiotic pairing problems, this sterility is often limited to the heterogametic sex (see below). This is not expected if sterility results from the disruption of chromosome pairing, which in most cases should afflict both males and females. Indeed, in some species such as *Drosophila*, intraspecific chromosomal rearrangements typically sterilize only females (male *Drosophila* have no recombination), while hybrid sterility is far more severe in males (Ashburner 1989).

Third, even in species hybrids whose chromosomes fail to pair during meiosis, we do not know whether this failure is caused by differences in chromosome arrangement or differences in genes. As Dobzhansky (1937) emphasized, both rearrangements *and* gene mutations can disrupt meiosis within species and so, presumably, within species hybrids. The best attempt to disentangle these causes remains the first: *Drosophila pseudoobscura* and *D. persimilis*, which differ by at least four inversions, produce sterile hybrid males (Dobzhansky 1937). Meiotic pairing in hybrids is abnormal and univalents are common. Dobzhansky (1933) showed that islands of tetraploid spermatocytes are often found in hybrid testes. All chromosomes in these 4N cells have a homologous pairing partner, and hence should show improved pairing if pairing problems in hybrid cells reflect structural differences between chromosomes. However, univalents are just as common in the hybrid tetraploid as in diploid spermatocytes. The hybrid meiotic problems must therefore have a genetic and not a chromosomal basis. This test has apparently not been repeated in any other hybridization.

A further problem with chromosomal speciation is that it depends critically on the semisterility of hybrids who are heterozygous for chromosome rearrangements. It is not widely appreciated, however, that heterozygous rearrangements theoretically expected to be deleterious (e.g., fusions and pericentric inversions) in reality often enjoy normal fitness, probably because segregation is regular or recombination is prevented (see discussion in Coyne *et al.* 1997). Any putative case of chromosomal speciation requires proof that different rearrangements actually cause semisterility in heterozygotes, and almost no studies have met this standard.

Finally, direct genetic analyses over the last decade have shown conclusively that postzygotic isolation in animals is typically caused by *genes*, not by large chromosome rearrangements. Many of these genes have been well mapped, and several have been genetically characterized or even cloned (Wittbrodt *et al.* 1989; Orr 1992; Perez *et al.* 1993). Our main task, therefore, is to understand how the evolution of genic differences can produce hybrid sterility and inviability.

The Dobzhansky–Muller model

Understanding the evolution of postzygotic isolation is difficult, because the phenotypes we are hoping to explain—the inviability and sterility of hybrids—seem maladaptive. The difficulty is best seen by considering the simplest possible model for the evolution of postzygotic isolation: change at a single gene. One species has genotype *AA*, the other *aa*, while *Aa* hybrids are completely sterile. Regardless of whether the common ancestor was *AA* or *aa*, fixation of the alternative allele cannot occur because the first mutant individual has genotype *Aa* and so is sterile. Using the metaphor of adaptive landscapes, it is hard to see how two related species can come to reside on different adaptive peaks unless one lineage passed through an adaptive valley.

This problem was finally solved by Bateson (1909), Dobzhansky (1937) and Muller (1942), who noted that, if postzygotic isolation is based on incompatibilities between *two or more genes*, hybrid sterility and inviability can evolve unimpeded by natural selection. If, for example, the ancestral species had genotype *aabb*, a new mutation at one locus (allele *A*) could be fixed by selection or drift in one isolated population because the *Aabb* and *AAbb* genotypes are perfectly fit. Similarly, a new allele (*B*) at the other locus could be fixed in a different population since *aaBb* and *aaBB* genotypes are also fit. But it is entirely possible that when the *AAbb* and *aaBB* populations come into contact, the resulting *AaBb* hybrids could be sterile or inviable. The *A* and *B* alleles have never been 'tested' together within a genome, and so may not function properly in hybrids.

Alleles showing this pattern of epistasis are called 'complementary genes.' Such genes need not, of course, have drastic effects on hybrid fitness; any particular incompatibility might lower hybrid fitness by only a small amount. It should also be noted that the Dobzhansky–Muller model is agnostic about the evolutionary causes of substitutions that ultimately produce hybrid sterility or inviability: purely adaptive or purely neutral evolution within populations can give rise to complementary genes and thus to postzygotic isolation.

The Dobzhansky–Muller model is the basis for almost all modern work in the genetics of postzygotic isolation. There is now overwhelming evidence that hybrid sterility and inviability do indeed result from such between-locus incompatibilities (reviewed in Orr 1997). Curiously, there have been few *theoretical* studies of the Dobzhansky–Muller model. Recent analyses, however, predict that the evolution of postzygotic isolation should show several regularities.

Patterns in the genetics of postzygotic isolation

Long before any formal studies of the Dobzhansky–Muller model, Muller (1942) predicted that the alleles causing postzygotic isolation must act asymmetrically. To see this, consider the two-locus scenario sketched above. Although the *A* and *B* alleles might be incompatible in hybrids, their allelomorphs *a* and *b* must be compatible. This is because the *aabb* genotype must represent an ancestral state in the evolution of the two species.

There is now good evidence that genic incompatibilities do in fact act asymmetrically. The best data come from Wu and Beckenbach's (1983) study of male sterility in *Drosophila pseudoobscura–D. persimilis* hybrids. When an X-linked region from one species caused sterility upon introgression into the other species' genome, they found that the reciprocal introgression had no such effect.

This observation has now been confirmed in other *Drosophila* hybridizations (e.g., Orr and Coyne 1989).

The second pattern expected under the Dobzhansky–Muller model was pointed out more recently. If hybrid sterility and inviability are caused by the accumulation of complementary genes, the severity of postzygotic isolation, as well as the number of genes involved, should 'snowball' much faster than linearly with time (Orr 1995; see also Menotti-Raymond *et al.* 1997). This follows from the fact that any new substitution in one species is potentially incompatible with the alleles at *every locus* that has previously diverged in the other species. (In our discussion above, the *B* substitution is potentially incompatible with the previous *A* substitution in the other species.) Later substitutions are therefore more likely to cause hybrid incompatibilities than earlier ones. Consequently, the cumulative number of hybrid incompatibilities increases much faster than linearly with the number of substitutions, K. If all incompatibilities involve pairs of loci, the expected number of Dobzhansky–Muller incompatibilities increases as K^2, or (assuming a rough molecular clock) as the square of the time since species diverged (Orr 1995). If incompatibilities sometimes involve interactions between more than two loci (see below), the number of hybrid incompatibilities will rise even faster. This snowballing effect requires that we interpret genetic studies of postzygotic isolation with caution. Because the genetics of hybrid sterility and inviability will quickly grow complicated as species diverge, it is easy to over-estimate the number of genes required to cause strong reproductive isolation (see Orr 1995 and below).

The limited data we possess are consistent with a snowballing effect, although they do not prove it (see Orr 1995 for a discussion). The simplest prediction of the snowballing hypothesis is that the number of mapped genes causing hybrid sterility or inviability should increase quickly with molecular genetic distance between species. But given the enormous difficulties inherent in accurately mapping and counting 'speciation genes,' it may be some time before such direct contrasts are possible.

Third, while we have assumed that hybrid incompatibilities involve *pairs* of genes, analysis of the Dobzhansky–Muller model shows that more complex hybrid incompatibilities, involving interactions among three or more of genes, should be common. The reason is not intuitively obvious, but is easily demonstrated mathematically (Cabot *et al.* 1994; Orr 1995). Certain paths to the evolution of new species are barred because they would require passing through intermediate genotypes that are sterile or inviable. It is easy to show, however, that the *proportion* of all imaginable paths to speciation allowed by selection increases with the complexity of hybrid incompatibilities (Orr 1995). Thus, for the same reason that two-gene speciation is 'easier' than single-gene speciation, so three-gene is easier than two-gene, and so on.

The evidence for complex incompatibilities is now overwhelming. They have been described in the *Drosophila obscura* group (Muller 1942; Orr unpublished), the *Drosophila virilis* group (Orr and Coyne 1989), the *Drosophila repleta* group (Carvajal *et al.* 1996), and the *Drosophila melanogaster* group (Cabot *et al.* 1994;

Davis *et al.* 1994). Indeed, such interactions could prove more common than the two-locus interactions discussed at length by Bateson, Dobzhansky, and Muller.

Theoretical analysis of the Dobzhansky–Muller model also has yielded counter-intuitive results about the effect of population subdivision on speciation. Many evolutionists have maintained, for example, that speciation is most likely in taxa subdivided into small populations. At least for postzygotic isolation, this idea is demonstrably false. Orr and Orr (1996) showed that if the substitutions ultimately causing Dobzhansky–Muller incompatibilities are driven by natural selection—as seems likely (Christie and Macnair 1984)—the waiting time to speciation grows *longer* as a species of a given size is splintered into ever smaller populations. If, on the other hand, the substitutions causing hybrid problems are originally neutral, population subdivision has little effect on the time to speciation. In no case is the accumulation of hybrid incompatibilities greatly accelerated by small population size. Unfortunately, we have little empirical data bearing on this issue. Although it might seem that the effect of population size on speciation rates could be estimated by comparing the rate of evolution of postzygotic isolation on islands versus continents, this comparison is confounded by the likelihood of stronger selection in novel island habitats, which itself might drive rapid speciation.

Finally, much of the theoretical work on the evolution of postzygotic isolation has been devoted to explaining one of the most striking patterns characterizing the evolution of hybrid sterility and inviability—Haldane's rule. Because this large and confusing literature has recently been reviewed elsewhere (e.g., Wu *et al.* 1996; Orr 1997), we will not attempt a thorough discussion here. Instead, we briefly consider the data from nature and sketch the leading hypotheses offered to explain them.

Haldane's rule

In 1922, J. B. S. Haldane noted that, if only one hybrid sex is sterile or inviable, it is nearly always the heterogametic (XY) sex. More recent and far more extensive reviews (Coyne 1992c) show that 'Haldane's rule' is obeyed in all animal groups that have been surveyed, e.g., *Drosophila*, mammals, Orthoptera, birds, and Lepidoptera (the latter two groups have heterogametic females). Indeed, it is likely that Haldane's rule characterizes postzygotic isolation in all animals having chromosomal sex determination. Moreover, these surveys show that Haldane's rule is *consistently* obeyed. In *Drosophila*, for instance, 114 species crosses produce hybrids who are sterile in only one gender. In 112 of these, it is the males who are sterile (Coyne 1992c). Comparative work in *Drosophila* also shows that Haldane's rule represents an early stage in the evolution of postzygotic isolation: hybrid male sterility or inviability arises quite quickly, while female effects appear only much later (Coyne and Orr 1989b, 1997).

For obvious reasons, Haldane's rule has received a great deal of attention: it represents one of the strongest patterns in evolutionary biology, and perhaps the *only* pattern characterizing speciation. In addition, the rule implies that there is some fundamental similarity in the genetic events causing speciation in all animals.

Although many hypotheses have been offered to explain Haldane's rule, most have been falsified. We will not consider these failed explanations here (for reviews see: Orr 1997; Wu *et al.* 1996; Coyne and Orr 1989a; Coyne *et al.* 1991; Coyne 1992c). Instead, we briefly review the three explanations of Haldane's rule that remain viable. There is strong evidence for two of these, and suggestive evidence for the third. In a field that has historically been rife with disagreement, a surprisingly good consensus has emerged that some combination of these hypotheses explains Haldane's rule (Wu *et al.* 1996; True *et al.* 1996; Orr 1997).

The first hypothesis, the 'dominance theory,' posits that Haldane's rule reflects the recessivity of X-linked genes causing hybrid problems. This idea was first suggested by Muller (1942), and his verbal theory was later formalized by Orr (1993b) and Turelli and Orr (1995). The mathematical work shows that heterogametic hybrids suffer greater sterility and inviability than homogametic hybrids whenever the alleles causing hybrid incompatibilities are, on average, partially recessive ($\bar{d} < \frac{1}{2}$; this parameter incorporates both the effects of dominance *per se* and any correlation between dominance and severity of hybrid effects; see Turelli and Orr 1995). The reason is straightforward. Although XY hybrids suffer the full hemizygous effect of all X-linked alleles causing hybrid problems (dominant and recessive), XX hybrids suffer twice as many X-linked incompatibilities (they carry twice as many Xs). These two forces balance when $\bar{d} = \frac{1}{2}$. But if $\bar{d} < \frac{1}{2}$, the expression of recessives in XY hybrids outweighs the greater number of incompatibilities in XX hybrids, and Haldane's rule results. Obviously, the dominance theory can account not only for Haldane's rule, but also for the well-known large effect of the X chromosome on hybrid sterility and inviability (Wu and Davis 1993; Turelli and Orr 1995).

There is now strong evidence that dominance explains Haldane's rule for hybrid inviability. In particular, the dominance theory predicts that, in *Drosophila* hybridizations obeying Haldane's rule for inviability, hybrid females who are forced to be homozygous for their X chromosome should be as inviable as F_1 hybrid males. (Such 'unbalanced' females possess an F_1 male-like genotype in which all recessive X-linked genes are fully expressed.) In both of the species crosses in which this test has been performed, unbalanced females are, as expected, completely inviable (Orr 1993a; Wu and Davis 1993). Similarly, there is evidence from haplodiploid species that hybrid backcross males (who are haploid) suffer more severe inviability than their diploid sisters (Breeuwer and Werren 1995).

There is also weaker indirect evidence that the alleles causing hybrid sterility act as partial recessives: Hollocher and Wu (1996) and True *et al.* (1996) found that while most heterozygous introgressions from one *Drosophila* species into another are reasonably fertile, many *homozygous* introgressions are sterile. It therefore seems likely that dominance contributes to Haldane's rule for both hybrid inviability and hybrid sterility. Last, it is worth noting that the dominance theory—unlike several alternatives—should hold in *all* animal taxa, regardless of which sex is heterogametic (Orr and Turelli 1996).

The second hypothesis posits that Haldane's rule reflects the faster evolution of genes ultimately causing hybrid male than female sterility (Wu and Davis 1993;

Wu *et al.* 1996). Wu and his colleagues offer two explanations for this 'faster male' evolution: (1) In hybrids, spermatogenesis may be disrupted far more easily than oogenesis, and (2) sexual selection may cause genes expressed in males to evolve faster than those expressed in females. While there is now good evidence for faster-male evolution (see below), this theory cannot be the sole explanation of Haldane's rule. First, it cannot explain Haldane's rule for sterility in those taxa having heterogametic *females*, e.g., birds and butterflies. After all, spermatogenesis and sexual selection involve males *per se*, while Haldane's rule pertains to heterogametic hybrids, male or not. Second, the faster-male theory cannot account for Haldane's rule for inviability in *any* taxa. Because there is strong evidence that genes causing lethality are almost always expressed in both sexes (reviewed in Orr 1997), it seems unlikely that hybrid 'male lethals' can evolve faster than 'female lethals.' Finally, it is not obvious that sexual selection would inevitably lead to faster substitution of 'male' than 'female' alleles. One can easily imagine, for instance, forms of sexual selection in which each substitution affecting a male character is matched by a substitution affecting female preference for that character. If sexual selection causes faster evolution of male sterility, one may need to consider processes like male–male competition in addition to male–female coevolution.

Despite these caveats, there is now good evidence—at least in *Drosophila*—that alleles causing sterility of hybrid males accumulate much faster than those affecting females (True *et al.* 1996; Hollocher and Wu 1996). (Unfortunately, both of these studies analysed the same species pair; analogous data from other species are badly needed.) While we cannot be sure of the mechanism involved, it certainly appears that 'faster-male' evolution plays an important role in Haldane's rule for sterility in taxa with heterogametic males.

The last hypothesis, the 'faster-X' theory, posits that Haldane's rule reflects the more rapid divergence of X-linked than autosomal loci (Charlesworth *et al.* 1987; Coyne and Orr 1989b). Charlesworth *et al.* (1987) showed that, if the alleles ultimately causing postzygotic isolation were originally fixed by natural selection, X-linked genes will evolve faster than autosomal genes if favorable mutations are on average partially recessive ($\bar{h} < \frac{1}{2}$). (It must be emphasized that this theory requires only that the favorable effects of mutations on their 'normal' conspecific genetic background are partially recessive; nothing is assumed about the dominance of alleles in hybrids. Conversely, the dominance theory requires only that the alleles causing hybrid problems act as partial recessives *in hybrids*; nothing is assumed about the dominance of these alleles on their normal conspecific genetic background.) Under various scenarios, this faster evolution of X-linked genes can indirectly give rise to Haldane's rule (Orr 1997).

There is some evidence that X-linked hybrid steriles and lethals do in fact evolve faster than their autosomal analogs. In their genome-wide survey of speciation genes in the *Drosophila simulans–D. mauritiana* hybridization, True *et al.* (1996) found a significantly higher density of hybrid male steriles on the X chromosome than on the autosomes. Hollocher and Wu (1996), however, found no such difference in a much smaller experiment. Thus, while faster-X evolution may contribute to Haldane's rule, the evidence for it is considerably weaker than that for both the dominance and faster-male theories. (The faster-X theory also suffers several other shortcomings described elsewhere [Orr 1997].) Definitive tests of the faster-X theory must await new experiments that, following True *et al.*, allow direct comparison of the number of hybrid steriles and lethals on the X vs. autosomes.

In sum, there is now strong evidence for both the dominance and faster-male theories of Haldane's rule. Future work must include better estimates of the dominance of hybrid steriles, better tests of the faster-X theory, and, most important, genetic analyses of Haldane's rule in taxa having heterogametic females. Although both the dominance and faster-male theories make clear predictions about the genetics of postzygotic isolation in these groups (Orr 1997), we have virtually no direct genetic data from these critical taxa. Last, it is important to determine if Haldane's rule extends beyond animals, particularly to species of plants with heteromorphic sex chromosomes.

Hybrid rescue mutations

No discussion of postzygotic isolation would be complete without mentioning a recent and remarkable discovery—'hybrid rescue' mutations. These are alleles that, when introduced singly into *Drosophila* hybrids, rescue the viability or fertility of normally inviable or sterile individuals (Watanabe 1979; Takamura and Watanabe 1980; Hutter and Ashburner 1987; Hutter *et al.* 1990; Sawamura *et al.* 1993a,b,c; Davis *et al.* 1996). All rescue mutations studied to date involve the *D. melanogaster–D. simulans* hybridization. In one direction of this cross, hybrid males die as late larvae; in the other, hybrid females die as embryos. All surviving hybrids are completely sterile (Sturtevant 1920). Several mutations are known that rescue the larval inviability: *Lethal hybrid rescue* (*Lhr*) from *D. simulans* (Watanabe 1979), and *Hybrid male rescue* (*Hmr*) and *In(1)AB* from *D. melanogaster* (Hutter and Ashburner 1987; Hutter *et al.* 1990). Despite some early confusion, it now appears that these rescue mutations have no effect on hybrid *embryonic* inviability, which is instead rescued by a different set of mutations: *Zygotic hybrid rescue* (*Zhr*) from *D. melanogaster* (Sawamura *et al.* 1993c), and *maternal hybrid rescue* (*mhr*) from *D. simulans* (Sawamura *et al.* 1993a). The fact that larval and embryonic lethality are rescued by non-overlapping sets of mutations strongly suggests that these forms of isolation have different developmental bases (Sawamura *et al.* 1993b).

Recently, Davis, *et al.* (1996) described a mutation that rescues, albeit weakly, the *fertility* of *D. melanogaster–D. simulans* hybrid females. Although little is known about this allele, its discovery suggests that it may be possible to bring all of the genetic and molecular technology available in *D. melanogaster* to bear on speciation. (This species has previously been of limited use in the genetics of speciation because it produces no fertile progeny when crossed to any other species.)

The discovery of hybrid rescue genes has several important implications. First, it suggests that postzygotic isolation may have a simple genetic basis. It seems quite unlikely that a single mutation could restore the viability of hybrids if lethality were caused by many different developmental problems. But a simple developmental basis implies in turn a simple *genetic* basis: if many genes were involved, it seems unlikely that they all would affect the same developmental pathway. These inferences have been roughly confirmed by more recent work on the developmental basis of larval inviability in *D. melanogaster–D. simulans* group hybrids, which shows that lethal hybrids suffer a profound mitotic defect that can be reversed by introduction of hybrid rescue mutations (Orr *et al.* 1997). This suggests (but does not prove) that hybrid inviability in this case results from a single developmental defect—failure to condense chromosomes during mitosis. [Such a suggestion may seem contradicted by recent introgression experiments showing that *many* different chromosome regions cause postzygotic isolation when introgressed from one *Drosophila* species into another (True *et al.* 1996; Hollocher and Wu 1996). But the overwhelming majority of these introgressions have discernible effects only when homozygous and so would not play any role in F_1 hybrid inviability or sterility (Hollocher and Wu 1996).]

The discovery of hybrid rescue genes may also provide an important short-cut to the cloning and characterization of speciation genes. It is possible that rescue mutations are alleles of the genes that normally cause hybrid inviability or sterility (Hutter and Ashburner 1987). If so, characterization of these mutations might quickly lead to the molecular isolation of speciation genes. Alternatively, rescue mutations might be second-site suppressors, i.e., mutations at some second set of loci that suppress the problems caused by a different, primary set of genes.

The role of endosymbionts in speciation

Recent work on postzygotic isolation has pointed to a novel cause of hybrid incompatibilities—cytoplasmic incompatibility (CI) resulting from infection by cellular endosymbionts. Although CI has now been found in at least five orders of insects (Stevens and Wade 1990; Hoffmann and Turelli 1997), most work has focused on two model systems, the fly *Drosophila simulans* (Hoffmann *et al.* 1986; Turelli and Hoffmann 1995) and the small parasitic wasp *Nasonia* (Perrot-Minnot *et al.* 1996; Werren 1998).

In both systems, hybrid embryonic lethality results when males from infected populations or species are crossed to females from uninfected populations or species. Moreover, in both cases the infective agent is the *rickettsia*-like endosymbiont *Wolbachia* (Turelli and Hoffmann 1995; Werren, 1997). Remarkably, antibiotic treatment of infected lines cures *Wolbachia* infections, allowing fully compatible crosses. Because infected-by-infected crosses are compatible—while the same crosses using genetically identical but cured females are incompatible—the presence of *Wolbachia* in females must confer immunity to the effects of fertilization by sperm from infected males. The basis of this immunity is unknown.

Because CI only occurs when 'naive' uninfected cytoplasm is fertilized by sperm from infected males, CI is typically unidirectional. Recently, however, cases have been found in both *Drosophila* (O'Neill and Karr 1990) and *Nasonia* (Breeuwer and Werren 1995; Perrot-Minnot *et al.* 1996) in which two different populations or species are infected by different varieties of *Wolbachia*. Crosses between individuals carrying these different types of *Wolbachia* are *bidirectionally* incompatible, so that postzygotic isolation is complete. This represents a remarkable mode of speciation, for *no* changes are required in an host's genome.

Although *Wolbachia* infections will surely prove common—and speciation workers should routinely test for them—there are reasons for questioning whether CI will prove an important cause of speciation. First, *Wolbachia* cannot explain Haldane's rule, which is ubiquitous among animals and characterizes (at least in *Drosophila*) an early and nearly-obligate step in the evolution of postzygotic isolation (Coyne and Orr 1989b, 1997). Although CI results in dead embryos, we know of no cases in which *Wolbachia* causes lethality of one sex only. Similarly, *Wolbachia* seems unlikely to be a common cause of hybrid sterility: among the many genetic analyses of hybrid sterility, only a single case involves an endosymbiont (Somerson, *et al.* 1984).

1.6 Conclusions

The increasing prominence of work on speciation reflects real progress: a number of important questions have been resolved by experiment and a number of patterns explained by theory. This progress, in turn, derives from several fundamental but rarely recognized changes in our approach to speciation. First, the field has grown increasingly *genetical*. As a consequence, a large body of grand but notoriously slippery questions (How important are peak shifts in speciation? Is sympatric speciation common?) have been replaced by a collection of simpler questions (Is the Dobzhansky–Muller model correct? What is the cause of Haldane's rule?). While it would be fatuous to claim that these new questions are more important than the old, there is no doubt that they are more tractable. Second, the connection between theory and experiment has grown increasingly close. While speciation once seemed riddled with amorphous and untestable verbal theories, the last decade of work has produced a body of mathematical theory yielding clear and testable predictions about the basis of reproductive isolation. Last, but most important, many of these predictions have been tested.

Despite this progress, many questions about speciation remain unanswered. Throughout this review we have tried to highlight those questions that seem to us both important and tractable. Most fall into two broad sets. The first concerns speciation in taxa that have been relatively ignored: Does reinforcement occur in plants? Do pre- and postzygotic isolation evolve at about the same rate in most taxa as in *Drosophila*? Do plants with heteromorphic sex chromosomes obey Haldane's rule? Do hybrid male steriles still evolve faster than female steriles in taxa having heterogametic females? How distinct are asexual taxa in sympatry?

The second set of questions concerns the evolution of prezygotic isolation, which has received less attention than postzygotic isolation. How complex is the genetic basis of sexual isolation? How common is reinforcement? Why is sexual isolation so often asymmetric? What is the connection between adaptive radiation and sexual isolation?

It may seem that trading yesterday's grand verbal speculations for today's smaller, more technical studies risks a permanent neglect of the larger questions about speciation. We believe, however, that more focused pursuits of tractable questions will ultimately produce better answers to these bigger questions. Just as no mature theory of population genetics was possible until we understood the mechanism of inheritance, so no mature view of speciation seems possible until we understand the origins and mechanics of reproductive isolation.

Acknowledgments

Our work is supported by grants from the National Institutes of Health to JAC and to HAO and from the David and Lucile Packard Foundation to HAO. We thank M. Turelli for helpful comments and discussion of the 'snowball effect,' and A. Magurran for calling our attention to the Haskins and Haskins paper. We are grateful for the hospitality of the Holiday Inn in Boulder, Colorado, whose swimming pool provided a venue for the genesis of this paper.

References

Arnold, S. J., Verrell, P. A., and Tilley, S. G. 1996 The evolution of asymmetry in sexual isolation: a model and a test case. *Evolution* **50**, 1024–1033.

Ashburner, M. 1989 *Drosophila: a laboratory handbook*. New York: Cold Spring Harbor Laboratory Press.

Barraclough, T. G., Harvey, P. H., and Nee, S. 1995 Sexual selection and taxonomic diversity in passerine birds. *Proc. R. Soc. London Ser. B* **259**, 211-215.

Barton, N. H. and Charlesworth, B. 1984 Genetic revolutions, founder effects, and speciation. *Annu. Rev. Ecol. Syst.* **15**, 133–164.

Bateson, W. 1909 Heredity and variation in modern lights. In: *Darwin and modern science* (ed. A. C. Seward), pp. 85–101. Cambridge: Cambridge University Press.

Baum, D. and Shaw, K. L. 1995 Genealogical perspectives on the species problem. *Monogr. Syst. Bot.* **53**, 289–303.

Bella, J. L., Butlin, R. K., Ferris, C., and Hewitt, G. M. 1992 Asymmetrical homogamy and unequal sex ratio from reciprocal mating-order crosses between *Chorthippus parallelus* subspecies. *Heredity* **68**, 345–352.

Bradshaw, H. D., Wilbert, S. M., Otto, K. G., and Schemske, D. W. 1995 Genetic mapping of floral traits associated with reproductive isolation in monkeyflowers (*Mimulus*). *Nature* **376**, 762–765.

Breeuwer, J. A. J. and Werren, J. H. 1995 Hybrid breakdown between two haplodiploid species: the role of nuclear and cytoplasmic genes. *Evolution* **49**, 705–717.

Butlin, R. 1987 Speciation by reinforcement. *Trends Ecol. Evol.* **2**, 8–13.

Cabot, E. L., Davis, A. W., Johnson, N. A., and Wu, C.-I. 1994 Genetics of reproductive isolation in the *Drosophila simulans* clade: complex epistasis underlying hybrid male sterility. *Genetics* **137**, 175–189.

Carvajal, A. R., Gandarela, M. R., and Naveira, H. F. 1996 A three-locus system of interspecific incompatibility underlies male inviability in hybrids betwen *Drosophila buzzattii* and *D. koepferi*. *Genetica* **98**, 1–19.

Charlesworth, B., Coyne, J. A., and Barton, N. 1987 The relative rates of evolution of sex chromosomes and autosomes. *Am. Nat.* **130**, 113–146.

Christie, P. and Macnair, M. R. 1984 Complementary lethal factors in two North American populations of the yellow monkey flower. *J. Hered.* **75**, 510–511.

Christie, P. and Macnair, M. R. 1987 The distribution of postmating reproductive isolating genes in populations of the yellow monkey flower, *Mimulus guttatus. Evolution* **41**, 571–578.

Cobb, M., Burnet, B., Blizard, R., and Jallon, J.-M. 1989 Courtship in *Drosophila sechellia*: Its structure, functional aspects, and relationship to those of other members of the *Drosophila melanogaster* species subgroup. *J. Insect Behav.* **2**, 63–89.

Cohan, F. M. 1998. Genetic structure of bacterial populations. In: Evolutionary genetics from molecules to morphology, (ed. R. Singh and C. Krimbas), Cambridge: Cambridge University Press. (in press).

Coulthart, M. D. and Singh, R. S. 1988 High level of divergence of male reproductive tract proteins between *Drosophila melanogaster* and its sibling species, *D. simulans. Mol. Biol. Evol.* **5**, 182–191.

Coyne, J. A. 1983 Genetic basis of differences in genital morphology among three sibling species of *Drosophila. Evolution* **37**, 1101–1118.

Coyne, J. A. 1989 The genetics of sexual isolation between two sibling species, *Drosophila simulans* and *Drosophila mauritiana. Proc. Natl Acad. Sci. USA* **86**, 5464–5468.

Coyne, J. A. 1992a Much ado about species. *Nature* **357**, 289–290.

Coyne, J. A. 1992b Genetics of sexual isolation in females of the *Drosophila simulans* species complex. *Genet. Res. Camb.* **60**, 25–31.

Coyne, J. A. 1992c Genetics and speciation. *Nature* **355**, 511–515.

Coyne, J. A. 1993a Recognizing species. *Nature* **364**, 298.

Coyne, J. A. 1993b The genetics of an isolating mechanism between two sibling species of *Drosophila. Evolution* **47**, 778–788.

Coyne, J. A. 1994 Ernst Mayr and the origin of species. *Evolution* **48**, 19–30.

Coyne, J. A. 1996a Genetics of differences in pheromonal hydrocarbons between *Drosophila melanogaster* and *D. simulans. Genetics* **143**, 353-364.

Coyne, J. A. 1996b Genetics of sexual isolation in male hybrids of *Drosophila* simulans and *D. mauritiana. Genet. Res.* **68**, 211–220.

Coyne, J. A. and Charlesworth, B. 1986 Location of an X-linked factor causing male sterility in hybrids of *Drosophila simulans* and *D. mauritiana. Heredity* **57**, 243–246.

Coyne, J. A. and Charlesworth, B. 1989 Genetic analysis of X-linked sterility in hybrids between three sibling species of *Drosophila. Heredity* **62**, 97–106.

Coyne, J. and Charlesworth, B. 1997 Genetics of a pheromonal difference affecting sexual isolation between *Drosophila mauritiana* and *D. sechellia. Genetics* **145**, 1015–1030.

Coyne, J. A. and Kreitman, M. 1986 Evolutionary genetics of two sibling species, *Drosophila simulans* and *D. sechellia. Evolution* **40**, 673–691.

Coyne, J. A. and Orr, H. A. 1989a Two rules of speciation. In: *Speciation and its consequences.* (ed. D. Otte and J. Endler.), pp. 180–207. Sunderland, MA: Sinauer Associates.

Coyne, J. A. and Orr, H. A. 1989b Patterns of speciation in *Drosophila. Evolution* **43**, 362–381.

Coyne, J. A. and Orr, H. A. 1997 Patterns of speciation in *Drosophila* revisited. *Evolution* **51**, 295–303.

Coyne, J. A., Orr, H. A., and Futuyma, D. J. 1988 Do we need a new species concept? *Syst. Zool.* **37**, 190–200.

Coyne, J. A., Charlesworth, B., and Orr, H. A. 1991 Haldane's rule revisited. *Evolution* **45**, 1710–1713.

Coyne, J. A., Barton, N. H., and Turelli, M. 1997 A critique of Sewall Wright's shifting balance theory of evolution. *Evolution* **51**, 643–671.

Cracraft, J. 1989 Speciation and its ontology: the empirical consequences of alternative species concepts for understanding patterns and processes of differentiation. In *Speciation and its consequences* (ed. D. Otte and J. A. Endler), pp. 28–59. Sunderland, MA: Sinauer Associates.

Darwin, C. R. 1859 *On the origin of species.* London: J. Murray,

David, J., Bocquet, C., Lemeunier, F., and Tsacas, L. 1974 Hybridation d'une nouvelle espéce *Drosophila mauritiana* avec *D. melanogaster* et *D. simulans. Ann. Genét.* **17**, 235–241.

Davies, N., Aiello, A., Mallet, J., Pomiankowski, A., and Silberglied, R. E. 1997 Speciation in two neotropical butterflies: extending Haldane's rule. *Proc. R. Soc. London Ser. B* **264**, 845–851.

Davis, A. W. and Wu, C.-I. 1996 The broom of the sorcerer's apprentice: the fine structure of a chromosomal region causing reproductive isolation between two sibling species of *Drosophila. Genetics* **143**, 1287–1298.

Davis, A. W., Noonburg, E. G., and Wu, C. I. 1994 Evidence for complex genetic interactions between conspecific chromosomes underlying hybrid female sterility in the *Drosophila simulans* clade. *Genetics* **137**, 191–199.

Davis, A. W., Roote, J., Morley, T., Sawamura, K., Herrmann, S., and Ashburner, M. 1996 Rescue of hybrid sterility in crosses between *D. melanogaster* and *D. simulans. Nature* **380**, 157–159.

Dobzhansky, T. 1933 On the sterility of the interracial hybrids in *Drosophila pseudoobscura. Proc. Natl Acad. Sci USA* **19**, 397–403.

Dobzhansky, T. 1935 A critique of the species concept in biology. *Phil. Sci.* **2**, 344–345.

Dobzhansky, T. 1937 *Genetics and the origin of species.* New York: Columbia University Press.

Eberhard, W. G. 1985 *Sexual selection and animal genitalia.* Cambridge, MA: Harvard University Press.

Eberhard, W. G. 1996 *Female control: sexual selection by cryptic female choice.* Princeton, NJ: Princeton University Press.

Ehrman, L. 1965 Direct observation of sexual isolation between allopatric and between sympatric strains of the different *Drosophila paulistorum* races. *Evolution* **19**, 459–464.

Endler, J. A. and Houde, A. E. 1995 Geographic variation in female preferences for male traits in *Poecilia reticulata. Evolution* **49**, 456–468.

Fenster, C. B., Diggle, P. K., Barrett, S. C. H., and Ritland, K. 1995 The genetics of floral development differentiating two species of *Mimulus* (Scrophulariaceae). *Heredity* **74**, 258–266.

Fenster, C. B. and Ritland, K. 1994 Quantitative genetics of mating system divergence in the yellow monkeyflower species complex. *Heredity* **73**, 422-435.

Fisher, R. A. 1930 *The genetical theory of natural selection.* Oxford: Oxford University Press.

Gilliard, E. T. 1956 Bower ornamentation versus plumage characters in bower-birds. *Auk* **73**, 450–451.

Grant, B. R. and Grant, P. R. 1996 Cultural inheritance of song and its role in the evolution of Darwin's finches. *Evolution* **50**, 2471–2487.

Grant, V. 1981 *Plant speciation.* 2nd edn. New York: Columbia University Press.

Gregory, P. G. and Howard, D. J. 1994 A postinsemination barrier to fertilization isolates two closely related ground crickets. *Evolution* **48**, 705–710.

Haldane, J. B. S. 1922 Sex-ratio and unisexual sterility in hybrid animals. *J. Genet.* **12**, 101–109.

Haskins, C.P. and Haskins, E.F. 1949 The role of sexual selection as an isolating mechanism in three species of poeciliid fishes. *Evolution* **3**, 160–169.

Heikkinen, E. and Lumme, J. 1991. Sterility of male and female hybrids of *Drosophila virilis* and *Drosophila lummei*. *Heredity* **67**, 1–11.

Hoffmann, A. A. and Turelli, M. 1997. Cytoplasmic incompatibility in insects. In *Influential passengers: inherited microorganisms and arthropod reproduction,* (ed. S. L. O'Neill, J. H. Werren, and A. A. Hoffmann), pp. 42–80. Oxford: Oxford University Press.

Hoffmann, A. A., Turelli, M., and Simmons, G. M. 1986 Unidirectional incompatibility between populations of *Drosophila simulans*. *Evolution* **40**, 692–701.

Hoikkala, A. and Lumme, J. 1984 Genetic control of the difference in male courtship sound between *D. virilis* and *D. lummei*. *Behav. Genet.* **14**, 827–845.

Hollocher, H. and Wu, C.-I. 1996 The genetics of reproductive isolation in the *Drosophila simulans* clade: X versus autosomal effects and male vs. female effects. *Genetics* **143**, 1243–1255.

Holman, E. W. 1987 Recognizability of sexual and asexual species of rotifers. *Syst. Zool.* **36**, 381–386.

Howard, D. J. 1993 Reinforcement: origin, dynamics, and fate of an evolutionary hypothesis. In *Hybrid zones and the evolutionary process* (ed. R. G. Harrison), pp. 46–69. Oxford: Oxford University Press.

Hutter, P. and Ashburner, M. 1987 Genetic rescue of inviable hybrids between *Drosophila melanogaster* and its sibling species. *Nature* **327**, 331–333.

Hutter, P., Roote, J., and Ashburner, M. 1990 A genetic basis for the inviability of hybrids between sibling species of *Drosophila*. *Genetics* **124**, 909–920.

Iwasa, Y. and Pomiankowski, A. 1995 Continual change in mate preferences. *Nature* **377**, 420–422.

Jones, C. D. 1998 The genetic basis of *Drosophila sechellia*'s resistance to a host plant toxin. *Genetics* **149**, 1899–1908.

Kaneshiro, K. Y. 1980 Sexual isolation, speciation, and the direction of evolution. *Evolution* **34**, 437–444.

Kelly, J. K. and Noor, M. A. F. 1996 Speciation by reinforcement: a model derived from studies of *Drosophila*. *Genetics* **143**, 1485–1497.

Khadem, M. and Krimbas, C. B. 1991 Studies of the species barrier between *Drosophila subobscura* and *D. madeirensis*. 1. The genetics of male hybrid sterility. *Heredity* **67**, 157–165.

King, M. 1993 *Species evolution*. Cambridge: Cambridge University Press.

Kirkpatrick, M. 1996 Good genes and direct selection in the evolution of mating preferences. *Evolution* **50**, 2125–2140.

Kirkpatrick, M. and Barton, N. 1995 *Déjà vu* all over again. *Nature* **377**, 388–389.

Lamnissou, K., Loukas, M., and Zouros, E. 1996 Incompatibilities between Y chromosome and autosomes are responsible for male hybrid sterility in crosses between *Drosophila virilis* and *Drosophila texana*. *Heredity* **76**, 603–609.

Lande, R. 1979 Effective deme sizes during long-term evolution estimated from rates of chromosomal rearrangement. *Evolution* **33**, 234–251.

Lande, R. 1981 Models of speciation by sexual selection on polygenic traits. *Proc. Natl Acad. Sci. USA* **78**, 3721–3725.

Lande, R. 1982 Rapid origin of sexual isolation and character divergence in a cline. *Evolution* **36**, 213–223.

Liou, L. W. and Price, T. D. 1994 Speciation by reinforcement of prezygotic isolation. *Evolution* **48**, 1451–1459.

Lofstedt, C., Hansson, B. S., Roelofs, W., and Bengtsson, B. O. 1989 No linkage between genes controlling female pheromone production and male pheromone response in the European corn borer, *Ostrinia nubialis* Hübner (Lepidoptera; Pyralideae). *Genetics* **123**, 553–556.

Macnair, M. R. and Christie, P. 1983 Reproductive isolation as a pleiotropic effect of copper tolerance in *Mimulus guttatus*. *Heredity* **50**, 295–302.

Macnair, M. R. and Cumbes, Q. J. 1989 The genetic architecture of interspecific variation in Mimulus. *Genetics* **122**, 211–222.

Marin, I. 1996 Genetic architecture of autosome-mediated hybrid male sterility in *Drosophila*. *Genetics* **142**, 1169–1180.

Masterson, J. 1994 Stomatal size in fossil plants: evidence for polyploidy in majority of angiosperms. *Science* **264**, 421–424.

Maynard-Smith, J. and Szathmáry, E. 1995 *The major transitions in evolution*. Oxford: W. H. Freeman/Spektrum.

Mayr, E. 1942 *Systematics and the origin of species*. New York: Columbia University Press.

Menotti-Raymond, M., David, V. A., and O'Brien, S. J. 1997 Pet cat hair implicates murder suspect. *Nature* **386**, 774.

Metz, E. C. and Palumbi, S. R. 1996 Positive selection and sequence rearrangements generate extensive polymorphism in the gamete recognition protein bindin. *Mol. Biol. Evol.* **13**, 397–406.

Mitra, S., Landel, H., and Pruett-Jones, S. 1996 Species richness covaries with mating system in birds. *Auk* **113**, 544–551.

Mitrofanov, V. G. and Sidorova, N. V. 1981 Genetics of the sex ratio anomaly in *Drosophila* hybrids of the *virilis* group. *Theor Appl Genet* **59**, 17–22.

Monti, L., Génermont, J., Malosse, C., and Lalanne-Cassou, B. 1997 A genetic analysis of some components of reproductive isolation between two closely related species, *Spodoptera latifascia* (Walker) and *S. descoinsi* (Lalanne-Cassou and Silvain) (Lepidoptera: Noctuidae.). *J. Evol. Biol* **10**, 121–143.

Muller, H. J. 1942 Isolating mechanisms, evolution, and temperature. *Biol. Symp.* **6**, 71–125.

Naveira, H. and Fontdevila, A. 1986 The evolutionary history of *Drosophila buzzatii*. The genetic basis of sterility in hybrids between *D. buzzattii* and its sibling *D. serido* from Argentina. *Genetics* **114**, 841–857.

Naveira, H. and Fontdevila, A. 1991 The evolutionary history of *Drosophila buzzatii*. XXI. Cumulative action of multiple sterility factors on spermatogenesis in hybrids of *D. buzzatii* and *D. koepferae*. *Heredity* **67**, 57–72.

Nei, M. 1972 Genetic distances between populations. *Am. Nat.* **106**, 282–292.

Nei, M. 1976 Mathematical models of speciation and genetic distance. In *Population genetics and ecology* (ed. S. Karlin and E. Nevo), pp. 723–766. New York: Academic Press, Inc.

Noor, M. 1995 Speciation driven by natural selection in *Drosophila*. *Nature* **375**, 674–675.

Noor, M. A. F. 1997a How often does sympatry affect sexual isolation in *Drosophila*? *Am. Nat.* **149**, 1156–1163.

Noor, M. A. F. 1997a Genetics of sexual isolation and courtship dysfunction in male hybrids of *Drosophila pseudoobscura* and *D. persimilis*. *Evolution* **51**, 809–815.

O'Neill, S. L. and Karr, T. L. 1990 Bidirectional incompatibility between conspecific populations of *Drosophila simulans*. *Nature* **348**, 178–180.

Orr, H. A. 1987 Genetics of male and female sterility in hybrids of *Drosophila pseudoobscura* and *D. persimilis*. *Genetics* **116**, 555–563.

Orr, H. A. 1989a Genetics of sterility in hybrids between two subspecies of *Drosophila*. *Evolution* **43**, 180–189.

Orr, H. A. 1989b Localization of genes causing postzygotic isolation in two hybridizations involving *Drosophila pseudoobscura*. *Heredity* **63**, 231–237.

Orr, H. A. 1992 Mapping and characterization of a 'speciation gene' in *Drosophila*. *Genet.l Res.* **59**, 73–80.

Orr, H. A. 1993a Haldane's rule has multiple genetic causes. *Nature* **361**, 532–533.

Orr, H. A. 1993b A mathematical model of Haldane's rule. *Evolution* **47**, 1606–1611.

Orr, H. A. 1995 The population genetics of speciation: the evolution of hybrid incompatibilities. *Genetics* **139**, 1805–1813.

Orr, H. A. 1997 Haldane's rule. *Annu.l Rev. Ecol. Syst.* **28**, 195–218.

Orr, H. A. and Coyne, J. A. 1989 The genetics of postzygotic isolation in the *Drosophila virilis* group. *Genetics* **121**, 527–537.

Orr, H. A. and. Coyne, J. A. 1992 The genetics of adaptation: a reassessment. *Am. Nat.* **140**, 725–742.

Orr, H. A. and Orr, L. H. 1996 Waiting for speciation: the effect of population subdivision on the time to speciation. *Evolution* **50**, 1742–1749.

Orr, H. A. and Turelli, M. 1996 Dominance and Haldane's rule. *Genetics* **143**, 613–616.

Orr, H. A., L. D. Madden, Coyne, J. A., Goodwin, R., and Hawley, R. S. 1997 The developmental genetics of hybrid inviability: a mitotic defect in *Drosophila* hybrids. *Genetics* **145**, 1031–1040.

Pantazidis, A. C., Galanopoulos, V. K., and Zouros, E. 1993 An autosomal factor from *Drosophila arizonae* restores normal spermatogenesis in *Drosophila mojavensis* males carrying the *D. arizonae* Y chromosome. *Genetics* **134**, 309–318.

Patterson, J. T. 1946 A new type of isolating mechanism in *Drosophila*. *Proc. Natl Acad. Sci. USA* **32**, 202–208.

Perez, D. E., Wu, C.-I., Johnson, N. A., and Wu, M.-L. 1993 Genetics of reproductive isolation in the *Drosophila simulans* clade: DNA-marker assisted mapping and characterization of a hybrid-male sterility gene, *Odysseus* (*Ods*). *Genetics* **134**, 261–275.

Perrot-Minnot, M.-J., Guo, L. R., and Werren, J. H. 1996 Single and double infections with Wolbachia in the parasitic wasp *Nasonia vitripennis*: effects on compatibility. *Genetics* **143**, 961–972.

Pontecorvo, G. 1943 Viability interactions between chromosomes of *Drosophila melanogaster* and *Drosophila simulans*. *J. Genet.* **45**, 51–66.

Prager, E. R. and Wilson, A. C. 1975 Slow evolutionary loss of the potential for interspecific hybridization in birds: a manifestation of slow regulatory evolution. *Proc. Natl Acad. Sci. USA* **72**, 200–204.

Prazmo, W. 1965 Cytogenetic studies on the genus *Aquilegia*. III. Inheritance of the traits distinguishing different complexes in the genus *Aquilegia*. *Acta Soc. Bot. Poloniae* **34**, 403–437.

Price, C. S. C. 1997 Conspecific sperm precedence in *Drosophila*. *Nature* **388**, 663–666.

Ramsey, J. M. and Schemske, D. W. 1998 The dynamics of polyploid formation and establishment in flowering plants. *Annu. Rev. Ecol. Syst.* (in press).

Rice, W. R. 1996 Sexually antagonistic male adaptation triggered by experimental arrest of female evolution. *Nature* **381**, 232–234.

Rice, W. R. and Hostert, E. E. 1993 Laboratory experiments on speciation: what have we learned in 40 years? *Evolution* **47**, 1637–1653.

Rieseberg, L. H., Desrochers, A. M., and Youn, S. J. 1995. Interspecific pollen competition as a reproductive barrier between sympatric species of *Helianthus* (Asteraceae). *Am. J. Bot.* **82**, 515–519.

Rieseberg, L. H. 1998 Genetic mapping as a tool for studying speciation. In: *Molecular systematics of plants*, 2nd edn (ed. D. E. Soltis, P. S. Soltis, and J. J. Doyle). New York: Chapman and Hall Inc. (in press).

Ringo, J. M. 1977 Why 300 species of Hawaiian *Drosophila*? The sexual selection hypothesis. *Evolution* **31**, 694–696.

Roberts, M. S. and Cohan, F. M. 1995 Recombination and migration rates in natural populations of *Bacillus subtilis* and *Bacillus mojavensis*. *Evolution* **49**, 1081–1094.

Roberts, M. S., Nakamura K. L., and Cohan, F. M. 1996 *Bacillus vallismortis* sp. nov., a close relative of *Bacillus subtilis*, isolated from soil in Death Valley, California. *Int. J. Syst. Bacteriol.* 46, 470–475.

Roelofs, W., Glover, T., X.-H. Tang, Sreng, I., Robbins, P., Eckenrode, C., Löfstedt, C., Hansson, B. S., and Bengtson, B. 1987 Sex pheromone production and perception in European corn borer moths is determined by both autosomal and sex-linked genes. *Proc. Natl Acad. Sci. USA* **84**, 7585–7589.

Ruiz, A. and Heed, W. B. 1988 Host-plant specificity in the cactophilic *Drosophila mulleri* species complex. *J. Anim. Ecol.* **57**, 237–249.

Saetre, G.-P., Moum, T., Bures, S., Král, M., Adamjan, M., and Moreno, J. 1997 A sexually selected character displacement in flycatchers reinforces premating isolation. *Nature* **387**, 589–592.

Sawamura, K., Taira, T., and Watanabe, T. K. 1993a Hybrid lethal systems in the *Drosophila melanogaster* species complex. I. The *maternal hybrid rescue* (*mhr*) gene of *Drosophila simulans*. *Genetics* **133**, 299–305.

Sawamura, K., Watanabe, T. K., and Yamamoto, M.-T. 1993b Hybrid lethal systems in the *Drosophila melanogaster* species complex. *Genetica* **88**, 175–185.

Sawamura, K., Yamamoto, M.-T., and Watanabe, T. K. 1993c Hybrid lethal systems in the *Drosophila melanogaster* species complex. II. The Zygotic hybrid rescue (*Zhr*) gene of *D. melanogaster*. *Genetics* **133**, 307–313.

Schäfer, U. 1978 Sterility in *Drosophila hydei* × *D. neohydei* hybrids. *Genetica* **49**, 205–214.

Schäfer, U. 1979 Viability in *Drosophila hydei* × *D. neohydei* hybrids and its regulation by genes located in the sex heterochromatin. *Biol. Zentralbl.* 98, 153–161.

Schliewen, U. K., Tautz, D., and Pääbo, S. 1994 Sympatric speciation suggested by monophyly of crater lake cichlids. *Nature* **368**, 629–632.

Schluter, D. 1996 Ecological causes of adaptive radiation. *Am. Nat.* **148** (Suppl.), S40–S64.

Schluter, D. 1998 Ecological causes of speciation. In: *Endless forms: species and speciation* (ed. D. J. Howard and S. H. Berlocher). Oxford: Oxford University Press (in press).

Schluter, D. and Price, T. 1993 Honesty, perception, and population divergence in sexually selected traits. *Proc. R. Soc. London Ser. B* **253**, 117–122.

Shaw, K. L. 1996 Polygenic inheritance of a behavioral phenotype: interspecific genetics of song in the Hawaiian cricket genus *Laupala*. *Evolution* **50**, 256–266.

Somerson, N. L., Ehrman, L., Kocka, J. P., and Gottlieb, F. J. 1984 Streptococcal L-forms isolated from *Drosophila paulistorum* semispecies cause sterility in male progeny. *Proc. Natl Acad. Sci. USA* **81**: 282–285.

Spencer, H. G., McArdle, B. H., and Lambert, D. M. 1986 A theoretical investigation of speciation by reinforcement. *Am. Nat.* **128**, 241–262.

Stratton, G. E. and Uetz, G. W. 1986 The inheritance of courtship behavior and its role as a reproductive isolating mechanism in two species of *Schizocosa* wolf spiders (Araneae: Lycosidea). *Evolution* **40**, 129–141.

Stevens, L. and Wade, M. J. 1990 Cytoplasmically inherited reproductive incompatibility in *Tribolium* flour beetles: the rate of spread and effect on population size. *Genetics* **124**, 367–372.

Sturtevant, A. H. 1920 Genetic studies on *Drosophila simulans*. I. Introduction. Hybrids with *Drosophila melanogaster*. *Genetics* **5**, 488–500.

Takamura, T. and Watanabe, T. K. 1980 Further studies on the lethal hybrid rescue (Lhr) gene of *Drosophila simulans*. *Jpn. J. Genet.* **55**, 405–408.

Tan, C. C. 1946 Genetics of sexual isolation between *Drosophila pseudoobscura* and *Drosophila persimilis*. *Genetics* **31**, 558–573.

Templeton, A. R. 1977 Analysis of head shape differences between two interfertile species of Hawaiian *Drosophila*. *Evolution* **31**, 630–641.

Templeton, A. R. 1981 Mechanisms of speciation—a population genetics approach. *Annu. Rev. Ecol. Syst.* **12**, 23–48.

Tomaru, M., Matsubayashi, H., and Oguma, Y. 1995. Heterospecific inter-pulse intervals of courtship song eleicit female rejection in *Drosophila biauraria*. *Anim. Behav.* **50**, 905–914.

Tomaru, M. and Oguma, Y. 1994. Genetic basis and evolution of species-specific courtship song in the *Drosophila auraria* complex. *Genet. Res. Camb.* **63**, 11–17.

True, J. R., Weir, B. S., and Laurie, C. C. 1996 A genome-wide survey of hybrid incompatibility factors by the introgression of marked segments of *Drosophila mauritiana* chromosomes into *Drosophila simulans*. *Genetics* 144, 819–837.

True, J. R., Liu, J., Stam, L. F., Zeng, Z.-B., and Laurie, C. C. 1997 Quantitative genetic analysis of divergence in male secondary sexual traits between *Drosophila simulans* and *D. mauritiana*. *Evolution* 51, 816–832.

Tsaur, S.-C. and Wu, C.-I. 1997 Positive selection and the molecular evolution of a gene of male reproduction, *Acp26Aa* of *Drosophila*. *Mol. Biol. Evol.* **14**, 544–549.

Turelli, M. and Hoffmann, A. A. 1995 Cytoplasmic incompatibility in *Drosophila simulans*: dynamics and parameter estimates from natural populations. *Genetics* **140**, 1319–1338.

Turelli, M. and Orr, H. A. 1995 The dominance theory of Haldane's rule. *Genetics* **140**, 389–402.

Turner, G. F. and Burrows, M. T. 1995 A model of sympatric speciation by sexual selection. *Proc. R. Soc. London Ser. B* **260**, 287–292.

Val, F. C. 1977 Genetic analysis of the morphological differences between two interfertile species of Hawaiian *Drosophila*. *Evolution* **31**, 611–629.

Vigneault, G. and Zouros, E. 1986 The genetics of asymmetrical male sterility in *Drosophila mojavensis* and *Drosophila arizonensis* hybrids: interaction between the Y chromosome and autosomes. *Evolution* **40**, 1160–1170.

Wade, M. J., Patterson, H., Chang, N. W., and Johnson, N. 1994 Postcopulatory, prezygotic isolation in flour beetles. *Heredity* **72**, 163–167.

Walsh, J. B. 1982 Rate of accumulation of reproductive isolation by chromosome rearrangements. *Am. Nat.* **120**, 510–532.

Wasserman, A. O. 1957 Factors affecting interbreeding in sympatric species of spadefoots (genus *Scaphiopus*). *Evolution* **11**, 320–338.

Wasserman, M. and Koepfer, H. R. 1977 Character displacement for sexual isolation between *Drosophila mojavensis* and *Drosophila arizonensis*. *Evolution* **31**, 812–823.

Watanabe, T. K. 1979 A gene that rescues the lethal hybrids between *Drosophila melanogaster* and *D. simulans*. *Jpn. J. Genet.* **54**, 325–331.

Watanabe, T. K. and Kawanishi, M. 1979 Mating preference and the direction of evolution in *Drosophila*. *Science* **205**, 906–907.

Werren, J. H. 1998 Wolbachia and speciation. In *Endless forms: species and speciation* (ed. D. Howard and S. Berlocher, eds.). Oxford: Oxford University Press.

White, M. J. D. 1969 Chromosomal rearrangements and speciation in animals. *Annu. Rev. Genet.* **3**, 75–98.

White, M. J. D. 1978 *Modes of speciation*. San Francisco: W. H. Freeman and Co.

Wittbrodt, J., Adam, D., Malitschek, B., Maueler, W., Raulf, F., Telling, A., Robertson,

S. M., and Schartl, M. 1989 Novel putative receptor tyrsosine kinase encoded by the melanoma-inducing *Tu* locus in *Xiphophorus*. *Nature* **341**, 415–421.

Wu, C.-I. and Beckenbach, A. T. 1983 Evidence for extensive genetic differentiation between the sex-ratio and the standard arrangement of *Drosophila pseudoobscura* and *D. persimilis* and identification of hybrid sterility factors. *Genetics* **105**, 71–86.

Wu, C.-I. and Davis, A. W. 1993 Evolution of postmating reproductive isolation: the composite nature of Haldane's rule and its genetic bases. *Am. Nat.* **142**, 187–212.

Wu, C. I., Johnson, N., and Palopali, M. F. 1996 Haldane's rule and its legacy: why are there so many sterile males? *Trends Ecol. Evol.* **11**, 281–284.

Zink, R. M. and McKitrick, M. C. 1995 The debate over species concepts and its implications for ornithology. *Auk* **112**, 701–719.

Zouros, E. 1973 Genic differentiation associated with the early stages of speciation in the mulleri subgroup of *Drosophila*. *Evolution* **27**, 601–621.

Zouros, E. 1981 The chromosomal basis of sexual isolation in two sibling species of *Drosophila*: *D. arizonensis* and *D. mojavensis*. *Genetics* **97**, 703–718.

2

Genetic diversity: do marker genes tell us the whole story?

T. Tregenza and R. K. Butlin

2.1 The units of diversity

To understand the variety of life and, more urgently, to decide how best to conserve it, biological diversity must be measured. Such an endeavour has two prerequisites: first, that we decide what we mean by diversity, and secondly that we are able to quantify the differences between organisms, concentrating on those differences we believe to be most relevant.

Because species are the units ecologists are best able to count (Magurran 1988), they are frequently used as a practical measure of diversity. This ought to mean that the still unresolved search for a universal species concept (Claridge *et al.* 1997) should be placed at the centre of ecological debates about biodiversity. However, practising ecologists usually cannot devote the time needed to tackle this thorny issue: they necessarily make pragmatic use of the species defined by systematists.

There are essentially three classes of species concepts which can be applied to real situations. These are based either on diagnosability, reproductive isolation or ecological distinctiveness. When measuring alpha-diversity (the number of species within a community), which of these concepts is used is unlikely to affect estimates of diversity, except with a minority of species which present problems either of asexual reproduction or of hybridisation. Clones within asexual 'species' may have different ecological roles (Vrijenhoek 1994), so if it is ecological distinctiveness which interests us, these clones may be more appropriate units than diagnosable species. Hybridisation may be sufficiently common to cause severe problems with species concepts based on reproductive isolation, or on diagnosability, depending on one's perspective: for example, hybridisation in a local flora has been interpreted very differently by Mayr (1992) and by others (e.g. Gornall 1997). It may well blur distinctions between ecological units.

More complex measures of biological diversity, such as beta-diversity (changes in species composition along an environmental gradient) and gamma-diversity (variation in species composition within similar habits at different sites) (Whittaker 1960) introduce another intractable problem with species concepts: how to deal with divergent allopatric populations. Concepts based on reproductive isolation really have no satisfactory answer. Alternatives based on diagnosability may offer a practical solution that is satisfactory for the systematist, but it is doubtful whether diagnosability can be used as a guide to ecological distinctiveness. An approach arising from conservation biology is to recognise 'evolutionarily significant units' as entities that are sufficiently distinct in non-trivial characters to be worthy of independent conservation efforts (Vogler and DeSalle 1994). This amounts to a species concept in its own right which aims at the common goal of recognising lineages with independent evolutionarily potential (Mayden 1997) but which, in common with its competitors, fails to provide an objective yardstick to measure 'distinctiveness'.

2.2 Genetic diversity

While species are units of ecological diversity, alleles are the units of genetic diversity. One could argue that they are the fundamental units of biological diversity since individuals of different species simply harbour more distinct alleles than individuals of the same species. Given that biodiversity cannot be measured in practice by counting alleles, one must find other features which can be used to indicate genetic diversity. One approach is to use neutral genetic markers to consider how genetic variation is partitioned within and between species. It may then be possible to use species diversity as a surrogate for total diversity, but only if the species is defined in a way that partitions neutral genetic variation consistently, and if neutral markers reliably reflect genetic variation of adaptive significance. Therefore, one must consider how much species differ genetically from one another, how much genetic variation there is within species, how consistent these two measures are across taxa, and the extent to which different classes of genetic variation show the same patterns. If these parameters vary widely, species are not equivalent units of biodiversity and must be weighted accordingly.

Weighting of ecological measures of species diversity on the basis of their genetic distance has recently been considered by several authors (May 1990; Humphries *et al.* 1995). This, too, has been mainly in the context of conservation, with the aim being to maximise the character diversity that is maintained in a limited area or within limited resources. In general, the result of weighting is to give more emphasis to more distinct taxa. Comparable weighting for the diversity within species has also been considered in the conservation context. Part of the motivation for defining 'evolutionarily significant units' is the recognition that some species, as currently delineated, contain several genetically distinct populations whereas others do not. At the level of genetic variation within populations, two distinct positions can be taken: either that genetically homogeneous populations are the most endangered and require the greatest intervention, or that genetically variable populations have the greatest biodiversity value, and the greatest evolutionary potential, and so should be most valued. Either way, reliable measure of genetic diversity within populations are needed.

In this chapter, we first consider the patterns of genetic divergence between taxonomic species and genetic variation within species, as revealed by allozyme electrophoresis. We then ask how well these measures reflect divergence or variability of quantitative traits. This leads to a particular consideration of divergence in traits that contribute to prezygotic reproductive isolation because of the importance of these traits in the origin, and thus the nature, of species, and because they appear to show even less dependence on general genetic divergence than other trait groups.

2.3 Allozymes

Allozymes are variant forms of proteins, usually soluble enzymes, determined by alternative alleles at a genetic locus. They are detected by separation in a gel on the basis of their mobility in an electric field (electrophoresis; see Avise 1994 for further details). They provide by far the largest coherent body of data on genetic variation within and between species. Not only is a large amount of allozyme data available in the literature, but much of it has been accumulated in databases. For example, Thorpe (1982) gathered data on Nei's genetic identity (a measure of the sharing of alleles and similarity of allele frequencies) from 106 studies (typically involving several species each), while Nevo and co-workers (Nevo *et al.* 1984) and Ward, Skibinski and Woodwark (Ward *et al.* 1992; Skibinski *et al.* 1993) assembled heterozygosity estimates for 1111 species and 3728 populations of more than 1500 species, respectively. The latter database includes genetic distance information as well as heterozygosities, and data are recorded locus by locus. This allows comparisons to be made across proteins as well as across populations, and allows taxonomic comparisons to be corrected for variation in the sample of proteins studied. The patterns revealed by these large surveys are striking. Primarily, and reassuringly, genetic identity is strongly related to the taxonomic hierarchy (Thorpe 1982) (Fig. 1); Nei genetic identities among populations within species are nearly always greater than $I = 0.8$ and 80% of values are greater than $I = 0.95$, identities between congeneric species typically range from $I = 0.35$ to 0.85. While identity values between species in different genera, and between congeneric species, overlap broadly, there is remarkably little overlap between identities within and among species. This is true despite the fact that Thorpe's survey included taxa ranging from seaweeds to mammals (with an admitted concentration on North American vertebrates). Only one major group, the birds, was omitted from the figure because 'their speciation processes seem to differ fundamentally from those of most other organisms' (Thorpe 1982, p. 150). Identities among bird species tended to fall in the range of within-species comparisons from other taxa.

If allozyme differences accumulate in a clock-like fashion, then Thorpe's data imply, for many taxa, that speciation typically occurs after a consistent time interval, most probably because it is associated with a general accumulation of genetic differences. Given that, in practice, most species are defined on the basis of morphological distinctiveness, this may simply mean that morphological differences accumulate in a roughly clock-wise manner along with allozyme differences. In birds, for whatever reason, morphological divergence is accelerated. However, this is not a very satisfying hypothesis because it does not explain the distinct gap in the distribution of identity values around $I = 0.8$ (Fig. 2.1).

A possible explanation for the discontinuity in genetic distances between and within species is that it results from Thorpe's deliberate omission of 'studies where there is taxonomic doubt (on grounds other than electrophoresis)' (Thorpe 1982, p. 150). However, a protein by protein analysis (Skibinski et al. 1993) also shows a marked contrast between intra- and interspecific genetic identities. Across seven proteins that are represented in a large majority of allozyme surveys, intraspecific identities all average more than $I = 0.96$ in vertebrates and $I = 0.91$ in invertebrates, while mean identities between species are all below $I = 0.85$ for vertebrates and $I = 0.72$ for invertebrates. This suggests that there is a genuine discontinuity in the number of shared allozymes and their allele frequencies within and between species.

An alternative way in which allozymes can be used as markers for genetic diversity is to examine variation within a population species. This can be expressed as an average expected heterozygosity, H, which provides a measure of allelic diversity across loci dependent on the numbers of alleles and their frequencies. Using heterozygosity expected under Hardy–Weinberg equilibrium, rather than observed heterozygosity, allows comparisons of genetic diversity across species

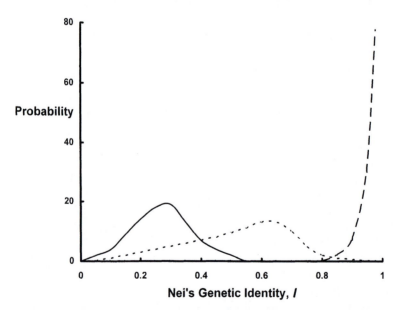

Fig. 2.1. Distributions of Nei's genetic identity (I) values between species in different genera, between congeneric species, and between populations within species. Redrawn from data in Thorpe (1982). ——, genera; - - -, species; – – –, population.

that are not influenced by factors such as inbreeding. Values of H vary across species from zero to about $H = 0.3$ in vertebrates or $H = 0.6$ in invertebrates (Ward *et al.* 1992). There is some variation in typical levels of variation across taxonomic groups from $H = 0.06$ in birds to $H = 0.16$ in molluscs but the variation is, in Gillespie's opinion 'not large enough to dispel the impression that there is a remarkable uniformity across taxa' (Gillespie 1991, p. 45). The remaining variation among species within higher taxa is substantial, suggesting that species are not equivalent units in terms of the genetic variability they harbour. A significant part of this variation in H is dependent on differences in population size among species (see below, and Box 2.1) but the remainder is unexplained.

Box 2.1. Neutral versus selective explanations for allozyme heterozygosity

Essentially, there are three classes of model: (a) strictly neutral drift; (b) slow removal of mildly deleterious alleles; or (c) maintenance by balancing selection. Under each of these models, heterozygosity is expected to be strongly related to effective population size, but the predicted levels of heterozygosity may be quite different (Nei and Graur 1984). The data gathered by Nei and Graur on heterozygosity and population size for 72 species do show a highly significant positive correlation between them, although the range of heterozygosity values is much less than is predicted from the range in population sizes (Fig. 2.2). Gillespie (1991) points out that the correlation is mainly due to *Drosophila* species at the high end (when most other data are from mammals) and carnivores at the low end. Within higher taxa, the major effect is probably the low heterozygosity in a few species that have been through very tight bottlenecks (such as the northern elephant seal, Hoelzel *et al.* 1993), with only a weak relationship otherwise.

Nei and Graur (1984) considered that the neutral model provided the best fit to the data in their survey. Skibinski *et al.* (1993) also concluded that the majority of allozyme variation could be explained under the neutral model. Their earlier analysis (Ward *et al.* 1992) showed substantial systematic variation in heterozygosity across proteins. For example, heterozygosity declines with increase in the number of subunits in a protein and increases with subunit molecular weight. The mean heterozygosity for a protein in vertebrates is strongly correlated with its mean heterozygosity in invertebrates. These patterns suggest selective constraints. However, they observed (Skibinski *et al.* 1993) a positive relationship between heterozygosity and genetic distance, both within and between populations and species. This is expected under the neutral theory and indicates that the variation among proteins is mostly due to variation in their neutral mutation rates.

On the other hand, various strands of evidence against the strictly neutral hypothesis have been gathered by Gillespie (1991). Until recently, only two general types of evidence were available: pattern tests and single locus tests. Pattern tests, such as those employed by Nei and Graur (1984) or Skibinski *et al.* (1993), suffer from overlap between the predictions of the three classes of model. Single locus tests have shown convincingly that *some* allozyme loci are under selection. For example, Watt (1991; Watt *et al.* 1996) has provided extensive evidence for differences in molecular function among phosophoglucose isomerase allozymes in *Colias* butterflies, and has related these to organismal performance and fitness. But single locus studies cannot be extended to allozymes in general and a few selectively maintained polymorphisms are consistent with the neutral model as a general explanation. Indeed, Watt himself has argued (1995) that a general 'selectionist-neutralist' debate is fruitless: some variants are selected, others are neutral.

DNA sequence data now provide an additional approach to the issue on the basis that silent-site polymorphism is likely to be strictly neutral and, therefore, can be used as a baseline for comparison (Kreitman and Akashi 1995). Silent-site variation is greatest in the region of the *Drosophila Adh* gene close to the codon responsible for a ubiquitous electrophoretic polymorphism for 'fast' and 'slow' alleles, supporting the interpretation of balancing selection on this locus (Hudson *et al.* 1987). More generally, if allozyme and silent-site polymorphisms are both determined by drift, they will be influenced by population size in the same way. However, there are significant departures from this pattern (Kreitman and Akashi 1995). Allozyme heterozygosity in *Drosophila melanogaster* is similar to, or greater than, heterozygosity in *D. simulans* but *simulans* has much more silent-site variation than does *melanogaster* (Aquandro 1991), and *D. melanogaster* has similar allozyme variability to humans but much greater silent-site variation (Li and Sadler 1991). If the silent-site variation reflects differences in effective population size, then the relative uniformity of allozyme heterozygosity across these three species, and generally, indicates pervasive balancing selection. Comparisons between heterozygosity and genetic distance show that amino-acid replacement substitutions are more common than expected, relative to silent-site substitutions, between species when compared with their proportions within species (McDonald and Kreitman 1991). This suggests divergence driven by selection and disagrees with the conclusion from largescale allozyme comparisons of heterozygosity and genetic distance, unless one concludes that the *Adh* locus in *Drosophila* is one of the exceptional examples of selection acknowledged by Skibinski *et al.* (1993).

2.4 How does allozyme variation correlate with trait variation?

A key question then, is to what extent variation among species or populations in allozyme heterozygosity provides an index of variation in the morphological, behavioural or life-history characters that are generally of more interest to the ecologist or conservation biologist. There is often an assumption that the two are positively correlated, based on the expectation that both are influenced by effective population size: larger populations have a greater input of variation by mutation

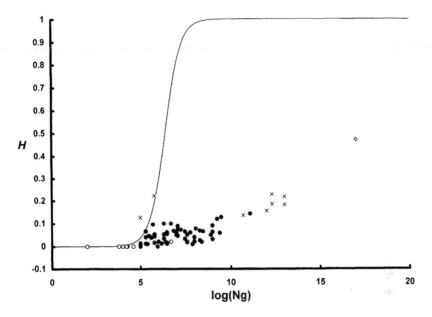

Fig. 2.2. A compilation of measurements of allozyme heterozygosity (*H*) in relation to population size (here expressed as the product of population size and generation time on a logarithmic scale). The neutral expectation is shown for comparison. Redrawn from data in Nei and Grau (1984) where further details of the data, and derivation of the neutral expectation, can be found. —, neutral expectation; ●, others; o, carnivores; ×, *Drosophila*; ◇ *Escherichia coli*.

and slower loss of variation by genetic drift. However, the data available to support this assumption are few (Mallet 1996). Lande (1988) predicts that the effects of population size will be rather different between neutral markers and quantitative traits, mainly because the higher mutational input to characters influenced by many loci enables them to maintain significant genetic variability at low population sizes and recover variability rapidly after population bottlenecks.

There is a significant positive relationship between population size and allozyme heterozygosity within several plant species with endangered populations (ranging in size from 1 to 10^5 individuals) (Young *et al.* 1996). This presumably results from loss of variation during contraction of the smaller populations. The question is whether there is a comparable reduction in quantitative variation in such populations, and, if so, whether this impairs their ability to adapt to changing environmental conditions (Storfer 1996). This is not only a pressing conservation issue but also a general question related to divergence between species and the process of speciation.

For allozymes, the impact of population size depends on the vexed question of neutral versus selective explanations for heterozygosity (see Box 2.1). The current balance of evidence favours a selective explanation for allozyme polymorphism rather than a strictly neutral explanation. Probably there is a mixture of balancing selection and selection against mildly deleterious alleles. If one assumes that the majority of quantitative traits in natural populations are under weak stabilising selection and are influenced by several to many loci (Lande 1976; Barton and Turelli 1989), then there is a broad similarity between the forces influencing both trait groups. Loci influencing quantitative traits under stabilising selection are also typically under weak balancing selection or have mildly deleterious alleles. Indeed, allozyme loci and quantitative loci are not necessarily mutually exclusive categories. The relationship between allozyme heterozygosity and population size should be reflected in a comparable relationship for heritable variation in quantitative traits (Houle 1989). The correlation may be similarly dominated by low values in populations with low effective population sizes as a result of recent bottlenecks, with Lande's (1988) proviso about the more rapid recovery of variability in quantitative traits because of the large numbers of loci involved. This relationship is not expected to extend to those quantitative traits that have a history of persistent directional selection, or are under particularly strong stabilising selection. Such traits are normally expected to have low heritability, regardless of population size.

Heritable variation in quantitative traits is, in fact, a ubiquitous feature of natural populations, and shows the expected relationship with intensity of selection, albeit with much scatter (Roff and Mousseau 1987). However, Houle (1989) found no good empirical support for the predicted correlation of genetic variance in quantitative traits with population size, although if total phenotypic variation could be used as an indicator of genetic variation (assuming environmental variance to be relatively constant) a few studies showed trends in the right direction. A direct comparison between allozyme heterozygosity and genetic variation in quantitative traits does not show a strong relationship either (Box 2.2).

Another way in which allozymes have been suggested to reflect an important general feature of genetic variability is through an association between individual heterozygosity and fitness. While numerous studies have reported correlations between allozyme heterozygosity and either growth rate or fluctuating asymmetry, the reasons for these relationships are unclear. It may be that allozyme loci are frequently in linkage disequilibrium with chromosomal regions containing deleterious recessive alleles, or it could be that individual allozyme loci have large direct effects on fitness (Gillespie 1991). Either way, these associations typically only explain a very small proportion of variance in the fitness measure (Britten 1996).

Overall, allozymes suggest that broad taxonomic distinctions concur with broad patterns of genetic variation. However, at the species level they may fail to provide a useful measure of general genetic diversity, presumably because often they are not selectively neutral and may be constrained by protein structure. Furthermore, there is little evidence for correlations between allozyme variation and variation in quantitative traits making allozyme heterozygosity a poor surrogate for other classes of genetic variation.

Box 2.2. Allozyme variation in relation to quantitative traits with different selection histories

Secondary sexual characters are a major class of traits with a history of directional selection and it has recently been argued that the operation of sexual selection on such characters can favour modifiers that enhance their additive genetic variation. A survey of data in the literature has supported this prediction (Pomiankowski and Møller 1995), although some doubt its theoretical basis (Rowe and Houle 1996). The process outlined by Pomiankowski and Møller is not necessarily specific to secondary sexual traits under female choice, it could apply to other traits that have been under directional selection. The expectation, then, is that genetic variation in quantitative traits with a history of weak stabilising selection will be correlated with allozyme heterozygosity, through their mutual dependence on effective population size, but the relationship will be weaker, or absent, from traits under directional selection.

In testing these predictions, it is clearly preferable to use the coefficient of additive genetic variation (CV_A), rather than the heritability, as the measure of genetic variability since it is not influenced by the level of non-genetic (or non-additive) variation (Houle 1992). Since Pomiankowski and Møller (1995) have accumulated data on CV_A for sets of characters currently under directional and stabilising sexual or non-sexual components of selection, we have added data on heterozygosity to their comparisons. Note that most characters are presumably under net stabilising selection, with any directional sexual selection balanced by other components of natural selection (Rowe and Houle 1996). However, exaggerated secondary sexual traits that are currently under directional sexual selection are likely to have a recent history of net directional selection. Despite the large number of allozyme studies in the literature, heterozygosity estimates are available for only 16 of the 32 species included in Pomiankowski and Møller's table 2. For this small dataset, the correlation between CV_A and heterozygosity is close to zero for characters under directional sexual selection ($r = -0.04$, $n = 11$) and negative for characters under stabilising selection ($r = -0.36$, $n = 9$; $r = -0.21$ for the combined data). It is, of course, possible that a larger survey would reveal a significant positive relationship in one or both classes but there is little support here for the use of allozyme data to infer the general evolutionary potential of a population or species. If there really is a negative correlation, which could, perhaps, result from weak selection in expanding populations versus strong selection in large or contracting populations, then allozyme data would be seriously misleading.

2.5 Sequence data

DNA sequence based methods may soon overtake allozymes in the volume of data available for comparisons within and between species. They have not done so yet, partly because early work used a variety of indirect approaches such as restriction fragment length polymorphisms (RFLPs) making it difficult to accumulate comparable data from across studies. Ultimately, direct sequencing must provide the most powerful description of genetic variation because it can eliminate the inferential steps needed with indirect methods like allozymes or RFLPs and can, potentially, avoid bias in the representation of different types of loci (although biases are certainly present in the currently available data, particularly in the choices of loci to examine). Sequences of coding regions have the added advantage of internal comparisons between silent-site, presumably largely neutral, variation and non-silent variation which is, at least potentially, under selection. Although allozyme data will remain the largest comparative resource for some time, comparisons with DNA sequence data can already help in their interpretation (see Box 2.1).

2.6 Genetic distance and genetic difference

Thorpe's (1982) survey of genetic distances between species showed a broad general agreement with the taxonomic hierarchy. Phenotypic divergence also undoubtedly increases as taxonomic relatedness decreases, although this pattern has rarely been documented systematically. The study of three plant families by Gilmartin (1980) demonstrates the point, which many take for granted, that the rate of increase in phenetic distance with taxonomic distance varies markedly among higher taxa. Probably, the same points could be made for ecological similarity. Allozyme and morphological variation are partitioned in broadly similar ways within and among populations but with differences in detail. For example, in *Phlox drummondii* (Schwaegerle *et al.* 1982) 94% of allozyme variation is within populations compared with 73% of heritable variation in quantitative traits. The underlying common variables are, of course, time of divergence and population size. But they are reflected imperfectly in each of these types of variation and so the correlations between them may be weak.

Attempting to define diversity simply in terms of the distribution of unique alleles ignores the fact that speciation is more than simply accumulation of phenetic, or even genetic, differences. Speciation involves the acquisition of evolutionary independence through the development of barriers to gene exchange. Genetic isolation then feeds back to preserve and promote differentiation. Thus, the evolution of characters contributing to reproductive isolation is of particular importance to the generation and maintenance of biological diversity. Since we are generally interested in the future diversity of ecosystems, we need to know whether reproductive isolation accumulates with general genetic divergence as indicated by allozyme data. This question has been addressed by a number of studies, both between and within species.

2.7 Does reproductive isolation accumulate with genetic divergence?

As we have discussed, allozyme data provide an imperfect surrogate for genetic divergence, but remain the best source of information for many groups. The large number of studies of species within the genus *Drosophila*, both of genetic distance and of reproductive isolation, has allowed Coyne and Orr (1989, 1997) to investigate the relationship between allozyme divergence and both pre- and postzygotic reproductive isolation. Postzygotic isolation (failure to produce viable offspring in interspecific crosses or production of sterile offspring), does show a very strong positive correlation with genetic distance between species. This suggests that genetic differences at loci responsible for isolation accumulate in much the same steady fashion as allozyme differences, dominated by drift or under weak selection. The loci involved are known to be numerous and to produce sterility or inviability through epistatic interactions (Wu and Palopoli 1995), as predicted by Dobzhansky (1937) and Muller (1942). However, little else can be said about them because their only phenotype is sterility or inviability in crosses and this is clearly not the effect which caused their divergence. At least in *Drosophila*, postzygotic isolation is a side-effect of slow general genetic divergence which may be caused by drift or by gradual adaptive evolution involving many small substitutions.

Coyne and Orr's analyses demonstrate a very different pattern for premating isolation due to assortative mating. In many of the species pairs for which they gathered data, strong premating isolation occurs despite low genetic distance ($D < 0.5$) and weak postzygotic isolation. This pattern is especially true for sympatric species pairs which show significantly higher premating isolation than do allopatric species pairs at low genetic distances. As a result, although there is a general increase in premating isolation with allozyme genetic distance, the relationship is much weaker than for postzygotic isolation. The implication is that, at least in sympatry, premating isolation evolves under strong selection. This selection may be direct selection for isolation as a result of producing dysfunctional hybrid offspring (reinforcement: Butlin 1987; Butlin and Tregenza 1997) but could also be an indirect effect of selection on mating signals and responses for other reasons. For example, sympatric species are more likely to differ in habitat associations than closely related allopatric species and selection may favour mating behaviour that is adapted to these habitats.

Although there is no comparable large-scale comparative study, there is evidence that mating signals do not diverge by steady accumulation of small differences. Study of the acoustic signals of males of the *Drosophila willistoni* group by Ritchie and Gleason (1995) is a case in point. Despite low levels of morphological divergence, the species have very distinctive songs. *D. equinoxialis* has the most divergent song structure but is not the most distantly related on the basis of allozymes (data quoted in Coyne an Orr 1989) whereas the distantly related ($D = 1.27$) *D. paulistorum* and *D. tropicalis* have rather similar songs, perhaps reflecting their allopatry. Using sequence data to measure genetic distance, Ritchie and Gleason (in prep.) find no significant correlation with song divergence (Mantel test, $r = -0.27$). While genetic distance is a good predictor of postzygotic isolation ($r = 0.68$, $P < 0.001$), it is a relatively poor predictor of prezygotic isolation ($r = 0.44$, $P = 0.035$).

A comparative study of mating signal evolution would ideally consider both signals and preferences in a group of species with known phylogenetic relationships. To some extent this type of study has been carried out in the recent tests of the 'sensory exploitation' hypothesis in *Physalaemus* toads (Ryan and Rand 1993), water mites (Proctor 1992) and swordtail fish (Basolo 1996) but these studies have concentrated on the origins of preferences and traits rather than quantitative changes. Measurements of carrier frequency spectra in the songs of a group of 20 related grasshopper species have recently been compared with sensitivity spectra of the receptor organs in the same species (Meyer and Elsner 1996). There is a strong correlation across species between the position of the low-frequency peak in the male song and the maximum sensitivity of the 'a-cells' that are responsible for registering the tympanal response to such frequencies ($r = 0.75$, $P < 0.001$). Unfortunately, grasshopper phylogeny is not well known and the current classification is probably not a reliable reflection of true relationships (D. R. Ragge, pers. comm.) so that phylogenetic interpretation of this correlation is not possible. However, it is striking that the pair *Chorthippus parallelus/montanus* and the trio *C. biguttulus/brunneus/mollis*, which are undoubtedly closely related, are very divergent in this characteristic of song and correspondingly in preference. It appears that song and preference characteristics co-evolve but can be modified rapidly and reversibly, a pattern that may be expected from intersexual selection (West-Eberhard 1983). As with the *Drosophila willistoni* group, these grasshoppers are very similar morphologically. They also show little genetic divergence (Butlin and Hewitt 1987; Mason *et al.* 1995). Mating signal evolution, and thus prezygotic isolation, has evolved independently of general genetic divergence.

Between species comparisons, while valuable and informative, are limited because they cannot adequately separate divergence that occurs during speciation from divergence after speciation. Therefore, it is also important to study differentiation among populations within species in relation to their genetic divergence. For example, there is strong assortative mating among some population pairs in the salamander *Desmognathus ochrophaeus* which is related to the degree of allozyme divergence (Tilley *et al.* 1990). Both ethological isolation and genetic distance are strongly related to geographic distance suggesting that they both reflect accumulation of genetic differentiation in allopatry, as for the allopatric between-species comparisons in *Drosophila* but again in contrast to the sympatric species pairs. Isolation is asymmetric at intermediate levels of divergence in accordance with the expectation from overlapping signal and preference distributions (Arnold *et al.* 1996). Another large study, with the túngara frog *Physalaemus pustulosus*, shows similar strong correlations between geographic separation of samples and signal trait divergence, in this case using directly measured features of acoustic signals (Ryan *et al.* 1996). Again, genetic distance is also strongly correlated with geographic distance, but geographic distance remains a better predictor of signal divergence than genetic distance.

In contrast to these studies, investigation of genetic and phenotypic divergence in the meadow grasshopper *Chorthippus parallelus* suggests that quantitative traits can vary extensively, independent of genetic variation and of each other (Butlin and Tregenza 1998). Seven populations from around Europe were sampled and F_1 individuals reared under standard conditions in the laboratory. A population phylogeny, based on nuclear DNA sequence (Cooper *et al.* 1993, 1995) allows comparison of the pattern of genetic variation with variation of other traits. There are correlations between genetic divergence and morphological (but not size) differences between pairs of populations, while signal traits (male song and cuticular contact pheromone composition) show little sign of any association. Additionally, reproductive isolation measured directly using assortative mating experiments, fails to show any correlation with genetic divergence, although it is correlated with both morphology and cuticular composition. Overall, it appears that in *C. parallelus*, at least some aspects of morphological divergence are associated with genetic divergence, but that signal traits have changed independently, and have apparently influenced prezygotic isolation along the way. Signal trait divergence in *C. parallelus* clearly cannot be explained simply as a by-product of genetic divergence.

A possible explanation for independent divergence in signal traits is that they are subject to sexual selection. This possibility has been examined in the brown planthopper, *Nilaparvata lugens*, where male 'songs' (produced by shaking the plant stem) and female responses have both been shown to vary between populations. Sequence data indicate that this divergence has occurred relatively recently. However, in contrast to the pattern expected if sexual selection were driving this divergence, signals have diverged more than preferences between populations. A quantitative genetic analysis of male and female signal differences between populations suggests that from one to 16 loci contribute to the divergence (Butlin 1996). However, the maximum effect of a single allelic substitution in the male signal is still small relative to the range of signals acceptable to an individual female, suggesting that rapid signal divergence in *Nilaparvata lugens* is most likely to be the result of genetic drift. This may be aided by the population structure of brown planthopper which, especially on cultivated rice, involves repeated colonisations by small numbers of individuals followed by rapid population growth, although this does not explain conservatism in other characters. Similar conclusions might be reached for rapid song divergence in the *Laupala* crickets of Hawaii (Shaw 1996). It may be possible for signal characters to drift because of the simultaneous drift in female preferences, as suggested by Lande's (1981) model of the Fisher runaway process: unlike characters under stabilising selection due to adaptation to the external environment, signal/response systems generate their own optima. This process may also have operated in the salamanders discussed above, but more slowly.

2.8 Conclusions

Species are not equivalent units of biological diversity. They range widely in their genetic variability, both within and among populations, and in their genetic similarity to other species. By far the largest database available at present to describe the patterns of genetic variation within and among species is the massive accumulation of studies using allozymes. Unfortunately, allozyme heterozygosity within species and allozyme genetic distances between species are generally only a poor guide to other types of genetic variation.

The evidence at present suggests that allozyme variation is not strictly neutral: some polymorphisms are maintained by strong selection and many more are influenced by weak selection. Nevertheless, heterozygosity within populations is correlated with effective population size and genetic distance between populations or species does accumulate in roughly clock-like fashion. General phenetic distance and postzygotic reproductive isolation also appear to increase steadily with time of independent evolution and thus correlate with allozyme-based genetic distance. This is probably because both trait groups are influenced by many genes that are individually under weak selection at most, like allozymes. If speciation were primarily a result of these traits, genetic distance between species and variation within species might be more consistent. But, speciation is actually dominated by the evolution of prezygotic isolation. The characters involved in prezygotic isolation frequently experience rather different patterns of selection: signal traits may experience periods of strong directional selection, or be under stabilising selection but with a shifting optimum dependent primarily on the preferences of the other sex. They may evolve rapidly and in unpredictable directions, and can result in rapid speciation at low levels of general genetic, including allozyme, divergence. This is especially true in sympatry; in allopatry prezygotic isolation may reflect general divergence more closely as in the salamander or túngara frog examples or in *Drosophila*, but this does not seem to be true of brown planthoppers or meadow grasshoppers. Perhaps this discrepancy is a result of the shorter time scales involved in the planthopper and grasshopper examples.

Within species, signal characters under directional selection are expected to show different patterns of variation within populations from other traits, and genetic variation in signal traits is not expected to be correlated with allozyme heterozygosity even if there is such a correlation for genetic variation in other traits. Populations may vary markedly in signal traits, and resulting assortative mating, without relation to patterns of genetic relationship among populations.

From the point of view of biodiversity, the outcome of speciation depends on the ecological differentiation of the resulting species. If speciation is driven mainly by prezygotic isolation, then it is likely to be de-coupled from ecological differentiation. Many species pairs will initially be allopatric or parapatric and contribute to beta or gamma, rather than alpha diversity. In contrast, ecological separation may drive the origin of prezygotic isolation in some cases (Schluter

1996). How ecological differentiation relates to allozyme divergence is an unanswered question, but it is likely that any connection will be weak. Understanding biodiversity, and preserving it, requires a knowledge of genetic variability in these critical character sets—mating signal systems and ecologically relevant traits—within and between populations and species. It is not enough either to count species, or to measure variation in marker loci, be they allozymes or the latest sequence based markers.

Acknowledgements

We are very grateful to Mike Ritchie for allowing us to use his unpublished data and for his comments on a draft of the manuscript, and to all members of the Ecology and Evolution Programme in Leeds for useful discussions. Vicky Pritchard and Adam Trickett have contributed to the work on grasshoppers and planthoppers which has been supported by the Royal Society, N.E.R.C. and B.B.S.R.C.

References

Aquandro, C. F. 1991. Molecular population genetics of *Drosophila*. In: *Molecular approaches to fundamental and applied entomology* (ed. J. Oakeshott and M. J. Whitten), pp. 222–266. New York: Springer-Verlag.

Arnold, S. J., Verrell, P. A. and Tilley, S. G. 1996. The evolution of asymmetry in sexual isolation: a model and a test case. *Evolution* **50**, 1024–1033.

Avise, J. C. 1994. *Molecular markers, natural history and evolution*. London: Chapman and Hall.

Barton, N. H. and Turelli, M. 1989. Evolutionary quantitative genetics: how much do we know? *Annu. Rev. Genet.* **23**, 337–370.

Basolo, A. L. 1996. The phylogenetic distribution of a female preference. *Syst. Biol.* **45**, 290–307.

Britten, H. B. 1996. Meta-analyses of the association between multilocus heterozygosity and fitness. *Evolution* **50**, 2158–2164.

Butlin, R. K. 1987. Speciation by reinforcement. *Trends Ecol. Evol.* **2**, 8–13.

Butlin, R. K. 1993. The variability of mating signals and preferences in the brown planthopper, *Nilaparvata lugens*. *J. Insect Behav.* **6**, 125–140.

Butlin, R. K. 1996. Co-ordination of the sexual signalling system and the genetic basis of differentiation between populations in the brown planthopper, *Nilaparvata lugens*. *Heredity* **77**, 369–377.

Butlin, R. K. 1997. What do hybrid zones in general, and the *Chorthippus parallelus* zone in particular, tell us about speciation? In: *Endless forms: species and speciation* (ed. D. J. Howard and S. Berlocher). New York: Oxford University Press (in press).

Butlin, R. K. and Hewitt, G. M. 1985. A hybrid zone between *Chorthippus parallelus parallelus* and *Chorthippus parallelus erythropus* (Orthoptera: Acrididae): behavioural characters. *Biol. J. Linnean Soc.* **26**, 287–299.

Butlin, R. K. and Hewitt, G. M. 1987. Genetic divergence in the *Chorthippus parallelus* species group (Orthoptera: Acrididae). *Biol. J. Linnean Soc.* **31**, 301–310.

Butlin, R. K. and Tregenza, T. 1997. Is speciation no accident? *Nature* **387**, 551–553.
Butlin, R. K. and Tregenza, T. 1998. Levels of genetic polymorphism: marker loci versus quantitative traits. *Phil. Trans. R. Soc. London Ser. B* 353, 187–198.
Claridge, M. F. 1985. Acoustic signals in the Homoptera. *Annu. Rev. Entomol.* **30**, 297–317.
Claridge, M. F., Dawahk, H. A. and Wilson, M. R. 1997. *Species: The units of biodiversity*. London: Chapman and Hall.
Claridge, M. F., Den Hollander, J. and Morgan, J. C. 1985a. The status of weed-associated populations of the brown planthopper, *Nilaparvata lugens* (Stål)—host race or biological species? *Zool. J. Linnean Soc.* **84**, 77–90.
Claridge, M. F., Den Hollander, J. and Morgan, J. C. 1985b. Variation in courtship signals and hybridisation between geographically definable populations of the rice brown planthopper, *Nilaparvata lugens* (Stål). *Biol. J. Linnean Soc.* **24**, 35–49.
Claridge, M. F., Den Hollander, J. and Morgan, J. C. 1988. Variation in hostplant relations and courtship signals of weed-associated populations of the brown planthopper, *Nilaparvata lugens* (Stål), from Australia and Asia: a test of the recognition species concept. *Biol. J. Linnean Soc.* **35**, 79–93.
Cooper, S. J. B. and Hewitt, G. M. 1993. Nuclear DNA sequence divergence between parapatric subspecies of the grasshopper *Chorthippus parallelus*. *Insect Mol. Biol.* **2**, 185–194.
Cooper, S. J. B., Ibrahim, K. M. and Hewitt, G. M. 1995. Postglacial expansion and genome subdivision in the European grasshopper *Chorthippus parallelus*. *Mol. Ecol.* **4**, 49–60.
Coyne, J. A. and Orr, H. A. 1989. Patterns of speciation in *Drosophila*. *Evolution* **43**, 362–381.
Coyne, J. A. and Orr, H. A. 1997. 'Patterns of speciation in *Drosophila*' revisited. *Evolution* **51**, 295–303.
Dagley, J. R., Butlin, R. K. and Hewitt, G. M. 1994. Divergence in morphology and mating signals, and assortative mating among populations of *Chorthippus parallelus* (Orthoptera: Acrididae). *Evolution* **48**, 1202–1210.
Dobzhansky, Th. 1937. *Genetics and the origin of species*. New York: Columbia University Press.
Ferris, C., Rubio, J. M., Serrano, L., Gosalvez, J. and Hewitt, G. M. 1993. One way introgression of a subspecific sex chromosome marker in a hybrid zone. *Heredity* **71**, 119–129.
Gilbert, D. G. and Starmer, W. T. 1985. Statistics of sexual isolation. *Evolution* **39**, 1380–1383.
Gillespie, J. H. 1991. *The causes of molecular evolution*. New York: Oxford University Press.
Gilmartin, A. J. 1980. Variations within populations and classification. II. Patterns of variation within Asclepiadaceae and Umbelliferae. *Taxon* **29**, 199–212.
Gornal, R. J. 1997. Practical aspects of the species in plants. In *Species: The units of biodiversity* (ed. M. F. Claridge, H. A. Dawah and M. R. Wilson), pp. 171–190. London: Chapman and Hall.
Hewitt, G. M. 1993. Postglacial distribution and species substructure: lessons from pollen, insects and hybrid zones. In *Evolutionary Patterns and Processes* (ed. D. R. Lees and D. Edwards), pp. 97–123. London: Academic Press.
Hewitt, G. M. 1996. Some genetic consequences of ice ages, and their role in divergence and speciation. *Biol. J. Linnean Soc.* **48**, 247–276.
Hoelzel, A. R., Halley, J., O'Brien, S. J., Campagna, C., Arnbom, T. Le Boeuf, B., Ralls, K. and Dover, G. A. 1993. Elephant seal genetic variation and the use of simulation models to investigate historical population bottlenecks. *J. Hered.* **84**, 443–449.

Houle, D. 1989. The maintenance of polygenic variation in finite populations. *Evolution* **43**, 1767–1780.

Houle, D. 1992. Comparing evolvability and variability of quantitative traits. *Genetics* **130**, 195–204.

Hudson, R. R., Kreitman, M. and Aguadé, M. 1987. A test of neutral molecular evolution based on nucleotide data. *Genetics* **116**, 153–159.

Humphries, C. J., Williams, P. H. and Vane-Wright, R. I. 1995. Measuring biodiversity value for conservation. *Rev. Ecol. Syst.* **26**, 93–111.

Jones, P. L. 1994. Molecular studies on *Nilaparvata*. Ph.D. Thesis, University of Wales.

Jones, P. L., Gacesa, P. and Butlin, R. K. 1996. Systematics of brown planthopper and related species using nuclear and mitochondrial DNA. In *The ecology of agricultural pests: biochemical approaches* (ed. W. O. C. Symondson and J. E. Liddell), pp. 133–148. London: Chapman and Hall.

Kelly-Stebbins, A. F. and Hewitt, G. M. 1972. The laboratory breeding of British Gomphocerine grasshoppers (Acrididae: Orthoptera). *Acrida* **1**, 233–245.

Kreitman, M. and Akashi, H. 1995. Molecular evidence for natural selection. *Annu. Rev. Ecol. Syst.* **26**, 403–422.

Lande, R. 1976. The maintenance of genetic variation by mutation in a polygenic character with linked loci. *Genet. Res.* **26**, 221–235.

Lande, R. 1981. The minimum number of genes contributing to quantitative variation between and within populations. *Genetics* **99**, 541–553.

Lande, R. 1988. Genetics and demography in biological conservation. *Science* **241**, 1455–1460.

Li, W.-H. and Sadler, L. A. 1991. Low nucleotide diversity in man. *Genetics* **129**, 513–523.

Magurran, A. E. 1988. *Ecological diversity and its measurement*. London: Croom Helm.

Mallett, J. 1996. The genetics of biological diversity: from varieties to species. In *Biodiversity: A biology of numbers and difference* (ed. K. J. Gaston), pp. 13–53. Oxford: Blackwell Science.

Mason, D., Butlin, R. K. and Gacesa, P. 1995. Mitochondrial DNA variation in the *Chorthippus biguttulus* group. *Mol. Ecol.* **4**, 121–126.

May, R. M. 1990. Taxonomy as destiny. *Nature* **347**, 129–130.

Mayden, R. L. 1997. A hierarchy of species concepts: the denouement in the sage of the species problem. In *Species: The units of biodiversity* (ed. M. F. Claridge, H. A. Dawah and M. R. Wilson), pp. 381–424. London: Chapman and Hall.

Mayr, E. 1992. A local flora and the biological species concept. *Am. J. Bot.* **79**, 222–238.

McDonald, J. H. and Kreitman, M. 1991. Adaptive evolution at the *Adh* locus in *Drosophila*. *Nature* **351**, 652–654.

Meyer, J. and Elsner, N. 1996. How well are frequency sensitivities of grasshopper ears tuned to species-specific song spectra? *J. Exp. Biol.* **199**, 1631–1642.

Muller, H. J. 1942. Isolating mechanisms, evolution and temperature. *Biol. Symp.* **6**, 71–125.

Neems, R. M. and Butlin, R. K. 1994. Variation in cuticular hydrocarbons across a hybrid zone in the grasshopper *Chorthippus parallelus*. *Proc. R. Soc. London Ser. B* **257**, 135–140.

Neems, R. M. and Butlin, R. K. 1995. Divergence in cuticular hydrocarbons between parapatric subspecies of the meadow grasshopper *Chorthippus parallelus* (Orthoptera: Acrididae). *Biol. J. Linnean Soc.* **54**, 139–149.

Neil, M. and Graur, D. 1984. Extent of protein polymorphism and the neutral mutation theory. *Evol. Biol.* **17**, 73–118.

Nevo, E., Beiles, A. and Ben-Shlomo, R. 1984. The evolutionary significance of genetic diversity: Ecological, demographic and life history correlates. In *Evolutionary dynamics of genetic diversity* (ed. G. S. Mani), pp. 13–213. Berlin: Springer-Verlag.

Pomiankowski, A. and Møller, A. P. 1995. A resolution of the lek paradox. *Proc. R. Soc. London Ser. B* **260**, 21–29.

Proctor, H. C. 1992. Sensory exploitation and the evolution of male mating behaviour: a cladistic test using water mites (Acari: *Parasitengona*). *Anim. Behav.* **44**, 745–752.

Ritchie, M. G. 1990. Does song contribute to assortative mating between subspecies of *Chorthippus parallelus* (Orthoptera: Acridadae)? *Anim. Behav.* **39**, 685–691.

Ritchie, M. G. and Gleason, J. M. 1995. Rapid evolution of courtship song pattern in *Drosophila willistoni* sibling species. *J. Evol. Biol.* **8**, 463–479.

Roff, D. A. and Mousseau, T. A. 1987. Quantitative genetics and fitness: lessons from *Drosophila. Heredity* **58**, 103–118.

Rowe, L. and Houle, D. 1996. The lek paradox and the capture of genetic variance by condition dependent traits. *Proc. R. Soc. London Ser. B* **263**, 1415–1421.

Ryan, M. J. and Rand, A. S. 1993. Sexual selection and signal evolution: the ghost of biases past. *Phil. Trans. R. Soc. London Ser. B* **340**, 187–195.

Ryan, M. J., Rand, A. S. and Weight, L. A. 1996. Allozyme and advertisement call variation in the túngara frog, *Physalaemus pustulosus. Evolution* **50**, 2435–2453.

Schluter, D. 1996. Ecological causes of adaptive radiation. *Am. Nat.* **148**, S40–S64.

Schwaegerle, K. E., Garbutt, K. and Bazzaz, F. A. 1986. Differentiation among nine populations of *Phlox*. I. Electrophoretic and quantitative variation. *Evolution* **40**, 506–517.

Shaw, K. L. 1996. Polygenic inheritance of a behavioural phenotype: interspecific genetics of song in the Hawaiian cricket genus *Laupala. Evolution* **50**, 256–266.

Skibinski, D. O. F., Woodwark, M. and Ward, R. D. 1993. A quantitative test of the neutral theory using pooled allozyme data. *Genetics* **135**, 233–248.

Storfer, A. 1996. Quantitative genetics: a promising approach for the assessment of genetic variation in endangered species. *Trends Ecol. Evol.* **11**, 343–348.

Templeton, A. R. 1981. Mechanisms of speciation: a population genetic approach. *Annu. Rev. Ecol. Syst.* **12**, 23–48.

Thorpe, J. P. 1982. The molecular clock hypothesis: Biochemical evolution, genetic differentiation and systematics. *Annu. Rev. Ecol. Syst.* **13**, 139–168.

Tilley, S. G., Verrell, P. A. and Arnold, S. J. 1990. Correspondence between sexual isolation and allozyme differentiation: A test in the salamander *Desmognathus ochrophaeus. Proc. Natl Acad. Sci. USA* **87**, 2715–2719.

Trickett, A. J. 1995. Sexual selection and the brown planthopper, *Nilaparvata lugens*. Ph.D. thesis, University of Leeds.

Vogler, A. P. and DeSalle, R. 1994. Diagnosing units of conservation management. *Cons. Biol.* **8**, 354–363.

Vrijenhoek, R. C. 1994. Unisexual fish: model systems for studying ecology and evolution. *Annu. Rev. Ecol. Syst.* **25**, 71–96.

Ward, R. D., Skibinski, D. O. F. and Woodwark, M. 1992. Protein heterozygosity, protein structure, and taxonomic differentiation. *Evol. Biol.* **26**, 73–160.

Watt, W. B. 1994. Allozymes in evolutionary genetics: self-imposed burden or extraordinary tool? *Genetics* **136**, 11–16.

Watt, W. B. 1995. Allozymes in evolutionary genetics: beyond the twin pitfalls of 'neutralism' and 'selectionism'. *Rev. Suisse Zool.* **102**, 869–882.

Watt, W. B., Donohue, K. and Carter, P. A. 1996. Adaptation at specific loci. VI. Divergence vs. parallelism of polymorphic allozymes in molecular function and the fitness-component effects among *Colias* species (Lepidoptera, Pieridae). *Mol. Biol. Evol.* **13**, 699–709.

West-Eberhard, M. J. 1983. Sexual selection, social competition, and speciation. *Q. Rev. Biol.* **58**, 155–183.

Whittaker, R. M. 1960. Vegetation of the Siskiyou Mountains, Oregon and California. *Ecol. Monogr.* **30** 279–338.

Wu, C.-I. and Palopoli, M. F. 1994. Genetics of postmating reproductive isolation in animals. *Annu. Rev. Genet.* **27**, 283–308.

Young, A., Boyle, T. and Brown, T. 1996. The population genetic consequences of habitat fragmentation for plants. *Trends Ecol. Evol.* **11**, 413–418.

3

Genetic diversity in natural populations

John Harwood and William Amos

3.1 Introduction

The Convention on Biological Diversity requires signatories to protect diversity at all levels from the genome to the ecosystem. This acknowledgement of the importance of genetic diversity within species is only right and proper, since this diversity is the base material on which all variety ultimately depends. Genetic diversity (whether measured by levels of heterozygosity or number of alleles per locus) is usually correlated with fitness (Mitton and Grant 1984; Allendorf and Leary 1986), and populations which are genetically diverse should be more able to respond to parasites, predators, and environmental change than their genetically impoverished counterparts. However, levels of genetic variability differ markedly between major taxonomic groups (Ward *et al.* 1992), and even between closely related species within the same taxon. Some species which are predicted to be genetically impoverished are not, while others harbour much less variability than expected. Similarly, there are exceptions to the rule that high levels of genetic variability equate to evolutionary health: some populations with low genetic diversity are increasing exponentially and others are declining, although they possess great variability.

Changes in the level of genetic diversity within a species or population are a consequence of the balance between the opposing processes of gain and loss. Diversity may be lost passively as a result of genetic drift, or actively through natural selection; it can be gained through mutation or gene flow from neighbouring populations or species. An important conclusion can be drawn from this simple summary: large or rapid changes in genetic variability are more likely to be the result of loss than gain, because the processes which increase variability are either slow (*de novo* mutations) or rare (influx from a neighbouring population or species).

In this chapter we briefly review the factors which are known to affect drift, selection, mutation and gene flow within populations. In the process, we examine the evidence that these factors can account for observed levels of diversity. Where there appear to be inconsistencies, we suggest additional mechanisms which may help to explain observed levels of diversity.

3.2 Genetic drift and the effect of population bottlenecks

The rate of loss of diversity from a population through genetic drift is inversely proportional to its variance effective size (N_e). As a result, all other things being equal, small populations carry less selectively neutral genetic diversity than equivalent large ones. The variance effective size is related to the total population size; but the nature of this relationship depends on inter-individual variation in offspring number, which may be affected by sex ratios, mating systems, generation time and spatial structuring. In certain special cases (for example when all individuals contribute equally to the next generation), the effective population size can be larger than the number of adult animals in the population. However, variance effective population size is usually between 0.25 and 1.0 times the number of adults (Nunney 1995). Under many circumstances, effective population size is close to one-half the number of adults (Nunney 1993). These calculations assume that matings occur at random between the individuals which are involved in reproduction. This is not always the case (see Sections 3.3 and 3.4).

In many cases, the expected positive relationship between genetic variability and population size is supported by empirical data (Tregenza and Butlin, Chapter 2; Young *et al.* 1996). However, there are a number of examples where variability is substantially less than would be expected from current population size (Prober and Brown 1994; Raijmann *et al.* 1994; Tregenza and Butlin, Chapter 2). These discrepancies are usually attributed to one or more severe reductions in size during the populations' histories, known as 'bottlenecks'. For example, eight of the 11 studies published in the journal *Conservation Biology* in 1996 and the first half of 1997 which included genetic diversity or variability in their title reported low levels of variability at some loci. In all cases this was attributed to an undocumented or poorly documented bottleneck in the population's history. Of the three studies which reported 'normal' levels of variability, one concluded that this was because the population in question had not gone through a bottleneck!

It is certainly true that 'bottlenecks' can have a dramatic effect on the rate of genetic drift because the effective size of a fluctuating population is given by its harmonic mean, which lies closer to the minimum than the mean value attained (Wright 1938), and because alleles are lost by chance during the process of reduction resulting in founder effects. However, few studies attempt to determine what size of bottleneck was necessary to explain the observed levels of variability. In practice, extreme bottlenecks may be required. The cheetah (*Acinonyx jubatus*) is thought to have lost '90–99%' of its allozyme variability (O'Brien 1994), yet its historical distribution covered all of Africa and much of the Middle East. To lose 99% heterozygosity by drift alone would require, for example, N_e to be reduced to two individuals for 16 generations.

Even in cases where the size of the historical bottleneck can be estimated, its explanatory power is often limited. The northern elephant seal (*Mirounga angustirostris*) also shows a near absence of genetic variability at certain loci. It was heavily exploited in the nineteenth century, culminating in a final large catch of 153 animals in 1854 (Townsend 1885). Seven more were killed in 1892, and small numbers were taken after this (Stewart *et al.* 1994). By 1922 the population had recovered to 350, and it currently exceeds 150 000 (Stewart *et al.* 1994). Hoelzel *et al.* (1993) estimate that the near absences of mitochondrial DNA variability in this species could have been generated by a bottleneck of less than 30 seals lasting for one generation. However, models of this kind do not appear to explain the extremely low levels of allozyme variability in the species (Hedrick 1996).

In many instances, low levels of genetic variability in some populations can be explained by mechanisms other than drift. For example, highly significant differences in heterozygosity exist between populations of the guppy (*Poecilia reticulata*) in upstream and downstream sections of the same river systems in Trinidad (Shaw *et al.* 1994). It is tempting to attribute this to drift, since downstream populations have a higher effective population size than upstream ones. However, a systematic loss of alleles is observed as sampling progresses upstream (Magurran *et al.* unpublished). This is more consistent with the erosion of diversity from upstream areas through unidirectional gene flow downstream than with drift.

A careful examination of the nature of the low levels of genetic diversity observed in populations which are believed to have passed through a bottleneck may help in identifying its causes. For example, it appears that the loss of allelic diversity is a more sensitive indicator of historical bottlenecks than reduction in heterozygosity (Leberg 1992; Brookes *et al.* 1997). An extreme bottleneck of two individuals will reduce heterozygosity by 25%, but will leave a maximum of four alleles at any given locus. The consequences of this are particularly evident in DNA fingerprints, which contain information from many hypervariable minisatellite markers. Each hypervariable locus may carry 100 or more alleles. As a result, the reduction in allelic diversity following a bottleneck can be readily detected. In a classic study of the Californian Channel Islands fox (*Urocyon littoralis*), Gilbert *et al.* (1990) showed that each island had a unique consensus banding pattern which was more or less shared by all individuals but which differed greatly between islands.

All of this discussion rests on the assumption that levels of genetic diversity which are described as 'low' in the literature are really low. Amos and Harwood (1998) have described how the decision-making process leading to the publication of scientific manuscripts can result in an inflated view of the amount of variability which is likely to be present in an undepleted population. For example, researchers studying breeding behaviour will search for genetic markers with high levels of variability and are likely to publish their results rapidly (Avise *et al.* 1989; Turner *et al.* 1992), while persistent failure to find variability is unlikely to be reported in the literature. The net result is an upward bias in the reported level of polymorphism, both at the level of individual markers and at the level of the species, and a perception that 'normal' levels of genetic variability are higher than they really are.

3.3 Selection

Even allowing for publication biases, it is clear that some species exhibit unusually low levels of variability. For example 4% of plants taxa (Hamrick and Godt 1989) and 5% of animal species examined (Nevo *et al.* 1984) show no allozyme polymorphism. In many case, this depletion cannot be explained by drift alone, and selective processes must have accelerated the rate of loss of variability above neutral expectations. Large-scale loss of variation through selection can occur as a result of inbreeding and its avoidance during bottlenecks, and through selective sweeps at larger population sizes.

Inbreeding depression is the name given to the reduction in fitness or viability resulting from the increased expression of deleterious recessive alleles at low population sizes because of the increased probability of matings between closely related individuals (Saccheri *et al.* 1996). Large populations are able to carry disproportionately more deleterious recessive alleles than smaller populations. If a population declines, the excess tends to be purged by selection (Barrett and Charlesworth 1991). In many species, behavioural mechanisms can minimize the effects of inbreeding, either by dispersal patterns which ensure that relatives tend not to meet when pairs are forming (Gilbert *et al.* 1991) or by some form of kin avoidance (Amos *et al.* 1993). Such behaviours are referred to as inbreeding avoidance.

Inbreeding depression and inbreeding avoidance filter the current gene pool in different ways: depression acts on individual genes and gene complexes while avoidance acts on whole sets of chromosomes. Both processes eventually result in greater levels of outbreeding, which may serve to maintain genetic diversity. However, some genetic diversity is inevitably lost as a result of inbreeding depression. The net effect of inbreeding will depend on what form of social organization operates and the magnitude of any population decline which occurs during purging.

As yet there is little evidence that either inbreeding depression or inbreeding avoidance play a major role in natural populations, although there is now good experimental field evidence that individuals from inbred populations have lower fitness than conspecifics from outbred populations (Chen 1993; Jimenez *et al.* 1994; Keller *et al.* 1994). The effects of inbreeding depression in controlled laboratory experiments may be quite strong. However, any purging is probably brief, and the consequences are rather slight if the inbreeding depression results from many genes, each of which exerts a small effect (Hedrick 1994).

If inbreeding depression is unlikely to account for the large-scale reduction in genetic diversity that is observed in some species, are there other mechanisms which can cause accelerated loss through natural selection? Meiotic drive is such a mechanism, since it is capable of generating change in many different parts of the genome, either simultaneously or in quick succession, through genome-wide selective sweeps. The way in which meiotic drive can have these effects, and can facilitate or even promote speciation, is described in more detail in Amos and

Harwood (1998). Here, we merely summarize the way in which it may explain the low levels of genetic diversity observed in some mammal species.

A genome-wide selective sweep requires simultaneous or sequential selection on many loci scattered across many or most chromosomes. There appear to be two possible ways in which these conditions might be met. Most obvious would be a Fisherian runaway process such as sexual selection, in which polygenic traits such as plumage patterns come under pressure for constant change. However, it is unclear how many chromosomes would be directly involved. Perhaps more interesting is the possibility of a molecular arms race forcing change in a key control gene forcing coevolution in widely scattered dependent genes.

Meiotic drive is the name given to the theoretical battle between the sex chromosomes in species where one sex is heterogametic (Crow 1979). In the male mammalian germ line, the Y chromosome only has a future in a male foetus and the X chromosome only ever in a female foetus. Thus, each sex chromosome is only half as fit as it could be if all progeny were males for the Y, or females for the X. Any mutation on one sex chromosome which debilitates the other will have a selective advantage, and this process creates a constant state of tension between the two chromosomes.

In mammals there are three X chromosomes for each Y, and it therefore seems reasonable to suppose that, in any such battle, the X will tend to be on the offensive and the Y on the defensive. A logical consequence is that the Y should shed any transcribed sequences which are not essential to its function and, indeed, the mammalian Y chromosome is degenerate (Morell 1994; Graves 1995), probably having fewer than 20 active genes. This degeneracy is usually explained as a consequence of the accumulation of deleterious mutations which begins as soon as the lack of recombination sets Müller's ratchet in motion (Fisher 1935; Charlesworth 1991). This process, while detectable (Rice 1994), appears to progress rather slowly (Allen and Ostrer 1994; Charlesworth *et al.* 1997). However, genes from the mammalian Y have been lost at an average rate which is one gene per 100 000 years. In fact, since all major mammalian radiations have reached similar levels of Y degeneration, it seems likely that most of the gene loss occurred early (Bull 1983), suggesting that early mammals may have shed tens of genes per speciation event.

Consider how the battle might run. Most of the time the two sex chromosomes are in unstable equilibrium and the sex ratio is 1 : 1. Any mutation which confers greater concealment to the Y or greater dominance to the X will increase the fitness of the chromosome on which it arose and will tend to sweep through to fixation. As the sweep progresses, the sex ratio will become distorted and, according to Fisherian principles, this will increase the selective advantage to any mutation on the opposing chromosome which redresses the sex ratio. However, the resulting change is unlikely to result in a perfect balance, and so a train of change and counter-change could follow.

A particularly interesting case would involve a non-silent substitution in the sex-determining gene SRY. SRY appears to control the expression of many male traits, presumably including secondary sexual characteristics. Consequently, a slight change in the SRY protein could affect the expression of a range of genes associated with male behaviour and appearance. Once fixed in the population, selection would then favour changes in the genes controlled by SRY so as to nullify or accommodate any detrimental effects associated with the original mutation. This will result in a genome-wide selective sweep which may cause incompatibilities between the sex chromosomes in the population unit where the event took place and their homologues in other population units. These changes could therefore facilitate speciation.

This model is highly speculative and applies only to species, such as mammals, with single gene sex determination. Nevertheless, some of the model's predictions (for example, that levels of Y-chromosome variability should be extremely low within a species, and that hybrid species should show sex ratio distortion, lowered sperm counts and infertility) are supported by empirical data (Amos and Harwood 1998). In addition, it demonstrates how periodic selective sweeps could result when an evolutionary arms race results in change on a control gene, which forces downstream genes to co-adapt.

In this sense, meiotic drive is a specific example of a more general model. Turning to the more general case, it has long been recognized that coevolution between genes within the genome has the potential to reduce hybrid fitness and hence to contribute to post-zygotic isolation (Dobzhansky 1937; Muller 1942). However, previous models have been based on the concept that mutations which become fixed in one genetic background may be detrimental in another. While logically compelling, this model does not seem powerful enough to explain the large numbers of strong, independent incompatibilities which, for example, separate closely related species of *Drosophila* (Davis and Wu 1996). More plausibly, these profound incompatabilities could have arisen by sequential rounds of change in a control gene followed by forced coevolution among the genes being controlled. In *Drosophila*, lack of sex ratio distortion among hybrids suggests that meiotic drive is not involved (Johnson and Wu 1992), a conclusion which is consistent with the lack of single gene sex determination and non-degenerate Y chromosome. Consequently, we would have to invoke related processes acting on other control genes. This is clearly an exciting area for future research.

3.4 Mutation and heterozygote instability

Ultimately, all genetic variability is generated by the process of mutation. Low levels of variability could therefore result from a reduction in the rate of mutation. This could, for example, occur following a change in some critical enzyme involved with DNA replication which either reduces the rate at which errors occur or improves the efficiency with which errors are corrected. Indeed, mutation rate evolution has been demonstrated in bacteria (Sniegowski *et al.* 1997). However, a change in mutation rate does not seem a plausible explanation for the low levels of genetic diversity observed in some mammal species, because their close relatives usually have relatively 'normal' levels of variability. Even in the unlikely event of the mutation rate falling suddenly to zero, the subsequent loss of variability through genetic drift will be relatively slow.

An alternative explanation for why small populations might appear genetically impoverished arises from recent work on microsatellite DNA sequences. This suggests that, rather than smaller populations having less variability, larger populations may have disproportionately more variability. The evidence for this is described more extensively elsewhere (Rubinsztein *et al.* 1995; Amos and Rubinsztein 1996; Amos and Harwood 1998), but is summarized briefly here.

Every base pair in the genome has a finite opportunity to mutate during DNA replication. However, extensive regions of heteroduplex DNA are formed between paired homologous chromosomes during meiosis. There is evidence that heterozygote sites are repaired in this state, providing a greater opportunity for mutation in heterozygotes than homozygotes. Since heterozygosity tends to increase with population size, large populations will gain genetic diversity more rapidly than smaller ones.

There is some direct evidence that this process does occur. In a human dataset comprising 19 informative mutations, significantly more mutations derived from the parent with the greater difference in length between his/her alleles (Amos and Rubinsztein 1996). Schug *et al.* (1997) found far fewer mutations than expected in homozygous inbred lines of *Drosophila melanogaster*. They estimated the average mutation rate was 10^{-1}–10^{-3} lower than that recorded in other species (Edwards *et al.* 1992; Weber and Wong 1993; Banchs *et al.* 1994).

The clearest evidence that heterozygosity can accelerate the rate of evolution will probably come from studies of microsatellites and other tandemly repeated sequences such as minisatellites. Biased mutation has been recorded in minisatellites (Jeffreys *et al.* 1994; Monckton *et al.* 1994), microsatellites (Weber and Wong 1993; Amos and Rubinsztein 1996; Amos *et al.* 1996; Crawford and Cuthbertson 1996; Primmer *et al.* 1996) and the tracts of triplet repeats responsible for a number of human diseases (Duyao *et al.* 1993; Rubinsztein *et al.* 1994). If mutation rate does vary with population size, expanding populations should harbour loci which are longer than their homologues in smaller populations. Preliminary evidence does conform to this prediction. Human microsatellites (Meyer *et al.* 1995; Rubinsztein *et al.* 1995), minisatellites (Gray and Jeffreys 1991) and triplet repeat disease alleles (Djian *et al.* 1996) all show a clear tendency to be longer than their homologues in chimpanzees, and microsatellites in the abundant barn swallow are consistently longer than their homologues in less abundant, related species (Ellegren *et al.* 1995).

Indirect evidence that heterozygote instability may not be restricted to tandemly repeated DNA sequences comes from genetic studies of hybrid zones, where rare alleles, known as hybrizymes (Barton *et al.* 1983; Woodruff 1989), which are not present in either pure population occur far more frequently than expected by chance. Hybrizymes could arise by recombination between two pre-existing alleles or be the result of an increased mutation rate in hybrids. Current evidence appears to support the mutation hypothesis (Hoffman and Brown 1995), and therefore also the heterozygote instability model because hybrid individuals should show higher levels of heterozygosity than either parent stock.

3.5 Gene flow and the effects of spatial and social structure

The spatial distribution of many species is fragmented, with patches of suitable habitat being separated by large areas of habitat which is either unsuitable or dangerous. The effects of this spatial structure on genetic diversity and evolution have long interested population geneticists. In the simplest case, a population which is sub-divided and where there is no migration between population units may lose allelic diversity more slowly than an undivided population, because different alleles will drift to fixation in each population unit. Even if there is migration, there can be local differentiation in neutral variation if the number of individuals which migrate between population units in each generation is less than a certain level (see review in Barton and Whitlock 1997).

The effect of sub-division on genetic diversity is much reduced if account is taken of the fact that small populations are more vulnerable to extinction than larger ones (McCauley 1991). This is because the greater inherent diversity of a sub-divided population is undermined by the loss of allelic diversity which occurs whenever a population unit goes extinct. If the population is to persist, there must be some finite probability that habitat patches which have been emptied by extinction can be recolonized. The resulting structure of spatially distributed population units which may disappear and then subsequently reappear is now generally known as a metapopulation.

A number of authors (Harrison 1991, 1994; Thomas 1994) have questioned whether the classic metapopulation model has general applicability. However, the populations of some taxonomic groups, such as butterflies (Thomas and Hanski 1997) and many plants (Giles and Goudet 1997), have a structure corresponding to Hanski and Simberloff's (1997) definition of a metapopulation as 'spatially structured into assemblages of local breeding populations and that migration among the local populations has some effect on long dynamics, including the possibility of population reestablishment following extinction'. The latter phenomenon is usually referred to as population 'turnover'.

The variance effective size of an entire metapopulation can be very much lower than the number of mature individuals in it (Hedrick and Gilpin 1997), and it can therefore suffer higher rates of genetic drift than an unstructured population of equivalent size . The extremely low variability observed at some loci in the cheetah may be explicable in this way (Gilpin 1991; Hedrick 1996). Hedrick (1996) has calculated that the observed levels of genetic variability in this species could be the result of a metapopulation structure with a variance effective population size of 200–2000 individuals. These calculations depend on specific assumptions about the way in which individuals move between population units and begs the question of what selective forces would promote a population structure which results in a rapid loss of genetic diversity. The way in which organisms or propagules move between patches is of critical importance in determining how spatial structure affects the stability and dynamics of populations (Rohani *et al.* 1997), as well as its effect on genetic diversity, and this is likely to be a particularly important area for research over the next decade.

Although, periodic extinction and recolonization may increase the rate at which genetic diversity is lost from a population, it usually results in an increase in the genetic variance among populations (Slatkin 1977; Wade and McCauley 1988; Whitlock and McCauley 1990). The magnitude of the effect depends on the way in which vacant sites are colonized. The two extreme cases are the propagule pool and migrant pool models of Slatkin (1977). In the propagule pool model, all colonists come from the same population unit, whereas in the migrant pool model colonists are drawn at random from all population units. The migrant pool model tends to reduce local differentiation compared with the propagule pool model. Whitlock and McCauley (1990) generalized Slatkin's propagule model and showed that the effect of colonization on population differentiation depends on the probability that colonizers have a common origin. Colonization favours differentiation unless the colonization process is very different from the process of migration between extant population units. Whitlock (1992) has shown that spatially variable migration patterns can also increase the degree of population differentiation.

Chesser and his colleagues (Chesser 1991; Sugg *et al.* 1996) have shown that the social organization of a species can result in non-random matings, for example where there is strong philopatry by one sex and where the other sex disperses. As a result of this process, variance effective population size can be at the upper end of the range described in Section 3.2. Where the philopatric sex forms stable social groups (as is the case in many mammals), rare alleles can become fixed within individual groups (Sugg *et al.* 1996). Since the same alleles are unlikely to become fixed in all groups, this process can increase the overall genetic diversity of the population and lead to local adaptation. This is directly analogous to the process which can increase diversity in spatially structured populations and, again, its effect depends critically on the way in which individuals move between population units.

3.6 Conclusions

In this chapter, we have taken a critical look at the extent to which the conventional mechanisms of genetic drift, gene flow, mutation and selection can explain interspecific and interpopulation variation in genetic variability. It seems clear to us that the scientific community is still some way from understanding the basis for this variability. The problem may be, in part, an artefact of the process of scientific publication but, nevertheless, we certainly need to know more about the way in which selection acts to accelerate the rate at which variability is lost from natural populations.

There is a strong tendency in the literature for genetic impoverishment to be attributed to population bottlenecks. However, the relationship between population size and observed levels of neutral variability is not as straightforward as is often assumed. Some relatively abundant species have low genetic diversity while some endangered species have no more or less variability than their more abundant relatives. In many cases where bottlenecks are assumed to have occurred there is little evidence that the species or population in question has spent long enough at sufficiently low numbers to account for the observed levels of variability. Alternative explanations for apparently low levels of variability include the effects of inbreeding and inbreeding avoidance, spatial structure, and the possibility that heterozygotes are more mutable than homozygotes. Some instances of extreme impoverishment in species whose close relatives have shown 'normal' levels of variability may be explained by natural selection resulting in a selective sweep, rather than passive loss through drift. Forced coevolution between a control gene and the loci it regulates could result in such a selective sweep; meiotic drive is one mechanism which could generate such coevolution.

References

Allen, B. S. and Ostrer, H. 1994 Conservation of human Y chromosome sequences among male great apes: implications for the evolution of Y chromosomes. *J. Mol. Evol.* **39**, 13–21.

Allendorf , W. and Leary, R. F. 1986 Heterozygosity and fitness in natural populations of animals. In *Conservation biology: the science of scarcity and diversity* (ed. M. E. Soule), pp. 57–76. Sunderland, MA: Sinauer Associates.

Amos, W. and Harwood, J. 1998. Factors affecting levels of genetic diversity in natural populations. *Phil. Trans. R. Soc. Lond.* **B 353**, 177–186.

Amos, W. and Rubinsztein, D. C. 1996 Microsatellites are subject to directional evolution. *Nature Genet.* **12**, 13–14.

Amos, W., Schlötterer, C. and Tautz, D. 1993 Social structure of pilot whales revealed by analytical DNA typing. *Science* **260**, 670–672.

Amos, W., Sawcer, S. J., Feakes, R. and Rubinsztein, D. C. 1996 Microsatellites show mutational bias and heterozygote instability. *Nature Genet.* **13**, 390–391.

Avise, J. C., Bowen, B. W. and Lamb, T. 1989 DNA fingerprints from hypervariable mitochondrial genotypes. *Mol. Biol. Evol.* **6**, 258–269.

Banchs, I., Bosch, A., Guimerà, J., Lázaro, C., Puig, A. and Estivill, X. 1994 New alleles at microsatellite loci in CEPH families mainly arise from somatic mutations in the lymphoblastoid cell lines. *Hum. Mutat.* **3**, 365–372.

Barrett, S. C. H. and Charlesworth, D. 1991 Effects of a change in the level of inbreeding on the genetic load. *Nature* **352**, 522–524.

Barton, N. and Whitlock, M. C. 1997 The evolution of metapopulations. in: *Metapopulation dynamics: ecology, genetics and evolution* (ed. I. Hanski and M. Gilpin), pp. 183–210. Academic Press, San Diego.

Barton, N. H., Halliday, R. B. and Hewitt, G. M. 1983 Rare electrophoretic variants in a hybrid zone. *Heredity* **50**, 139–146.

Brookes, M. I., Graneau, Y. A., King, P., Rose, O. C., Thomas, C. D. and Mallet, J. L. B. 1997 Genetic analysis of founder bottlenecks in the rare British butterfly *Plebejus argus*. *Conservation Biol.* **11**, 648–661.

Bull, J. J. 1983 *Evolution of sex determining mechanisms.* Menlo Park, CA: Benjamin Cummins,

Charlesworth, B. 1991 The evolution of sex chromosomes. *Science* **251**, 1030–1033.

Charlesworth, B., Charlesworth, D., Hnilicka, J., Yu, A. and Guttman, D. S. 1997 Lack of degeneration of loci on the neo-Y chromosome of *Drosophila americana americana. Genetics* **145**, 989–1002.

Chen, X. 1993 Comparison of inbreeding and outbreeding in hermaphroditic *Arianta arbustorum* (L.) (land snail). *Heredity* **71**, 456–461.

Chesser, R. K. 1991 Influence of gene flow and breeding tactics on gene diversity within populations. *Genetics* **129**, 573–583.

Crawford, A. M. and Cuthbertson, R. P. 1996 Mutations in sheep microsatellites. *Genome Res.* **6**, 876–879.

Crow, J. F. 1979 Genes that violate Mendel's rules. *Sci. Am.* **240**, 134–146.

Davis, A. W. and Wu, C.-I. 1996 The broom of the sorcerer's apprentice: the fine structure of a chromosomal region causing reproductive isolation between two sibling species of *Drosophila. Genetics* **143**, 1287–1298.

Djian, P., Hancock, J. M. and Chana, H. S. 1996 Codon repeats in genes associated with human diseases: Fewer repeats in the genes of nonhuman primates and nucleotide substitutions concentrated at sites of reiteration. *Proc. Natl Acad. Sci. USA* **93**, 417–421.

Dobzhansky, T. 1937 *Genetics and the origin of species*, p. 364. New York: Columbia University Press.

Duyao, M., Ambrose, C., Myers, R., Novelleto, A., Persichetti, F., Frontali, M., Folstein, S., Ross, C., Franz, M., Abbott, M., Gray, J., Conneally, P., Young, A. *et al.* 1993 Trinucleotide repeat length instability and age of onset in Huntington's disease. *Nature Genet.* **4**, 387–392.

Edwards, A., Hammond, H. A., Jin, L., Caskey, C. T. and Chakraborty, R. 1992 Genetic variation at five trimeric and tetrameric tandem repeat loci in four human population groups. *Genomics* **12**, 241–253.

Ellegren, H., Primmer, C. R. and Sheldon, B. C. 1995 Microsatellite evolution: directionality or bias in locus selection. *Nature Genet.* **11**, 360–362.

Fisher, R. A. 1935 The sheltering of lethals. *Am. Nat.* **69**, 446–455.

Gilbert, D. A., Lehman, N., O'Brien, S. J. and Wayne, R. K. 1990 Genetic fingerprinting reflects population differentiation in the California Channel Island fox. *Nature* **344**, 764–767.

Gilbert, D. A., Packer, C., Pusey, A. E., Stephens, J. C. and O'Brien, S. J. 1991 Analytical DNA fingerprinting in lions: parentage, genetic diversity and kinship. *J. Hered.* **82**, 378–386.

Giles, B. E. and Goudet, J. 1997 A case study of genetic structure in a plant metapopulation. In *Metapopulation dynamics: ecology, genetics, and evolution.* (ed. I. Hanski and M. Gilpin), pp. 429–454. New York: Academic Press.

Gilpin, M. 1991 The genetic effective size of a metapopulation. *Biol. J. Linn. Soc.* **42**, 165–175.

Graves, J. A. M. 1995 The origin and evolution of the mammalian Y chromosomes and Y-borne genes—an evolving understanding. *Bioessays* **17**, 311–320.

Gray, I. C. and Jeffreys, A. J. 1991 Evolutionary transience of hypervariable minisatellites in man and the primates. *Proc. R. Soc. London Ser. B* **243**, 241–253.

Hanski, I. and Simberloff, D. 1997 The metapopulation approach, its history, conceptual domain, and application to conservation. In *Metapopulation biology: ecology, genetics and evolution.* (ed. I. Haski and M. E. Gilpin), pp. 5–26. San Diego, CA: Academic Press.

Hamrick, J. L. and Godt, M. J. W. 1989 Allozyme diversity in plant species. In *Plant *population genetics, breeding and genetic resources* (ed. A. H. D. Brown, M. T. Clegg, A. L. Kahler and B. S. Weir).

Harrison, S. 1991 Local extinctions in a metapopulation context: an empirical evaluation. In *Metapopulation dynamics: empirical and theoretical investigations* (ed. M. E. Gilpin and I. Hanski), pp. 73–88. Academic Press, London..

Harrison, S. 1994 Metapopulations and conservation. In *Large-scale ecology and conservation biology* (ed. P. J. Edwards, R. M. May and N. R. Webb), pp. 111–128. Blackwell Sciences Ltd, Oxford.

Hedrick, P. W. 1994 Purging inbreeding depression and the probability of extinction: full-sib matings. *Heredity* **73**, 363–372.

Hedrick, P. W. 1996 Bottleneck(s) or metapopulation in cheetahs. *Conservation Biol.* **10**, 897–899.

Hedrick, P. W. and Gilpin, M. E. 1997 Genetic effective size of a metapopulation. In *Metapopulation dynamics: ecology, genetics, and evolution.* (ed. I. Hanski and M. Gilpin), pp. 166–181. New York: Academic Press.

Hoelzel, A. R., Halley, J. and O'Brien, S. J. 1993 Elephant seal genetic variation and the use of simulation models to investigate historical population bottlenecks. *J. Hered.* **84**, 443–449.

Hoffman, S. M. G. and Brown, W. M. 1995 The molecular mechanism underlying the rare allele phenomenon in a subspecific hybrid zone of the California field-mouse, *Peromyscus californicus. J. Mol. Evol.* **41**, 1165–1169.

Jeffreys, A. J., Tamaki, K., MacLeod, A., Monckton, D. G., Neil, D. L. and Armour, J. A. L. 1994 Complex gene conversion events in germline mutation at human minisatellites. *Nature Genet.* **6**, 136–145.

Jimenez, J. A., Hughes, K. A., Alaks, G., Graham, L. and Lacy, R. C. 1994 An experimental study of inbreeding depression in a natural habitat. *Science* **266**, 271–273.

Johnson, N. A. and Wu, C.-I. 1992 An empirical test of the meiotic drive models of hybrid sterility: sex ratio data from hybrids between *Drosophila simulans* and *Drosophila sechellia. Genetics* **130**, 507–511.

Keller, L. F., Arcese, P., Smith, J. N. M., Hochachka, W. M. and Stearns, S. C. 1994 Selection against inbred song sparrows during a natural population bottleneck. *Nature* **372**, 356–357.

Leberg, P. L. 1992 Effects of population bottlenecks on genetic diversity as measured by allozyme electrophoresis. *Evolution* **46**, 447–494.

McCauley, D. E. 1991 Genetic consequences of local population extinction and recolonization. *Trends Ecol. Evol.* **6**, 5–8.

Meyer, E., Wiegand, P., Rand, S. P., Kuhlmann, D., Brack, M. and Brinkmann, B. 1995 Microsatellite polymorphisms reveal phylogenetic relationships in primates. *J. Mol. Evol.* **41**, 10–14.

Mitton, J. B. and Grant, M. C. 1984 Associations among protein heterozygosity, growth rate, and developmental homeostasis. *Annu. Rev. Ecol. Syst.* **15**, 479–499.

Monckton, D. G., Neumann, R., Guram, T., Fretwell, N., Tamaki, K., MacLeod, A. and Jeffreys, A. J. 1994 Minisatellite mutation rate variation associated with a flanking DNA sequence polymorphism. *Nature Genet.* **8**, 162–170.

Morell, V. 1994 Rise and fall of the Y chromosome. *Science* **263**, 171–172.

Muller, H. J. 1942 Isolating mechanisms, evolution and temperature. *Biol. Symp.* **6**, 71–125.

Nevo, E., Bieles, A. and Schlomo, B. 1984 The evolutionary significance of genetic diversity: ecological, demographic and life history consequences. In *Evolutionary dynamics of genetic diversity* (ed. G. S. Mani).

Nunney, L. 1993 The influence of mating system and overlapping generations on effective population size. *Evolution* **47**, 1329–1341.

Nunney, L. 1995 Measuring the ratio of effective population size to adult numbers using genetic and ecological data. *Evolution* **49**, 389–392.

O'Brien, S. J. 1994 A role for molecular genetics in biological conservation. *Proc. Natl Acad. Sci. USA* **91**, 5748–5755.

Primmer, C., Ellegren, H., Saino, N. and Møller, A. P. 1996 Directional evolution in germline microsatellite mutations. *Nature Genet.* **13**, 391–393.

Prober, S. M. and Brown, A. H. D. 1994 Conservation of the grassy white box woodlands—population genetics and fragmentation of *Eucalyptus albens. Conservation Biol.* **8**, 1003–1013.

Raijmann, L. E. L., Vanleeuwen, N. C., Kersten, R., Oostermeijer, J. G. B., Dennijs, H. G. M. and Menken, S. B. J. 1994 Genetic variation and outcrossing rate in relation to population size in *Gentiana pneumonanthe* L. *Conservation Biol.* **8**, 1014–1026.

Rice, W. R. 1994 Degeneration of a nonrecombining chromosome. *Science* **263**, 230–232.

Rohani, P., May, R. M. and Hassell, M. P. 1997 Metapopulations and equilibrium stability—the effects of spatial structure. *J. Theor. Biol.* **181**, 97–109.

Rubinsztein, D. C., Amos, W., Leggo, J., Goodburn, S., Ramesar, R. S., Old, J., Bontrop, R., McMahon, R., Barton, D. E. and Ferguson-Smith, M. A. 1994 Mutational bias provides a model for the evolution of Huntington's disease and predicts a general increase in disease prevalence. *Nature Genet.* **7**, 525–530.

Rubinsztein, D. C., Amos, W., Leggo, J., Goodburn, S., Jain, S., Li, S. H., Margolis, R. L., Ross, C. A. and Ferguson-Smith, M. 1995 Microsatellites are generally longer in humans compared to their homologues in non-human primates: evidence for directional evolution at microsatellite loci. *Nature Genet.* **10**, 337–343.

Saccheri, I. J., Brakefield, P. M. and Nichols, R. A. 1996 Severe inbreeding depression and rapid fitness rebound in the butterfly *Bicyclus anynana* (Satyridae). *Evolution* **50**, 2000–2013.

Schug, M. D., Mackay, T. F. C. and Aquadro, C. F. 1997 Low mutation rates of microsatellites in *Drosophila melanogaster. Nature Genet.* **15**, 99–102.

Shaw, P. W., Carvalho, G. R., Magurran, A. E. and Seghers, B. H. 1994 Factors affecting the distribution of genetic variability in the guppy, *Poecilia reticulata. J. Fish Biol.* **45**, 875–888.

Slatkin, M. 1977 Gene flow and genetic drift in a species subject to frequent local extinctions. *Theor. Popul. Biol.* **12**, 253–262.

Sniegowski, P. D., Gerrish, P. J. and Lenski, R. E. 1997 Evolution of high mutation rates in experimental populations of *E. coli. Nature* **387**, 703–705.

Stewart, B. S., Yochem, P. K., Huber, H. R., DeLong, R. L., Jameson, R. J., Sydeman, W. J., Allen, S. G. and Le Boeuf, B. J. 1994 History and present status of the northern elephant seal population. in: *Elephant seals: population ecology, behaviour and physiology* (ed. B. J. Le Boeuf and R. M. Laws), pp. 29–48. Berkley and Los Angeles: University of California Press.

Sugg, D. W., Chesser, R. K., Dobson, F. S. and Hoogland, J. L. 1996 Population genetics meets behavioural ecology. *Trends Ecol. Evol.* **11**, 338–342.

Thomas, C. D. 1994 Extinction, colonization and metapopulations: environmental tracking by rare species. *Conservation Biol.* **8**, 373–378.

Thomas, C. D. and Hanski, I. 1997 Butterfly metapopulations. In *Metapopulation dynamics: ecology, genetics, and evolution.* (ed. I. Hanski and M. Gilpin), pp. 359–386. New York: Academic Press.

Townsend, C. H. 1885 An account of recent captures of the california sea-elephant, and statistics relating to the present abundance of the species. *Proc. US Nat. Mus.* **8**, 90–93.

Turner, B. J., Elder, J. F., Laughlin, T. F., Davis, W. P. and Taylor, D. S. 1992 Extreme clonal diversity and divergence in populations of selfing hermaphroditic fish. *Proc. Natl Acad. Sci. USA* **89**, 10643–10647.

Wade, M. J. and McCauley, D. E. 1988 Extinction and recolonization: Their effects on the genetic differentiation of local populations. *Evolution* **42**, 995–1005.

Ward, R. H., Skibinski, D. O. F. and Woodwark, M. 1992 Protein heterozygosity, protein structure, and taxonomic differentiation. *Evol. Biol.* **26**, 73–159.

Weber, J. L. and Wong, C. 1993 Mutation of human short tandem repeats. *Hum. Mol. Genet.* **2**, 1123–1128.

Whitlock, M. C. 1992 Nonequilibrium population structure in forked fungus beetle: Extinction, colonization, and the genetic variance among populations. *Am. Nat.* **139**, 952–970.

Whitlock, M. C. and McCauley, D. E. 1990 Some population genetic consequences of colony formation and extinction: Genetic correlation within founding groups. *Evolution* **44**, 1717–1724.

Woodruff, R. C. 1989 Genetic anomalies associated with *Cerion* hybrid zones: the origin and maintenance of new electrophoretic variants called hybrizymes. *Biol. J. Linn. Soc.* **36**, 281–294.

Wright, S. 1938 Size of population and breeding structure in relation to evolution. *Science* **87**, 430–431.

Young, A. G., Boyle, T. and Brown, T. 1996 The population genetic consequences of habitat fragmentation in plants. *Trends Ecol. Evol.* **11**, 413–418.

4

Sympatric morphs, populations and speciation in freshwater fish with emphasis on arctic charr

Skúli Skúlason, Sigurður S. Snorrason and Bjarni Jónsson

In recent years we have seen an increasing number of studies emphasizing both intraspecific variability among populations and individual differences within populations. Of particular interest in this respect are species which have distinct morphs that use different habitats and food resources (Wimberger 1994; Skúlason and Smith 1995; Smith and Skúlason 1996). Such resource polymorphism has been detected and studied in several vertebrate taxa: in birds (Sutherland 1987; Smith 1990, 1993); in amphibians (Collins and Cheek 1983; Pfennig 1992; Pfennig and Collins 1993; Maret and Collins 1997); and extensively in fish (e.g. Echelle and Kornfield 1984; Noakes *et al.* 1989; Schluter and McPhail 1993; Robinson and Wilson 1994). The ubiquity of resource polymorphism stimulates numerous questions regarding the ecological and genetic factors that might promote the evolution of resource polymorphism and how it may be maintained. The purpose of this chapter is to examine resource polymorphism in freshwater fish in an evolutionary context. After a general overview emphasizing the ecological scenarios and similarities of patterns across taxonomic groups, we will examine the importance of genetic and environmental factors in producing resource polymorphism. We then offer an evolutionary proposal leading to resource polymorphism, emphasizing the importance of behaviour. Finally, resource polymorphism will be examined in relation to population formation and speciation. We will focus on fish in the northern hemisphere, with the salmonid species arctic charr (*Salvelinus alpinus*) as our main example.

4.1 Resource polymorphism in freshwater fish

Freshwater fish constitute a large and diverse group. In some cases, especially in old, large lakes, closely related species form intralacustrine (often sympatric) species flocks (cf. Echelle and Kornfield 1984). In young systems intraspecific populations and morphs are frequently encountered. Resource polymorphism in fish is particularly common in arctic and subarctic areas where numerous lakes and rivers were formed as the icecap retreated at the end of the last glacial epoch some 10 000–15 000 years ago. These systems may still be considered to be in the colonization phase. Hence, they are usually species poor and invading species have been and are being presented with a diversity of uncontested habitats and food resources, thus promoting processes of character release and resource polymorphism (Robinson and Wilson 1994; Skúlason and Smith 1995). Such postglacial diversification has apparently occurred repeatedly and independently in several distinct groups of fish such as salmonids, sticklebacks, sunfish, smelt, and whitefish (see details in tables and references therein in Robinson and Wilson 1994; Skúlason and Smith 1995; and Smith and Skúlason 1996).

Intraspecific morphs differ in various respects; behaviour, life history characteristics such as adult body size, body shape and colour. Phenotypic differences among sympatric forms clearly correlate with the availability, number and discreteness of habitat and food resources in the respective lake or freshwater system (e.g. Hindar and Jonsson 1982; Malmquist et al. 1992; Robinson et al. 1993).

The selective factors that are proposed to give rise to resource polymorphism relate to resource diversity and in particular inter- and intraspecific competition. This has been extensively reviewed for fishes in the northern hemisphere in the context of character release and character displacement (Schluter and McPhail 1993; Robinson and Wilson 1994). It is well established that resource polymorphism can originate and be maintained by density and frequency dependent or disruptive selection (Rice 1984; Wilson 1989; Wood and Foote 1990; Pfennig 1992; Hori 1993; Smith 1993).

Resource polymorphism represents diversifying evolution on a fine scale producing distinct ecological groups. Because of shared origin, coexisting morphs provide unique opportunities to study the function of ecological, developmental and genetic factors in diversifying evolution. This is particularly important because these phenomena relate to fundamental questions of population divergence and speciation (Smith and Todd 1984; West-Eberhard 1986, 1989; Rice 1987; Diehl and Bush 1989; Wilson 1989, 1992; Rice and Hostert 1993; Bush 1994). In general, it is intriguing how evolutionarily young many polymorphic systems appear to be, as seen in freshwaters of the northern hemisphere.

4.2 Similarities of processes

The striking similarities in the pattern of diversity in different groups of fish suggests that common processes of evolution have been operating. For instance, in lakes (often landlocked) discrete open water (pelagic or limnetic) and benthic habitats are found, and in such lakes limnetic and benthic forms of fish often coexist (reviews in Schluter and McPhail 1993; Robinson and Wilson 1994). However, the differences that exist among species, and among localities within species, indicate that the magnitude or 'degree' of segregation varies. This may reflect historic differences, e.g. geological ages of systems and different colonization histories, and/or different generation time within and among species.

In threespine stickleback (*Gasterosteus aculeatus*) sympatric forms are generally regarded as separate species (McPhail 1984, 1994; Schluter and McPhail 1993) but in arctic charr gene flow among sympatric forms varies considerably across lakes (Hindar *et al.* 1986; Magnússon and Ferguson 1987; Skúlason *et al.* 1989a; Danzmann *et al.* 1991; Hartley *et al.* 1992; Gíslason 1998; see later discussion). We suggest that by careful comparisons within and among systems the relative importance and function of various ecological and evolutionary mechanisms promoting diversification can to some extent be unravelled.

4.3 Arctic charr and resource polymorphism: a focus on Iceland

Compared with other salmonids, arctic charr are extremely diverse both within and among localities. In many ways this variability exemplifies post-glacial diversification of freshwater fishes in the northern hemisphere (Savvaitova 1980; Behnke 1984; Noakes *et al.* 1989; Robinson and Wilson 1994; Snorrason *et al.* 1994; Skúlason and Smith 1995; Smith and Skúlason 1996). It has a circumpolar distribution and both anadromous and non-migratory populations exist, the latter often in landlocked lakes (Johnson 1980). In numerous cases sympatric forms are found using different habitats and food resources and differing in various phenotypic attributes, such as body growth, age and size at first sexual maturity, body coloration and body shape (e.g. Dörfel 1974; Hindar and Jonsson 1982; Behnke 1984; Snorrason *et al.* 1994; Reist *et al.* 1995; see Tables 4.1 and 4.2).

Arctic charr tolerates cold, unstable and non-productive environments better than do other salmonids (Johnson 1980; Noakes 1989). For instance, arctic charr juveniles grow faster in cold water than Atlantic salmon (*Salmo salar*) juveniles (Berg and Jonsson 1989). Arctic charr embryos can develop normally within the littoral gravel of lakes whereas brown trout (*Salmo trutta*) and Atlantic salmon require running water or springwater during incubation. Among the salmonids arctic charr may thus have had the best physiological and developmental pre-adaptations for successful colonization of cold and unstable freshwater systems formed in the wake of the latest glaciation.

Arctic charr is common in Iceland. The only other species of native freshwater fish on the island are brown trout, Atlantic salmon, threespine stickleback and European eel (*Anguilla anguilla*) (Skúlason *et al.* 1992). All five species can potentially migrate between sea and freshwater but only arctic charr, brown

Table 4.1 Phenotypic characteristics involved in resource polymorphism in arctic charr

Behaviour	Life history	Morphology	Spawning segregation
foraging	growth pattern	gill raker count	temporal
social	size at maturity	size and shape of mouth	spatial
migration	age at maturity	snout curvature	behavioural
movement patterns	egg quality	fin size and position	a combination of the above

Table 4.2 Most important differences among the sympatric arctic charr morphs in Thingvallavatn, Iceland

	Benthic charr blunt snout, subterminal mouth, pointed snout, terminal mouth stocky body, dark back and sides		Limnetic charr fusiform body	
	large benthivorous	small benthivorous	planktivorous charr	piscivorous charr
Morphology				
no. of gill rakers				
mean	25.9	24.3	27.4	26.8
Life-history				
adult size	25–40 cm	7–15 cm	11–20 cm	25–40 cm
maturity age	7–10 years	2–4 years	3–5 years	6–10 years
spawning time	mid July–mid-Aug.	Aug.–Dec.	late Sept.–mid Oct.	Sept.–Nov.
spawning site	littoral zone springs	littoral zone general	littoral zone general	littoral zone general
Habitat	benthic—littoral	benthic—among stones	pelagic zone	sub-littoral benthic
Food	snails	snails and insect larvae	planktonic crustaceans	threespined stickleback

trout and threespine stickleback have populations that spend the whole life cycle in freshwater, often because of migratory barriers. It is generally assumed that all freshwater fish populations in Iceland stem from postglacial invasions and the low number of species, compared with similar latitudes on the continents, is a consequence of the geographic isolation of the island and the young age of its freshwater systems (Skúlason *et al.* 1992).

Freshwater systems in Iceland are diverse and numerous. It is estimated that there are 1840 lakes more than 0.1 km^2 and about 9000 lakes more than 0.01 km^2 (Aðalsteinsson *et al.* 1989). Many of these lakes host arctic charr. The variability in physical and biological properties (e.g. productivity) of freshwater systems in Iceland is mostly determined by the age and nature of the volcanic bedrock that forms various parts of the island; for instance how porous it is and how easily it releases dissolvable minerals into the water (see Garðarsson 1979 for classification of Icelandic freshwater systems). In most volcanic lakes water enters lakes through underground springs. This has a stabilizing effect in terms of physical properties such as water flow and temperature and levels of dissolved minerals are high due to the young age of the surrounding bedrock. Therefore, these lakes are often very productive and ecologically diverse (Jónasson 1979, 1992). Furthermore, volcanic lakes offer benthic habitats characterized by uneroded lava substrates, fissures and crevices which add new dimensions to the physical complexity of these habitats. Such habitats may be fundamental in promoting new life styles for fish, e.g. novel feeding methods and modifications of the feeding apparatus, as well as providing diverse spawning and nursery habitats for fish (Skúlason *et al.* 1989a; Snorrason *et al.* 1989).

The diversity of Icelandic freshwater systems and the low level of interspecific competition has provided the scenario for the evolution of extensive diversity in Icelandic arctic charr (Snorrason *et al.* 1989; Skúlason *et al.* 1992). The best described case and the most complex is that of the landlocked lake Thingvallavatn which is situated in a rift valley in SW Iceland and is mostly surrounded by neovolcanic lava (Jónasson 1992). The lake harbours four morphs of arctic charr (Figure 4.1), two benthic specialists; small benthivorous-, large benthivorous charr, and two limnetic morphs; one piscivorous and a planktivore (Snorrason *et al.* 1989; Skúlason *et al.* 1989b). The morphs show distinct segregation in habitat use (Sandlund *et al.* 1987), diet (Malmquist *et al.* 1992), parasites (Frandsen *et al.* 1989), and differ extensively in size and various life history characters (Jonsson *et al.* 1988) and trophic morphology (Snorrason *et al.* 1989, 1994). The morphs are almost certainly locally adapted and conspecific (Magnússon and Ferguson 1987; Danzmann *et al.* 1991; Snorrason *et al.* 1994; Volpe and Ferguson 1996, Gíslason 1998). The differences in life history and morphology among the morphs apparently represent the whole range of reported diversity in the species with the benthic morphs representing a rare and an extreme local specialization (Jonsson *et al.* 1988; Snorrason *et al.* 1994). Furthermore, compared with other cases of sympatric morphs of arctic charr the ecological segregation of the morphs in Thingvallavatn is unusually distinct and stable, for instance diet segregation is maintained the year round (Malmquist *et al.* 1992), while in other

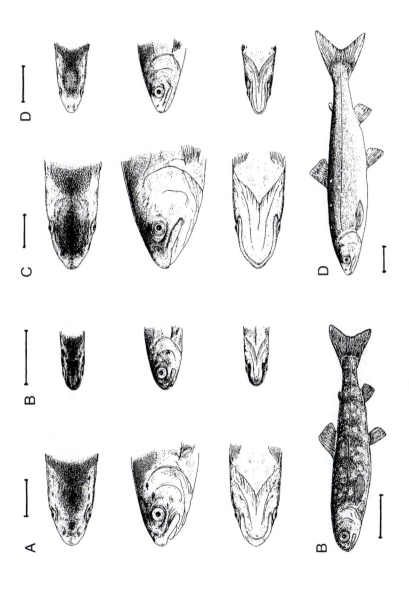

Fig. 4.1 Drawings of the four morphs of Arctic charr in Thingvallavatn, Iceland. (A) Large benthivorous morph. (B) Small benthivorous morph. (C) Planktivorous morphs. (D) Piscivorous charr. Scale bars are 25 mm. Reprinted from Snorrason et al. (1994) by permission of the publisher Academic Press Ltd, London

cases diet segregation is more flexible (Hindar and Jonsson 1982; Riget *et al.* 1986; Skúlason *et al.* 1992). Laboratory feeding trials with wild fish showed clear differences in foraging behaviour among morphs and these differences correlated with the diet specialization in Thingvallavatn (Malmquist 1992).

In most other Icelandic lakes arctic charr are opportunistic (flexible) in diet selection and typically prefer the prey which is most abundant at each time (Adalsteinsson 1979; Skúlason *et al.* 1992). In some lakes two or three morphs are found, differing in adult size and to some extent feeding habits but only in a few cases have observations indicated morphologically distinct differences, but in no case as extensive and clear as in Thingvallavatn (Skúlason *et al.* 1992; Jónsson and Skúlason, in press). Elsewhere, where sympatric morphs of arctic charr have been studied differences in habitat, food, growth and maturation pattern are common (e.g. Hindar and Jonsson 1982; Jonsson and Hindar 1982; Riget *et al.* 1986) but clear differences in body morphology are much less common (Dörfel 1974; Walker *et al.* 1988; Snorrason *et al.* 1989, 1994; Reist *et al.* 1995; Jónsson and Skúlason, in press). Thus, it appears that sympatric morphs of arctic charr differ in the level of specialization with the situation in Thingvallavatn representing a relatively high level of stability and specialization.

4.4 Phenotypic plasticity and genetic basis

Phenotypic plasticity has been defined as 'the change in the expressed phenotype of a genotype as a function of the environment' (Bradshaw 1965). Thus plasticity can also be defined as the ability of a single genotype to produce more than one alternative form of morphology, physiological state, and/or behaviour in response to environmental conditions (West-Eberhard 1989). Adaptive phenotypic plasticity responds to both artificial and natural selection (Scheiner 1993). Fish tend to be quite plastic in many phenotypic attributes (e.g. Liem and Kaufman 1984; Stearns and Crandall 1984; Meyer 1987, 1990; Noakes *et al.* 1989; Thorpe 1989; Wainwright 1991; Wimberger 1991, 1992, 1994; Travis 1994), but common garden and family breeding experiments (e.g. in aquaculture), also show direct genetic basis for important phenotypic attributes that differ within and among morphs and populations (Gjedrem 1983; Thorpe *et al.* 1983; Sutterlin and MacLean 1984; Nilsson 1990; Silverstein and Hershberger 1992; Crandell and Gall 1993). In some species and populations of vertebrates, resource polymorphism is primarily due to genetic polymorphism (Smith 1993; Hori 1993) but in some other species morphs may result primarily from phenotypic plasticity (Nordeng 1983; Meyer 1987; Pfennig 1992; Robinson and Wilson 1996).

Arctic charr are extremely plastic in behaviour, growth and maturation (Nordeng 1983; Vrijenhoek *et al.* 1987; Jobling *et al.* 1993; Brown *et al.* 1992). In rearing and transplantation experiments Nordeng (1983) demonstrated that progeny from intramorph crosses of three coexisting life history forms (differing in adult size and maturation pattern) of arctic charr in Norway produced all three adult forms, with the occasional individual representing all three forms at different times in its life. A common garden rearing experiment of the progeny of two morphs of arctic charr in Vangsvatnet, Norway, differing mainly in growth and adult size, body colour and to a lesser extent in trophic morphology, showed that adult size differences had a small genetic component, indicating considerable phenotypic plasticity in these characters. Coloration was a function of size, but differences in trophic morphology had a clear genetic component (Hindar and Jonsson 1993). Another similar experiment involving two sympatric size-morphs from Stora Rösjön, Sweden showed much clearer genetic contribution to both growth and maturation patterns than did both of the above rearing studies (Svedäng 1990). Laboratory rearing experiments with the progeny of the four morphs from Thingvallavatn showed a clear genetic component to differences in growth, age at first maturity, trophic morphology, body colour and foraging behaviour (Skúlason *et al.* 1989b, 1993, 1996). A study on laboratory reared progeny of benthic and pelagic morphs from Loch Rannoch in Scotland showed that substantial differences in social behaviour had a genetic basis (Mikheev *et al.* 1996).

A comparison of the available data on growth, maturity and behaviour for wild and laboratory reared morphs of arctic charr (above studies) shows that morphs and/or their progeny respond differently to variable environmental conditions, i.e. the morphs have different norms of reaction (Skúlason *et al.* 1996; Mikheev *et al.* 1996). For example, small benthivorous charr in Thingvallavatn appears to be less plastic than planktivorous charr (Skúlason *et al.* 1996) which is perhaps not surprising when the temporal stability of resources in their respective niches is compared (Malmquist *et al.* 1992; Snorrason *et al.* 1992, 1994). Phenotypic plasticity can also be related to variable feeding opportunities. Rearing experiments involving sympatric limnetic and benthic forms of threespine stickleback from British Columbia showed that greater plasticity in trophic morphology of the limnetic form was associated with a more variable diet, compared with the benthic form (Day *et al.* 1994).

A rearing experiment comparing 15 Icelandic allopatric populations of arctic charr revealed considerable phenotypic differences. The offspring of artificial matings of wild fish were reared in several fish farms differing in environmental factors. Offspring of different populations responded to some extent differently to different environments (i.e. had different norms of reaction, Adalsteinsson *et al.* 1992; Eythórsdóttir *et al.* 1993) suggesting varying levels of phenotypic plasticity and thus probably ecological flexibility among populations.

4.5 Evolutionary process: a proposal focusing on arctic charr

It is reasonable to expect phenotypic plasticity to increase in situations where the environment fluctuates (e.g. Newman 1992; Pfennig 1992; Travis 1994). Given the unstable environments salmonids have evolved in (e.g. Noakes 1989) particularly as these have been affected by glaciation (especially in the case of arctic charr) it is not unexpected that these fish display high levels of adaptive phenotypic plasticity and that this has been important in the evolution of different morphs, especially in the early stages of segregation (West-Eberhard 1986, 1989; Vrijenhoek *et al.* 1987; Skúlason *et al.* 1996).

A comparison of all the rearing experiments discussed in the last section shows that the relative importance of phenotypic plasticity and genetic contribution to phenotypic differences between morphs and populations of arctic charr differs within and among systems. Thus, the developmental switch mechanism producing different morphs appears to be controlled both by genetic and environmental factors (West-Eberhard 1989), and the relative importance of these effects depends most likely on past and present selective environments as well as developmental constraints or trade-offs, for instance related to the cost of plasticity (e.g. Newman 1992). This phenomenon needs to be studied much further, especially as it holds an important key to our understanding of the evolutionary processes of resource polymorphism and possibly speciation.

Examining the function of plasticity and genetic basis in morph formation in relation to variation in ecological specialization, e.g. Vangsvatnet versus Thing-vallavatn, it appears that the genetic contribution is greater and the plasticity of morphs probably less in cases where specialization is considerable and morpho-logical and behavioural differences relatively pronounced. Considering the evolu-tionary history of the morphs in Thingvallavatn in this manner, it can be suggested that following post-glacial colonization of the lake, the formation of morphs was primarily a consequence of phenotypic plasticity (cf. Nordeng 1983) in behaviour, life history and to some extent morphology; but as time passed and the discreteness and stability of niches in the lake increased, ecological segregation of morphs became stable and their phenotypic differences more clear, now having a noticeable genetic component (Skúlason *et al.* 1996). Implicit in this proposal is that the evolution of morphs, especially the apparent increase in their ecological specialization, has been characterized by some loss of plasticity in behaviour, life history and morphology (Table 4.3).

In general, the continuum from great flexibility to considerable specialization that exists in arctic charr in different lakes (Skúlason *et al.* 1992; Snorrason *et al.* 1994; Jónsson and Skúlason, in press) and its relationship with environmental and genetic effects on development, as the various rearing experiments indicate (Nordeng 1983; Skúlason *et al.* 1989b, 1993, 1996; Svedäng 1990; Hindar and Jonsson 1993) strongly suggests that an evolutionary process similar to the one suggested for Thingvallavatn may be ongoing in some polymorphic arctic charr systems, while flexibility and phenotypic plasticity continues to be favoured in other systems. This pattern could apply to other polymorphic species as well, and needs further studies and careful comparisons within and among freshwater systems (Table 4.3).

Phenotypic attributes, the importance of behaviour

Let us now consider the integrative role of various phenotypic attributes and their development in relation to the above model of segregation (Table 4.3). In vertebrates, morphology is well known to be relatively conservative to change while behaviour is much more flexible. Flexible foraging behaviour in fish can be of major adaptive importance, especially in environments where resources fluctuate in some important way (Dill 1983; Noakes 1989). The occupation by morphs of different habitats and their variable selection of food items is clearly a behavioural phenomenon and the flexibility in this behaviour differs among morphs and populations (Hindar and Jonsson 1982; Malmquist *et al.* 1992; Malmquist 1992; Skúlason *et al.* 1993). The fact that morphological differences among sympatric morphs of arctic charr are often vague despite differences in diet and habitat (Nordeng 1983; Skúlason *et al.* 1992; Hindar and Jonsson 1993; Jónsson and Skúlason, in press; Snorrason *et al.* 1994) underlines the primary importance of variable behaviour in their evolution, with morphological differences perhaps only having a clear function in cases where specialization is considerable (Snorrason *et al.* 1994).

Detailed studies on brook charr (*Salvelinus fontinalis*), a species closely related to arctic charr, have demonstrated intrapopulation diversity in foraging behaviour,

Table 4.3 A proposal on how the various cases of reported resource polymorphism in fish may be classified to one of four proposed phases of diversification that may lead to speciation (A–D). The proposed mechanisms operating in each phase are in order of importance. Note that although the character groups are also in order of importance they are not exclusively tied to a corresponding mechanism. We are simply proposing that for a given phase the mechanism and a character group in the same line are the most common combination to be expected. The actual phase various cases of diversification are found in depends on historic factors, ecological conditions, e.g. discreteness and stability of resources, opportunities for spatial and temporal segregation in spawning or behavioural factors promoting assortative mating.

Time	Phase	Controlling mechanism	Character group*	Examples	Selected references
	A	1. Phenotypic plasticity	behaviour	arctic charr	Nordeng 1983, Hindar and Jonsson 1993
		2. Genetic basis	life history	sunfish	Ehlinger and Wilson 1988, Robinson and Wilson 1994, 1996
			morphology	cichlids	Meyer 1989, 1990
	B	1. Genetic basis	life history	arctic charr	Skúlason *et al.* 1989b,1993
		2. Phenotypic plasticity	behaviour	cichlids	Hori 1993, Wimberger 1994
		3. Population segregation	morphology		
	C	1. Genetic basis	morphology	cichlids	Seehausen *et al.* 1997
		2. Population segregation	life history	sockeye salmon	Wood and Foote 1996
		3. Phenotypic plasticity	behaviour	arctic charr	D. Gíslason 1998
	D	Reproductive isolation		sticklebacks	McPhail 1994

*These characters primarily refer to resource based phenotypic differences.

primarily movement patterns, in recently emerged young (Grant and Noakes 1988). These differences are reflected in variable uses of microhabitat such as water column and current speed as well as diet selection (Grant and Noakes 1988; Noakes 1989; McLaughlin and Grant 1994; McLaughlin et al. 1994). This information has strong implications for the intramorph and, in particular, potential intermorph and population differences in foraging behaviour and diet selection in polymorphic fish species. Sympatric limnetic and benthic morphs of this species have recently been found in Quebec lakes, Canada (Bourke et al. 1997, in press).

Although no detailed behavioural studies have been done on wild juveniles of arctic charr, behavioural studies within populations or morphs and rearing experiments in the laboratory have shown dramatic variability in foraging and social behaviour (Brown et al. 1992; Jobling et al. 1993; Skúlason et al. 1993; Mikheev et al. 1996) which has many similarities to the behavioural variability in brook charr. It is thus a reasonable proposal that a population of arctic charr invading a fishless lake, e.g. at the end of the last glaciation, with several distinct habitats and food types, could very quickly (especially when intrapopulation competition is high) occupy the different niches primarily on the basis of behavioural differences. This process could be based both on behavioural plasticity of individuals and on genetic polymorphism in the population for the relevant behavioural characters (Table 4.3).

Variation in early foraging and social behaviour can have long-term life history consequences as demonstrated by studies on Atlantic salmon Salmo salar (Metcalfe et al. 1989, Metcalfe and Thorpe 1992; Thorpe et al. 1992; Metcalfe 1993). Thus, early behaviour can determine growth patterns which are closely correlated with other developmental events at later stages (e.g. threshold factors) such as 'decisions' to migrate, shift niches or become sexually mature (Thorpe et al. 1992; Metcalfe 1993; Forseth et al. 1994; Skúlason et al. 1996). In turn, body size and developmental condition can obviously have constraining effects on behaviour. Importantly, these studies suggest that the relevant behaviour patterns of individuals may be quite flexible early in life but become less flexible at later stages, even only a few weeks after first exogenous feeding (Metcalfe et al. 1989).

Thus, it can be visualized that variation in early behaviour can have immediate consequences for habitat and food selection with possible long-term consequences for growth patterns, ontogenetic niche shifts and/or maturation patterns (Forseth et al. 1994; Snorrason et al. 1994; Skúlason et al. 1996). In general, a group of arctic charr that has recently occupied a particular resource will experience certain developmental consequences related to its new niche, which will affect the adult phenotype. Examples show that the ability of morphs to exploit other resources (i.e. flexibility of morphs) varies among systems (see previous discussion).

Comparing studies of sympatric size-morphs of arctic charr (see Griffiths 1994) it is apparent that in general the morph that attains larger adult body size has a broader diet selection, e.g. based on seasonal switching in food types, while the diet of the morph with smaller adult body size is more restricted. This strongly suggests that morphs differ in behavioural flexibility with the small morph displaying greater ecological specialization than the larger one. A study of two morphs, differing greatly in adult size, in a small landlocked lake in Iceland supports the above conclusion. This study indicated that the differences in flexibility in diet selection between morphs was evident in young fish and independent of body size (Jónsson 1996). Thus, the differences in growth patterns and adult size often seen between sympatric morphs of arctic charr may originally be based on early variation in feeding behaviour and/or habitat selection.

We have already explained that the behavioural segregation of morphs can apparently occur without any detectable morphological differences. Nevertheless, even minor morphological variation may play a role and interact with behaviour in the early stages of segregation. McLaughlin and Grant (1994) found that the behaviour morphs of brook charr had minor morphological differences which were correlated with the different microhabitats they occupied (see also Swain and Holtby 1989 for coho salmon *Oncorhynchus kisutch*). Morphology can to some extent channel behaviour of morphs in a functional way (Liem and Kaufman 1984; Ehlinger and Wilson 1988; Meyer 1989; Wainwright 1991; Malmquist 1992; Skúlason *et al.* 1993). Morphology can also be indirectly influenced by behaviour, for instance trophic structures can respond to different diets (Meyer 1987; Wimberger 1991, 1992, 1994). Morphological differences can also result from allometric growth patterns, particularly when morphs differ in adult size (Humphries 1984), or age structure (Jónsson and Skúlason, in press). Morphological differences can thus, like differences in behaviour, be determined by environmental as well as genetic factors.

Like other polymorphic fish species, sympatric morphs of arctic charr are often either limnetic or benthic (Schluter and McPhail 1993; Robinson and Wilson 1994). In cases where morphological differences are evident, the limnetic morph has a fusiform body, pointed snout and terminal jaw structures, whereas the benthic morph has a chunkier body, blunt snout and sometimes a subterminal jaw shape (Skúlason *et al.* 1992; Snorrason *et al.* 1994). These differences result from a heterochronic shift in the development of morphology. The limnetic morphotype is the most usual morphology of salmonids, including arctic charr, but the benthic morphotype is a paedomorph of the more common limnetic morphology (Skúlason *et al.* 1989b). In other words, morphs with benthic features retain embryonic and juvenile characteristics in the adult phenotype. This seems to relate to the differential timing of skull ossification in late embryonic stages (S. Skúlason unpublished data; G. M. Eiríksson personal communication). Benthic morphs may also retain juvenile coloration in the adult stage (Hindar and Jonsson 1982; Snorrason *et al.* 1994). Furthermore, the benthic morphs in Thingvallavatn continue to occupy the littoral benthic nursery habitat while the limnetic morphs leave this habitat, probably early in life (Sandlund *et al.* 1987). The apparent change in the timing of developmental events in the evolution of morphs (i.e. heterochrony) could be based on a simple change in a regulatory gene mechanism (see Liem and Kaufman 1984; Snorrason *et al.* 1994; Skúlason and Smith 1995) potentially modified by other genes as well as environmental and maternal factors (Meyer 1987; West-Eberhard 1989; Collins *et al.* 1993).

In fish, differences in temperature during early development can produce different phenotypes, for example by fixing some components of the potential growth of muscle tissue (Johnston *et al.* 1996). Such developmental changes can subsequently channel alternative behavioural and morphological pathways. Increased egg size and yolk density are known to shorten the larval period and impose developmental constraints on the larvae (Balon 1990). Heterochronic change, correlated with differences in developmental temperatures and yolk densities may promote morphological segregation in sympatric arctic charr in the lake Vatnshlídarvatn, Iceland (Jónsson 1996; Jónsson and Skúlason, in press), where increase in egg size and egg density can be seen as an adaptation to counter high juvenile mortality (cf. Sargent *et al.* 1987). The fact that differences among arctic charr morphs in behaviour, life history and morphology may arise early in life (Balon 1984; Jonsson *et al.* 1988; Skúlason *et al.* 1989, 1993, 1996; Jónsson and Skúlason, in press) needs further consideration.

4.6 Population formation and speciation

In many cases resource polymorphism in freshwater fish is accompanied by the evolution of reproductive isolation (e.g. Skúlason and Smith 1995; Smith and Skúlason 1996; see other references discussed in these papers). Thus, through ecological and phenotypic specialisation (fixation) and reproductive isolation coexisting resource morphs can become new species (West-Eberhard 1989; see also Rice 1984, 1987; Diehl and Bush 1989; Wilson 1989; Rice and Hostert 1993; Bush 1994; Gíslason 1998). However, in order to understand this process better, the ecological and evolutionary relationship between resource polymorphism, population formation (limited gene flow among morphs) and speciation needs to be examined much more closely. Here, we provide an overview of some of the relevant information.

Molecular genetic studies on sympatric and allopatric forms of freshwater fish have in several cases suggested that sympatric morphs of the same species have repeatedly evolved locally and do not represent separate invasions of allopatric populations (e.g. Hindar *et al.* 1986; Foote *et al.* 1989; Bodaly *et al.* 1992; Taylor and Bentzen 1993; Hindar 1994). However, in some cases patterns of genetic diversity indicate allopatric origin of currently coexisting morphs (Bernatchez and Dodson 1990; Ferguson and Taggart 1991; Ferguson and Hynes 1995).

In general, the use of the terms 'allopatry' and 'sympatry' in the evolution of coexisting fish morphs or species is often somewhat unclear, or at least the distinction too vague. Naturally, many freshwater systems are, or have been in the past, clearly geographically isolated from each other, thus promoting allopatric population segregation. On the other hand population segregation within freshwater systems is based on much less obvious isolation factors. Freshwater systems differ in their structural complexity; for instance, some lakes offer distinct spawning habitats, promoting population segregation, while others do not (Smith and Todd 1984). Thus, in cases of intralacustrine population segregation it is necessary to describe the nature of isolation in each case carefully to avoid misunderstanding (Smith and Todd 1984).

Intralacustrine reproductive segregation of morphs can be based on different spawning locations (sometimes called 'microallopatry', e.g. because of strong philopatry or homing), different spawning seasons (allochrony) as well as mate selection on the spawning grounds (Kurenkov 1977; Dominey 1984; Smith and Todd 1984; Foote 1988; Foote and Larkin 1988; Skúlason *et al.* 1989a; Sigurjónsdóttir and Gunnarsson 1989; Meyer 1993; see Diehl and Bush 1989 for a discussion of 'extrinsic' and 'intrinsic' isolation, the importance of non-allopatric isolation and the perhaps misleading use of the term 'microallopatry', see Tables 4.1 and 4.2). These mechanisms may interact, for instance behavioural and allochronic isolation mechanisms may both operate in the same system (Skúlason *et al.* 1989a; Sigurjónsdóttir and Gunnarsson 1989). In addition to the prezygotic isolation mechanisms discussed above, postzygotic isolation mechanisms have also been identified. For instance, studies on two sympatric forms of the pacific salmon *Oncorhynchus nerka* which interbreed, strongly suggest that genetic segregation is primarily based on reduced fitness of hybrids (Wood and Foote 1990, 1996).

Considering prezygotic isolation among morphs, many similarities exist among cases. In order to understand details, however, the particular ecological situation in each case needs to be examined as well as the biological characteristics of the respective species, especially concerning life history and reproduction. In general taxa that show polymorphism or extensive radiation have large eggs and protect their egg or young (cf. Balon 1985), but these characteristics are likely to promote discrete spawning habits and reproductive isolation. In arctic charr, intralacustrine morphs may spawn in different locations and/or at different times of the year (e.g. Brenner 1980; Skúlason *et al.* 1989a; Skúlason *et al.* 1992; Jónsson 1996; Klemetsen *et al.* in press). Like other salmonids, arctic charr is a localized benthic spawner (Balon 1985) and sexually mature fish home to natal grounds to spawn. Thus, distinct spawning- and nursery habitats can develop among morphs. It is easy to envisage how such segregation can occur if available spawning grounds within a lake are discrete, e.g. areas with high flows of ground water and/or creeks and rivers connected to the lake (Skúlason *et al.* 1989a; Jónsson 1996; see for brown trout: Crozier and Ferguson 1986; Ferguson and Hynes 1995). Most often salmonids spawn in the autumn, and this timing is tied in with seasonal variation in temperature and food availability (e.g. for first feeding juveniles in spring). Sympatric morphs of arctic charr may diverge from this pattern and spawn earlier or later in the year, for instance when lakes offer spawning and nursery grounds which are less affected by season, such as areas with high flows of stable ground water at constant temperature (Brenner 1980; Skúlason *et al.* 1989a; Klemetsen *et al.* 1997). Similarly, seasonal changes in the availability of morph-specific food types may, through constraints on annual patterns of gonad production, lead to reproductive isolation by allochrony (Skúlason *et al.* 1989a).

Spawning behaviour and mate selection is usually sophisticated and often related to body size and colour, for instance in salmonids (Foote 1988). Because sympatric morphs of salmonids often differ in size at sexual maturity this may result in behavioural isolation (Sigurjónsdóttir and Gunnarsson 1989). In general, behavioural reproductive isolation among populations and closely related species of fish is likely to be common (e.g. Dominey 1984) and this needs much further attention, especially in relation to ideas of speciation (see discussion about cichlids later in this section).

The degree of genetic divergence between sympatric morphs is highly variable both within and among species. In some cases gene flow may be unimpeded, while in other cases, sympatric forms may appear partially or completely reproductively isolated (Hindar et al. 1986; Magnúson and Ferguson 1987; Foote et al. 1989; Danzmann et al. 1991; Ferguson and Taggart 1991; Bodaly et al. 1992; Hartley 1992; Hartley et al. 1992; Hindar 1994; McPhail 1994; Volpe and Ferguson 1996; Gíslason 1998).

The degree of gene flow may be related to the magnitude and persistence of ecological segregation of morphs as well as the age of the respective system (number of generations). As discussed earlier these ecological features also have bearing on the evolution of morph specialization and the relative importance of environmental and genetic factors on the development of alternative phenotypes. For example, in some lakes arctic charr morphs are primarily formed because of phenotypic plasticity while in some other lakes, for instance Thingvallavatn, morphs show high ecological and phenotypic divergence and there is evidence of reduced gene flow among them (Nordeng 1983; Skúlason et al. 1989a; Skúlason et al. 1992; Hindar and Jonsson 1993; Snorrason et al. 1994; Bjøru and Sandlund 1995; Volpe and Ferguson 1996; Gíslason 1998). Thus, phenotypic segregation of morphs can solely be based on non-genetic plastic responses of adaptive significance. However, if segregation persists and is associated with assortative mating we can expect different norms of reaction to evolve, e.g. in lakes where there are opportunities for temporal and/or spatial spawning segregation (Tables 4.1 and 4.2).

There is strong evidence suggesting that sympatric species of closely related freshwater fish have in some cases evolved from morphs or ecological forms that have become reproductively isolated (Echelle and Kornfield 1984; McPhail 1994; Skúlason and Smith 1995; Smith and Skúlason 1996; Schluter 1996). This provides an important and unique opportunity to study the importance of ecological factors in divergent selection, phenotypic segregation, and speciation (Schluter 1996). Gíslason (1998), studying five lakes with sympatric morphs of arctic charr, found a significant correlation between the degree of morphological and genetic divergence, indicating permissive coupling of these processes. In one small mountain lake the genetic analyses showed an advanced state of divergence with total reproductive isolation of morphs.

The explosive speciation of cichlids in the African great lakes may provide an excellent example of this. Studies on the numerous species of haplochromine cichlids in Lake Victoria suggest that mate choice of females for differently coloured males may maintain reproductive isolation between sympatric species and colour morphs (Seehausen et al. 1997; Galis and Metz 1998 see also Turner in this volume). The adaptive radiation of these cichlid species is clearly related to resource segregation, involving numerous adaptive specializations to a diversity of feeding niches. Thus, speciation processes in this system appear to be based on sexual selection, promoting and maintaining reproductive isolation among resource-based morphs and species. This is substantiated by a recent model of sympatric speciation focusing on mate choice of females (Payne and Krakauer 1997).

A convincing example of ecological speciation is found in cichlids from Cameroon, West Africa (Trewavas *et al.* 1972). Mitocondrial DNA analyses of cichlid species flocks endemic to two crater lakes strongly suggests that each lake contains a monophyletic group of species that originated sympatrically (Schliewen *et al.* 1994). Species within lakes are more closely related to each other than they are to riverine species or species from adjacent lakes. While resource polymorphism was not demonstrated in this instance, it is likely to have been an intermediate step given the ecological nature of species segregation in these lakes and that many lakes in East and West Africa contain polymorphic populations of cichlids (Meyer *et al.* 1990; Meyer 1993; Hori 1993).

It is our belief that future studies on resource polymorphism and its evolutionary significance will further reveal its potential commonness and importance in understanding how intraspecific variability may translate into species differences and that this has been underestimated as an evolutionary phenomenon (see also West-Eberhard 1986, 1989; Wilson 1992; Robinson and Wilson 1994; Skúlason and Smith 1995; Smith and Skúlason 1996)

Acknowledgements

We thank Stefán Óli Steingrímsson and Ólöf Ýrr Atladóttir for improving drafts of the manuscript and Pierre Magnan, Anders Klemetsen and Guðni Magnús Eiríksson for allowing us to cite unpublished material and papers in press.

References

Adalsteinsson, H. 1979 Size and food of arctic charr *Salvelinus alpinus* and stickleback *Gasterosteus aculeatus*. *Oikos* **32**, 228–231.

Aðalsteinsson, H., Rist, S., Hermannsson, S. and Pálsson S. 1989 Stöðuvötn á Íslandi. Skrá um vötn staerri en 0.1km² (Lakes in Iceland. Report on lakes larger than 0.1km²). Report by the Energy authority. (In Icelandic).

Adalsteinsson, S., Hilmarsdóttir, T., Svavarsson, E. and Pétursdóttir T. 1992 Comparison of 15 strains of arctic charr (*Salvelinus alpinus*) in Iceland. *Icel. Agr. Sci.* **6**, 135–142.

Balon, E. K. 1984 Life history of arctic charrs, an epigenetic explanation of their invading ability and evolution. In *Biology of the Arctic Charr* (ed. L. Johnson and B. L. Burns), pp. 109–141. Winnipeg: University of Manitoba Press.

Balon, E. K. 1985 *Early life histories of fishes: new developmental, ecological and evolutionary perspectives*. Developments in Env. Biol. Fish. 5, Dr. Junk, W. Publishers, Dordrect, 280 pp.

Balon, E. K. 1990 Epigenesis of an epigeneticist: the development of some alternative concepts on the early ontogeny and evolution of fishes. *Guelph. Ichthyol. Rev.* **1**, 1–48.

Behnke, R. J. 1984 Organizing the diversity of the Arctic char complex. In *Biology of the Arctic Charr* (ed. L. Johnson and B. L. Burns), pp. 3–21. Winnipeg: University of Manitoba Press.

Berg, O. K. and Jonsson, B. 1989 Migratory patterns of anadromous atlantic salmon, brown trout and arctic charr from the Vardnes river in Northern Norway. In *Proceedings of the salmonid migration and distribution symposium* (ed. E. Brannon and B. Jonsson), pp. 106–115. June 23–25, 1987 Second international symposium. School of Fisheries University of Washington, Seattle, Washington and Norwegian Institute for Nature Research Throndheim, Norway.

Bernatchez, L. and Dodson, J. J. 1990 Allopatric origin of sympatric populations of lake whitefish (*Coregonus clupeaformis*) as revealed by mitochondrial-DNA restriction analysis. *Evolution* 44, 1263–1271.

Bjøru, B. and Sandlund, O. T. 1995 Differences in morphology and ecology within a stunted arctic charr population. *Nordic J. Freshw. Res.* **71**, 163–172.

Bodaly, R. A., Clayton, J. W., Lindsey, C. C. and Vuorinen, J. 1992 Evolution of the lake whitefish (*Coregonus clupeaformis*) in North America during the Pleistocene: genetic differentiation between sympatric populations.*Can. J. Fish. Aquat. Sci.* 49, 769–779.

Bourke, P., Magnan P. & Rodriguez, M. A. 1997 Individual variations in habitat use and morphology in brook charr. *J. Fish Biol.* **51**, 783-794.

Bourke, P., Magnan P. & Rodriguez, M. A. Phenotypic responses of lacustrine brook charr in relation to the intensity of interspecific competition. *Evolutionary Ecology*. In press.

Bradshaw, A. D. 1965 Evolutionary significance of phenotypic plasticity in plants. *Adv. Genet.* **13**, 115–155.

Brenner, T. 1980 The arctic charr, *Salvelinus alpinus*, in the prealpine Attersee, Austria. In *Charrs: Salmonid fishes of the genus Salvelinus* (ed. E. K. Balon), pp. 765–772. The Hague: Dr. W. Junk Publishers.

Brown, G. E., Brown, J. A. and Srivastava, R. K. 1992 The effect of stocking density on the behavior of arctic charr *Salvelinus alpinus* L. *J. Fish Biol.* **41**, 995–963.

Bush, G. L. 1994 Sympatric speciation in animals: new wine in old bottles. *Trends Ecol. Evol.* 9, 285–288.

Collins, J. P. and Cheek, J. E. 1983 Effect of food and density on development of typical and cannibalistic salamander larvae in *Ambystoma tigrinum nebulosum. Am. Zool.* **23**, 77–84.

Collins, J. P., Zerba, K. E. and Sredl, M. J. 1993 Shaping intraspecific variation: development, ecology and the evolution of morphology and life history variation in tiger salamanders. *Genetica* **89**, 167–183.

Crandell, P. A. and Gall, G. A. E. 1993 The genetics of age and weight at sexual maturity based on individually tagged rainbow trout *Oncorhynchus mykiss. Aquaculture* **117**, 95–105.

Crozier, W. W. and Ferguson, A. 1986 Electrophoretic examination of the population structure of brown trout, *Salmo trutta* L., from the Lough Neagh catchment, Northern Ireland. *J. Fish Biol.* **28**, 459–477.

Danzmann, R. G., Ferguson, M. M., Skúlason, S., Snorrason, S. S. and Noakes, D. L. G. 1991 Mitochondrial DNA diversity among four sympatric morphs of Arctic char, *Salvelinus alpinus*, from Thingvallavatn, Iceland. *J. Fish. Biol.* 39, 649–659.

Day, T., Pritchard, J. and Schluter, D. 1994. A comparison of two sticklebacks. *Evolution* 48, 1723–1734.

Diehl, S. R. and Bush, G. L. 1989 The role of habitat preference in adaptation and speciation. In *Speciation and its consequences* (eds. D. Otte and J. A. Endler), pp. 345–365. Sunderland, MA: Sinauer Associates, Inc.

Dill, L. M. 1983 Adaptive flexibility in the foraging behavior of fishes. *Can. J. Fish. Aquat. Sci.* 40, 398–408.

Dominey, W. J. 1984 Effects of sexual selection and life history on speciation: species flocks in African cichlids and Hawaiian *Drosophila*. In *Evolution of fish species flocks.* (ed. A. A. Echelle and I. Kornfield), pp. 231–250. Orono, ME: University of Maine Press.

Dörfel, V. H. J. 1974 Untersuchungen zur problematik der saiblingspopulationen (*Salvelinus alpinus* L.) im Überlinger See (Bodensee). *Arch. Hydrobiol.* (Supplement) 47, 80–105.

Echelle, A. A. and Kornfield, I. (ed.) 1984 *Evolution of fish species flocks.* Orono, ME: University of Maine Press.

Ehlinger, T. S. and Wilson, D. S. 1988 Complex foraging polymorphism in bluegill sunfish. *Proc. Natl Acad. Sci. USA* 85, 1878–1882.

Eyþórsdóttir, E., Pétursdóttir, Þ. and Svavarsson, E. 1993 Samanburður á bleikjustofnum (Comparison of arctic charr populations). Ráðunautafundur (Icelandic report).

Ferguson, A. and Hynes, R. A. 1995 Molecular approaches to the study of genetic variation in Salmonid fishes. *Nordic J. Freshw. Res.* 71, 23–32.

Ferguson, J. and Taggart, J. B. 1991 Genetic differentiation among the sympatric brown trout (*Salmo trutta*) populations of Lough Melvin, Ireland. *Biol. J. Linn. Soc.* 43, 221–237.

Foote, C. J. 1988 Male mate choice dependent on male size in salmon. *Behaviour* 106, 63–80.

Foote, C. J. and Larkin, P. A. 1988 The role of male choice in the assortative mating of sockeye salmon and kokanee, the anadromous and non-anadromous forms of *Oncorhynchus nerka. Behaviour* 106, 43–62.

Foote, C. J., Wood, C. C. and Withler, R. E. 1989 Biochemical genetic comparison of sockeye and kokanee, the anadromous and non-anadromous forms of Oncorynchus nerka. *Can. J. Fish. Aquat. Sci.* 46, 149–158.

Forseth, T., Ugedal, O. and Jonsson, B. 1994 The energy budget, niche shift, reproduction and growth in a population of Arctic char, *Salvelinus alpinus. J. Anim. Ecol.* 63, 116–126.

Frandsen, F., Malmquist, H. J. and Snorrason, S. S. 1989 Ecological parasitology of polymorphic Arctic charr, *Salvelinus alpinus* (L.) in Thingvallavatn, Iceland. *J. Fish Biol.* 34, 281–297.

Galis, F. and Metz, J. A. J. 1998 Why are there so many cichlid species? *Trends Ecol. Evol.* 13, 1–2.

Garðarsson, A. 1979 Vistfraeðileg flokkun íslenskra vatna. (Ecological classification of Icelandic freshwater systems) *Týi* 9, 1–10. (In Icelandic).

Gíslason, D. 1998 Genetic and morphological variation in polymorphic arctic charr, *Salvelinus alpinus*, from Icelandic lakes. M. Sc. thesis, University of Guelph, Canada.

Gjedrem, T. 1983 Genetic variation in quantitative traits and selective breeding in fish and shellfish. *Aquaculture* 86, 51–72.

Grant, J. W. A. and Noakes, D. L. G. 1988 Aggression and foraging mode of young-of-the-year brook charr, *Salvelinus fontinalis* (Pisces, Salmonidae). *Behav. Ecol. Sociobiol.* 22, 435–445.

Griffiths, D. 1994 The size structure of lacustrine arctic charr (Pisces: Salmonidae) populations. *Biol. J. Linn. Soc.* 43, 221–237.

Hartley, S. E. 1992 Mitochondrial DNA analysis of Scottish populations of Arctic charr, *Salvelinus alpinus* (L.) *J. Fish. Biol.* 4, 219–224.

Hartley, S. E., McGowan, C., Greer, R. B and Walker, A. F. 1992 The genetics of sympatric Arctic charr [*Salvelinus alpinus* (L.)] populations from Loch Rannoch, Scotland. *J. Fish Biol.* **41**, 1021–1031.

Hindar, K. 1994 Alternative life histories and genetic conservation. In *Conservation genetics* (ed. V. Loeschcke, J. Tomink and S. K. Jain), pp. 323–336. Basel: Birkauser Verlag.

Hindar, K. and Jonsson, B. 1982 Habitat and food segregation of dwarf and normal arctic charr (*Salvelinus alpinus*) from Vangsvatnet Lake, western Norway. *Can. J. Fish. Aqatic. Sci.* 39, 1030–1045.

Hindar, K. and Jonsson, B. 1993 Ecological polymorphism in Arctic charr. *Biol. J. Linn. Soc.* 48, 63–74.

Hindar, K., Ryman, N. and Stahl, G. 1986 Genetic differentiation among local populations and morphotypes of Arctic charr, *Salvelinus alpinus*. *Biol. J. Linn. Soc.* 27, 269–285.

Hori, M. 1993 Frequency-dependent natural selection in the handedness of scale-eating cichlid fish. *Science* **260**, 216–219.

Humphries, J. M. 1984 Genetics of speciation in pupfishes from Laguna Chichancanab, Mexico. In *Evolution of fish species flocks.* (ed. A. A. Echelle and Kornfield I.), pp. 129–139. Orono, ME: University of Maine Press.

Jobling, M., Jørgensen, E. H., Arnesen, A. M. and Ringø, E. 1993 Feeding, growth and environmental requirements of arctic charr: a review of aquaculture potential *Aquaculture International* **1**, 20–46.

Johnson, L. 1980 The arctic charr, *Salvelinus alpinus*. In *Charrs: salmonid fishes of the genus* Salvelinus (ed. E. K. Balon), pp. 15–98. The Hague: Dr. W. Junk Publishers.

Johnston, I. A., Vieira, V. L. A. and Hill, J. 1996 Temperature and ontogeny in ectotherms: muscle phenotype in fish. In *Animals and temperature. Phenotypic and evolutionary adaptation* (ed. I. A. Johnston and A. F. Bennett), pp. 153–181. Cambridge University Press.

Jónasson, P. M. (ed.) 1979 *Lake Mývatn. Oikos* **32**.

Jónasson, P. M. (ed.) 1992 *Ecology of oligotrophic, subarctic Thingvallavatn. Oikos* **64**.

Jonsson, B. 1996 Polymorphic segregation in Arctic charr Salvelinus alpinus (L.) from lake Vatnshlídarvatn, Northern Iceland. Unpublished masters thesis, Oregon State University.

Jonsson, B. and Hindar, K. 1982 Reproductive strategy of dwarf and normal Arctic charr (*Salvelinus alpinus*) from Vangsvatnet Lake, Western Norway. *Can. J. Fish. Aquatic. Sci.* 39, 1404–1413.

Jónsson, B. and Skúlason, S. Polymorphic segregation in arctic charr *Salvelinus alpinus* (L.) from Vatnshlídarvatn, a small shallow Icelandic lake. *Biol. J. Linn. Soc.*, in press.

Jónsson, B., Skúlason, B., Snorrason, S. S., Sandlund, O. T., Malmquist, H. J., Jonasson, P. M., Gydemo, R. and Linden, T. 1988 Life history variation of polymorphic Arctic charr (Salvelinus alpinus) in Thingvallavatn, Iceland.*Can. J. Fish. Aquatic. Sci.* 45, 1537–1547.

Klemetsen, A., Amundsen, P. A., Knudsen, R. and Bjørn, H. 1997 A profundal, winter-spawning morph of arctic charr *Salvelinus alpinus* (L.) in Fjellfrøsvatn, northern Norway. *Nordic J. Freshw. Res.* **73**, 13–73.

Kurenkov, S. I. 1977 Two reproductively isolated groups of kokanee salmon, *Oncorhynchus nerka kennerlyi*, from Lake Kronotskiy. *J. Ichthyol.* 17, 526–534.

Liem, K. F. and Kaufman, L. S. 1984 Intraspecific macroevolution: Functional biology of the polymorphic Cichlid species *Cichlasoma minckleyi*. In *Evolution of fish species flocks.* (ed. A. A. Echelle and Kornfield I.), pp. 203–215. Orono, ME: University of Maine Press.

Magnusson, K. P. and Ferguson, M. M. 1987 Genetic analysis for four sympatric morphs of the Arctic charr, *Salvelinus alpinus*, from Thingvallavatn, Iceland. *Environ. Biol. Fish.* 20, 67–73.

Malmquist, H. J. 1992 Phenotype-specific feeding behaviour of two arctic charr *Salvelinus alpinus* morphs. *Oecologia* 92, 354–361.

Malmquist, H. J., Snorrason, S. S., Skúlason, S. Jonsson, B., Sandlund, O. T. and Jonasson, P. M. 1992 Diet differentiation in polymorphic Arctic charr in Thingvallavatn, Iceland. *J. Anim. Ecol.* 61, 21–35.

Maret, T. J. and Collins, J. P. 1997 Ecological origin of morphological diversity: a study of alternative trophic phenotypes in larval salamanders. *Evolution* 51, 898–905.

McLaughlin, R. L. and Grant, J. W. A. 1994 Morphological and behavioural differences among recently-emerged brook charr, *Salvelinus fontinalis*, foraging in slow-vs fast-running water, *Environ. Biol. Fish.* 39, 289–300.

McLaughlin, R. L., Grant, J. W. A. and Kramer, D. L. 1994 Foraging movements in relation to morphology, water-column use, and diet for recently emerged brook trout (*Salvelinus fontinalis*) in still-water pools. *Can. J. Fish. Aquat. Sci.* 51, 1–9.

McPail, J. D. 1984 Ecology and evolution of sympatric sticklebacks (*Gasterosteus*): Morphological and genetic evidence for a species pair in Enos Lake, British Columbia. *Can. J. Zool.* 62, 1402–1408.

McPhail, J. D. 1994 Speciation and the evolution of reproductive isolation in the stickle-backs (*Gasterosteus*) of south-western British Columbia. In *The evolutionary biology of the threespine stickleback* (ed. M. A. Bell and S. A. Foster), pp. 399–437. Oxford: Oxford University Press.

Metcalfe, N. B. 1993 Behavioral causes and consequences of life history variation in fish. *Mar. Behav. Physiol.* 23, 205–217.

Metcalfe, N. B. and Thorpe, J. E. 1992 Early predictors of life-history events: the link between first feeding date, dominance and seaward migration in Atlantic salmon, *Salmo salar* L. *J. Fish Biol.* 41 (Supplement B), 93–99.

Metcalfe, N. B., Huntingford, F. A., Thorpe, J. E. and Adams, C. E. 1989 The effects of social status on life-history variation in juvenile salmon. *Can. J. Zool.* 68, 2630-2636.

Meyer, A. 1987 Phenotypic plasticity and heterochrony in *Ciclasoma managuense* (Pisces, Cichlidae) and their implications for speciation in cichlid fishes. *Evolution* 41, 1357–1369.

Meyer, A. 1989 Cost of morphological specialization: feeding performance of the two morphs in the trophically polymorphic cichlid fish, *Chiclasoma citrinellum*. *Oecologia* 80, 431–436.

Meyer, A. 1990 Ecological and evolutionary consequences of the trophic polymorphism in *Cichlasoma citrinellum* (Pisces: Cichlidae). *Biol. J. Linn. Soc.* 39, 279–299.

Meyer, A. 1993 Phylogenetic relationships and evolutionary processes in East African cichlid fishes. *Trends Ecol. Evol.* 8, 279–284.

Meyer, A., Kocher, T. D., Basasibwaki, P. and Wilson, A. C. 1990 Monophyletic origin of Lake Victoria cichlid fishes suggested by mitochondrial DNA sequences. *Nature* 347, 550–553.

Mikheev, V. N., Adams, C. E., Huntingford, E. A. and Thorpe, J. E. 1996 Behavioural responses of benthic and pelagic Arctic charr to substratum heterogeneity. *J. Fish Biol.* 49, 494–500.

Newmann, R. A. 1992 Adaptive plasticity in amphibian metamorphosis. *Bio. Sci.* 42, 671–678.

Nilson, J. 1990 Heritability estimates of growth-related traits in Arctic char (*Salvelinus alpinus*). *Aquaculture* 84, 211–217.

Noakes, D. L. G. 1989 Early life history and behaviour of charrs. In *biology of charrs and masu salmon. Proceedings of the international symposium on charrs and masu salmon* (ed. H. Kawanabe, F. Yamazaki and D. L. G. Noakes), pp. 173–186. *Physiol. Ecol. Jpn. Spec.* Vol. I.

Noakes, D. L. G., Skúlason, S. and Snorrason, S. S. 1989 Alternative life-history styles in salmonid fishes with emphasis on Arctic char, (Salvelinus alpinus). In *Alternative life-history styles of animals* (ed. M. N. Bruton), pp. 329–346. Dordrecht: Kluwer Academic Publishers.

Nordeng, H. 1983 Solution to the 'charr problem' based on arctic char (*Salvelinus alpinus*) in Norway. *Can. J. Fish. Aquat. Sci.* 40, 1372–1387.

Payne, R. J. H. and Krakauer, D. C. 1997 Sexual selection , space and speciation. *Evolution* **51**, 1–9.

Pfennig, D. W. 1992 Polyphenism in spadefoot toad tadpoles as a locally adjusted evolutionarily stable strategy. *Evolution* **46**, 1408–1420.

Pfennig, D. W. and Collins, J. P. 1993 Kinship affects morphogenesis in cannibalistic salamanders. *Nature* **362,** 836–838

Reist, J. D., Gyselman, E., Babaluk, J. A., Johnson, J. D. and Wissink, R. 1995 Evidence for two morphotypes of Arctic char (*Salvelinus alpinus* (L.)) from Lake Hazen, Ellesmere Island, Northwest Territories, Canada. *Nordic. J. Freshw. Res.* **71**, 396–410.

Rice, W. R. 1984 Sex chromosomes and the evolution of sexual dimorphism. *Evolution* 38, 735–742.

Rice, W. R. 1987 Speciation via habitat specialization: the evolution of reproductive isolation as a correlated character. *Evol. Ecol.* 1, 301–314.

Rice, W. R. and Hostert, E. E. 1993 Laboratory experiments on speciation: What have we learned in 40 years? *Evolution* 47, 1637–1650.

Riget, A., Nygaard, K. H. and Christensen, B. 1986 Population structure, ecological segregation, and reproduction in a population of arctic char (*Salvelinus alpinus*). *Can. J. Fish. Aquat. Sci.* **43**, 985–992.

Robinson, B. W. and Wilson, D. S. 1994 Character release and displacement in fishes,a neglected literature. *Am. Nat.* **144**, 596–627.

Robinson, B. W. and Wilson, D. S. 1996 Genetic variation and phenotypic plasticity in a trophically polymorphic population of pumpkinseed sunfish (*Lepomis gibbosus*). *Evol. Ecol.* **10**, 631–652

Robinson, B. W., Wilson, D. S., Margosian, S. S. and Lotito, P. T. 1993 Ecolocical and morphological differentiation of pumpkinseed sunfish in lakes without bluegill sunfish. *Evol. Ecol.* 7, 451–464.

Sandlund, O. T., Jonsson, B., Malmquist, H. J., Gydemo, R., Lindem, T., Skúlason, S., Snorrason, S. S. and Jonasson, P. M. 1987 Habitat use ef arctic charr *Salvelinus alpinus* in Thingvallavatn, Iceland. *Environ. Biol. Fish.* 20, 263–274.

Sargent, R. C., Taylor, P. D. and Gross, M. R. 1987 Parental care and the evolution of egg size in fishes. *Am. Nat.* **129**, 32–46.

Savvaitova, K. A. 1980 Taxonomy and biogeography of charrs in the palearrctic. In *Charrs: Salmonid fishes of the genus Salvelinus.* (ed. E. K. Balon), pp. 281-294. The Hague: Dr. W. Junk Publishers.

Scheiner S. M. 1993 Genetics and evolution of phenotypic plasticity. *Annu. Rev. Ecol. Syst.* 24, 35–68.

Schliewen, U.K., Tautz, D. and Paabo, S. 1994 Sympatric speciation suggested by monophyly of crater lake cichlids. *Nature* 368, 629–632.

Schluter, D. 1996 Ecological speciation in postglacial fishes. *Phil. Trans. R. Soc. London* **351**, 807–814.

Schluter, D. and McPhail, J. D. 1993 Character displacement and replicate adaptive radiation. *Trends Ecol. Evol.* 8, 197–200.

Seehausen, O, van Alphen J. J. M. and Whitte, F. 1997 Cichlid fishes diversity threatened by eutrophication that curbs sexual selection. *Science* **277**, 1808–1811.

Sigurjónsdóttir, H. and Gunnarsson, K. 1989 Alternative mating tactics of arctic charr, *Salvelinus alpinus*, in Thingvallavatn, Iceland. *Environ. Biol. Fish.* 26, 159-176.

Silverstein, J. T. and Hershberger, W. K. 1992 Precocious maturation in coho salmon *Oncorhynchus kisutch*: estimation of the additive genetic variance. *J. Hered.* **83**, 283–286.

Skúlason, S. and Smith, T. B. 1995 Resource polymophisms in vertebrates, *Trends Ecol. Evol.* **10**, 366–370.

Skúlason, S., Snorrason, S. S. Noakes, D. L. G. and Ferguson, M. M. 1996 Genetic basis of life history variations among sympatric morphs of Arctic char *Salvelinus alpinus*. *Can. J. Fish. Aquat. Sci.* 53, 1807–1813.

Skúlason, S., Snorrason, S. S., Noakes, D. L. G., Ferguson, M. M. and Malmquist, H. J. 1989a Segregation in spawning and early life history among polymorphic Arctic charr, *Salvelinus alpinus*, in Thingvallavatn, Iceland. *J. Fish. Biol.* 35, 225- 232.

Skúlason, S., Noakes, D. L. G. and Snorrason, S. S. 1989b Ontogeny of trophic morphology in four sympatric morphs of arctic charr *Salvelinus alpinus* in Thingvalla-vatn, Iceland. *Biol. J. Linn. Soc.* 38, 281–301.

Skúlason, S., Antonsson, T., Guðbergsson, G., Malmquist, H. J. and Snorrason, S. S. 1992 Variability in Icelandic Arctic charr. *Icel. Agr. Sci.* **6**, 142–153.

Skúlason, S., Snorrason, S. S., Ota, D. and Noakes, D. L. G. 1993 Genetically based differences in foraging behaviour among sympatric morphs of arctic charr (Pisces; Salmonidae). *Anim. Behav.* 45, 1179–1192.

Smith, T. B. 1990 Comparative breeding biology of the two bill morphs of the black-bellied seedcracker (*Pyrenestes ostrinus*). *Auk* **107**, 153–160.

Smith, T. B. 1993 Disruptive selection and the genetic basis of bill size polymorphism in the African finch, *Pyrenestes*. *Nature* **363**, 618–620.

Smith, T. B. and Skúlason, S. 1996 Evolutionary significance of resource polymorphisms in fishes, amphibians, and birds. *Annu. Rev. Ecol. Syst.* **27**, 111–133.

Smith, G. R. and Todd, T. N. 1984 Evolution of fish species flocks in north temperate lakes. In *Evolution of fish species flocks* (ed. A. A. Echelle and I. Kornfield), pp. 47–68. Orono, ME: University of Maine Press.

Snorrason, S. S., Skúlason, S., Sandlund, O. T., Malmquist, H. J., Jonsson, B. and Jónasson, P. M. 1989 Shape polymorphism in Arctic char, *Salvelinus alpinus*, in Thingvallavatn, Iceland. *Physiol. Ecol. Jpn. Spec.* Vol. I. pp. 393–404.

Snorrason, S. S., Jónasson, P. M., Jonsson, B., Lindem, T., Malmquist, H. J. Sandlund, O. T. and Skúlason, S. 1992 Population dynamics of the planktivorous arctic charr *Salvelinus alpinus* ('murta') in Thingvallavatn. *Oikos* **64**, 352–364.

Snorrason, S. S., Skúlason, S., Jonsson, B., Malmquist, H. J. and Jonasson, P. M. 1994 Trophic specialization in Arctic charr *Salvelinus alpinus* (Pisces: Salmonidae): morpho-logical divergence and ontogenetic niche shifts. *Biol. J. Linn. Soc.* 52, 1–18.

Stearns, S. C. and Crandall, R. E. 1984 Plasticity for age and size at sexual maturity: a life-history response to unavoidable stress. In *Fish reproduction: strategies and tactics*. (ed. G. W. Potts and R. J. Wootton), pp. 14–33. Academic Press, London.

Sutherland, W. J. 1987 Why do animals specialize? *Nature* **325**, 483–484.

Sutterlin, A. M. and MacLean, D. 1984 Age at first maturity and the early expression of ocyte recruitment proxesses in two forms of Atlantic salmon (*Salmo salar*) and their hybrids. *Can. J. Fish. Aquat. Sci.* 41, 1139–1149.

Svedäng, H. 1990 Genetic basis of life-history variation of dwarf and normal Arctic charr, *Salvelinus alpinus* (L.) in Stora Rosjöen, central Sweden. *J. Fish Biol.* 36, 917–932.

Swain, D. P. and Holtby, L. B. 1989 Differences in morphology and behavior in juvenile coho salmon (*Oncorhynchus kisutch*) rearing in a lake and in its tributary stream. *J. Fish. Aquat. Sci.* 46, 1406–1414.

Taylor, E. B. and Bentzen, P. 1993 Evidence for multiple origins and sympatric divergence of trophic ecotypes of smelt (*Osmerus*) in northeastern North America. *Evolution* 47, 813–832.

Thorpe, J. E. 1989 Developmental variation in salmonid populations. *J. Fish. Biol.* 35 (Suppl. A.), 295–303.

Thorpe, J. E., Morgan, R. I. G., Talbot, C. and Miles, M. S. 1983 Inheritiance of developmental rates in Atlantic salmon, *Salmo salar* L,. *Aquaculture* 33, 119-128.

Thorpe, J. E., Metcalfe, N. B. and Huntingford, F. A. 1992 Behavioral influences on life-history variation in juvenile Atlantic salmon, *Salmo salar*. *Environ. Biol. Fish.* 33, 331–340.

Travis, J. 1994 Evaluating the adaptive role of morphological plasticity. In *Ecological morphology. Integrative organismal biology* (ed. P. C. Wainwright and S. M. Reilly), pp. 99–122. The University of Chicago Press.

Trewavas, E., Green, J. and Corbert, S. A. 1972 Ecological studies on crater lakes in West Cameroon, fishes of Barombi Mbo. *J. Zool. Lond.* 167, 41–95.

Volpe, J. P. and Ferguson, M. M. 1996 Molecular genetic examination of the polymorphic Arctic charr, *Salvelinus alpinus*, of Thingvallavatn, Iceland. *Mol. Ecol.* 5, 763–772.

Vrijenhoek, R. C., Marteinsdóttir, G. and Schenck, R. 1987 Genotypic and phenotypic aspects of niche diversification in fishes. In *Community and evolutionary ecology of North American stream fishes* (ed. W. Matthews and D. Heins), pp. 245–250. Norman, OK. University of Oklahoma Press.

Wainwright, P. C., Osenberg, C. W. and Mittlebach, G. G. 1991 Trophic polymorphism in the pumpkinseed sunfish (*Lepomis gibbosus* Linnaeus) effects of environment on ontogeny. *Funct. Ecol.* 5, 40–55.

Walker, A. F., Greer, R. B. and Gardner, A. S. 1988 Two ecologically distinct forms of arctic charr *Salvelinus alpinus* (L.) in Loch Rannoch. *Biol. Conserv.* 43, 43–61.

West-Eberhard, M. J. 1986 Alternative adaptations, speciation, and phylogeny (a review). *Proc. Natl Acad. Sci. USA* 83, 1388–1392.

West-Eberhard, M. J. 1989 Phenotypic plasticity and the origins of diversity. *Annu. Rev. Ecol. Syst.* 20, 249–278.

Wilson, D. S. 1989 The diversification of single gene pools by density- and frequency-dependent selection. In *Speciation and its consequences* (ed. D. Otte and J. Endler), pp. 366–385. Sunderland, MA: Sinauer Assoc.

Wilson, E. O. 1992 *The diversity of life*. New York: W. W. Norton and Company.

Wimberger, P. H. 1991 Plasticity of jaw and skull morphology in the neotropical cichlids. *Geophagus brasiliensis and G. steindachneri*. *Evolution* 45, 1545–63.

Wimberger, P. H. 1992 Plasticity of fish body shape, the effects of diet, development family and age in two species of *Geophagus* (Pisces: Cichlidae). *Biol. J. Linn. Soc.* 45, 197–218.

Wimberger, P. H. 1994 Trophic polymorphisms, plasticity, and speciation in vertebrates. In *Theory and application in fish feeding ecology* (ed. D. J. Stouder, K. L. Fresh and R. J. Feller), pp. 19–43. Columbia, SC: University of South Carolina Press.

Wood, C. C. and Foote, C. J. 1996 Evidence for sympatric genetic divergence of anadromous and non-anadromous morphs of Sockeye salmon (*Oncorhynchus nerka*). *Evolution* 50, 1265–1279.

Wood, C. C. and Foote, C. J. 1990 Genetic differences in early development and growth of sympatric sockeye salmon and Kokanee (*Oncorhynchus nerka*) and their hybrids. *Can. J. Fish. Aquat. Sci.* 47, 2250–2260.

5

Sexual selection and natural selection in bird speciation

Trevor Price

5.1 Introduction

Natural selection, sexual selection and random drift have all been postulated to play roles in the origin of species. Natural selection has the strongest empirical support, because rapid speciation is best documented in adaptive radiations: the more or less simultaneous divergence of numerous lines from a common ancestor to exploit new adaptive zones (Simpson 1953, p.223; Schluter 1998). Artificial selection experiments, mimicking natural selection, often produce reproductive isolation as a correlated response (Rice and Hostert 1993). Comparisons of sister clades with differing numbers of species have been used to identify traits which accelerate speciation rates. Heard and Hauser (1994) reviewed 47 proposed 'key innovations'. Thirty-five (74%) were traits enabling exploitation of new resources (such as phytophagous habit in insects; Mitter *et al.* 1988), 10 (22%) affected dispersal, and just two (4%) were related to sexual selection.

Since Heard and Hauser's review more examples of traits accelerating speciation have been proposed, and some have implicated sexual selection. In birds, clades with relatively many species have been found to be associated with polygamous mating (Mitra *et al.* 1996; Møller and Cuervo 1998), with the frequency of sexually dichromatic species (Barraclough *et al.* 1995), and with the presence of sexual dimorphism in feather ornamentation (Møller and Cuervo 1998). Polygamy and sexual dimorphism are both thought to be indicators of strong sexual selection, and sexual selection is thus invoked as an agent of speciation. Previously, many authors had suggested a role for sexual selection in speciation, based on comparative evidence demonstrating that closely related species often vary most strikingly in sexually selected traits (Sibley 1957; Kaneshiro 1980; West-Eberhard 1983; Dominey 1984; Ryan and Rand 1993a). One good example is that of the South American bellbirds (Figure 5.1, also see Snow, 1976). The four species differ in facial ornamentation (Figure 5.1). Snow (1976, p.88) argued that 'some very powerful and arbitrary selective force is continually acting on the males'. Snow (1976) and West-Eberhard (1983) suggested that this force was sexual selection.

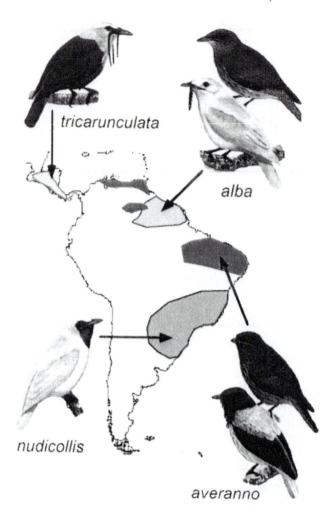

tricarunculata

alba

nudicollis

averanno

Fig. 5.1 The males of the four species of bellbirds, genus *Procnias,* and their distribution in Latin America. Females of two of the species are also shown. Figures and distributions are from Snow (1982).

The attractiveness of sexual selection as an agent of speciation derives from three main observations. First, closely related species often differ in traits which appear to have evolved by sexual selection. Second, sexually selected traits can exhibit almost infinite variety, as is shown for example by the ornaments of the bellbirds (West-Eberhard 1983). Natural selection often produces convergences and parallelisms, and if traits subject to parallel evolution are used in species recognition, several derived populations may continue to recognize each other, even as they do not recognize the ancestral type (Rice and Hostert 1993; Schluter and Nagel 1995). Third, because sexual selection requires communication between a signaller and a receiver, coevolution of the signalling trait and recognition system can result in members of some populations not being recognized by members of other populations (Kaneshiro 1980; Lande 1981). Pre-mating reproductive isolation automatically develops in parallel with the evolution of the sexually selected traits. This is demonstrated most clearly in models of Fisherian sexual selection by female choice (Lande 1981; Iwasa and Pomiankowski 1995; Payne and Krakauer 1997). Given certain preference functions, females at different positions along lines of equilibria will most often mate with males from their own population, and ignore members of other populations (Lande 1981; Ryan and Rand 1993a; Figure 5.2a).

Sexual selection offers an apparent mechanism for the generation of pre-mating isolation, but none for the generation of ecological differences. Natural selection generates ecological differences, but the way in which species recognition develops is left to be explained. In this review I consider how the two processes, and random drift, might interact. I address two questions:

1. What is the role for sexual selection in the establishment of pre-mating reproductive isolation among populations?
2. What is the role of sexual selection in adaptive radiations?

I draw on examples from birds, with some alternative taxa and speciation mechanisms considered in the discussion. I conclude, based on limited evidence,

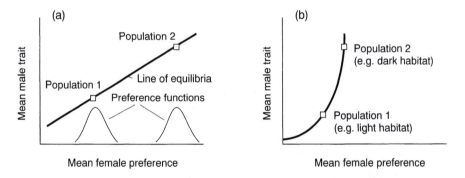

Fig. 5.2 (A) Under a Fisherian model of sexual selection, two populations can easily diverge to different points along a line of equilibria (Lande 1981). The female preference functions co-evolve with the male trait, and given the distribution of preferences illustrated, females would not recognize males of the other population as mates. Similar principles could apply when females have open-ended preferences for different traits (Lande 1981; Iwasa and Pomiankowski 1995). (B) One way by which sexually selected traits can evolve when they are indicators of male quality is if there is a cost to female search, and traits are more easily seen in some environments than others. The intensity of sexual selection on the male trait changes more rapidly than the female preference, which shows comparatively little evolution. In the example illustrated, the mean female preference and male trait lie along a curve in association with a habitat gradient (e.g. from lighter to darker; after Schluter and Price 1993).

that sexual selection is important in producing a great diversity of traits which are used in species recognition, but may have little effect on rates of ecological diversification in adaptive radiations.

5.2 Sexual selection and speciation

I equate speciation with the origin of reproductive isolation between diverging populations. Theories of sexual selection imply that correlated evolution of female preferences can automatically lead to reproductive isolation, as exemplified by the runaway (Fisherian) models. In this section I suggest two reasons why evolution of female preference as a consequence of a genetic correlation between male trait and preference is unlikely to play a major role in speciation. The reasons are first, most preferences are likely to be subject to some direct selection, and second, female preferences are often open-ended so that females from all populations prefer extreme males. Throughout the paper I concentrate on the evolution of female preferences for male sexually-selected traits, but given these conclusions, speciation may result as easily if traits are evolving via competition between males or by competition for resources other than mates (West-Eberhard 1983).

In the runaway models of sexual selection female preferences readily evolve and thereby amplify spatial patterns of variation in male traits (Lande 1982; Payne and Krakauer 1997). Divergence between populations in female preferences may be sufficient to cause speciation (Lande 1981; Iwasa and Pomiankowski 1995; Payne and Krakauer 1997). The models rely on the female preference evolving as a correlated response to evolution of the male trait (Lande 1981), and depend critically on the female preference being subject to no, or at least very little, direct selection (Kirkpatrick 1996). Numerous studies have demonstrated that this is seldom likely to be the case. Indeed females usually seem to base their choice on traits which indicate male quality (e.g. Price *et al.* 1993). When there are benefits to choice, the mean female preferences deterministically evolve to a single point (Kirkpatrick 1985; Price *et al.* 1993). Opportunities for runaway sexual selection (Kirkpatrick 1985, 1996) and divergence via genetic drift (Kirkpatrick 1985; Price *et al.* 1993) become severely limited.

Although there are still some proponents of the importance of Fisherian mechanisms (Iwasa and Pomiankowski 1995; Payne and Krakauer 1997), direct selection on female preferences seems more likely to guide their evolution. Often females prefer extreme development of the male trait (Ryan and Keddy-Hector 1992) perhaps because the males with the most extreme development also provide most benefits. The potential for rapid diversification in the male trait as well as the female preference seems to be restricted, because we might expect one kind of male to provide maximum benefits. However, there are many situations where variation in the environment can cause geographical variation in expression of a male trait.

The mean expression of the male trait is attributable to a balance between benefits (mating success) and costs (Kirkpatrick 1985; Price *et al.* 1993). Therefore, geographical variation in sexually selected traits will arise if there is geographical variation in the intensity of sexual or natural selection. For example, the intensity of natural selection on male traits may vary in association with resource abundances (Hill 1994; Møller *et al.* 1995) or predators (Houde and Endler 1990; Endler and Houde 1995). The intensity of sexual selection can also vary if the male trait is more easily seen, or is a better indicator of male quality, in some environments (Schluter and Price 1993). Under all these mechanisms substantial geographical variation in the male trait may be accompanied by only modest variation in female preference (Schluter and Price 1993, see Figure 5.2b). For example, in Hill's (1994) study of house finches, females from all populations showed strong preferences for bright males; variation in the male trait is attributed to the ease with which males can incorporate carotenoids into their diet.

If situations similar to the kind identified by Hill (1994) are common, geographical variation in male traits will not to lead to speciation, because females from all populations still prefer one kind of trait. One way by which populations can diverge in the male trait, and also to some extent in female preferences for those traits, is by the spread and fixation of new male phenotypes in different populations. There are likely to be many female preferences latent in a population, that are not expressed because the appropriate male phenotypes do not exist (Arak and Enquist 1993; Ryan and Rand 1993b). For example Burley *et al.* (1982) working with zebra finches, and Ryan and Rand (1993a,b) working with frogs have shown how males can be made more attractive to females by enhancing aspects of their signals (in Burley's case, by adding colour rings). Hidden preferences may arise as a side-effect of evolution of females under both natural and sexual selection (Ryan and Rand 1993b) or continually be gained and lost as a result of co-evolutionary arms races between female and male (Arak and Enquist 1993; Krakauer and Johnstone 1995). One consequence of hidden preferences is that new male mutations may be favoured (i.e. attract females) even if the males that carry them are of low quality. In the only model of this process, females evolve resistance to such mutations as they appear (Krakauer and Johnstone 1995). However, if the new male traits are, or become, better indicators of quality than the old ones (e.g. because they are more easily perceived in the environment), then they can rapidly replace the old trait (Schluter and Price 1993). Population divergence will readily occur as different male mutations in the different populations become fixed (Muller 1942; Sibley 1957). The bellbird variation (Figure 5.1) seems to be a good example of this sort of process.

If the male traits spread as a result of pre-existing biases in females and/or variation in the environment sexually selected traits may relatively easily evolve, but female preferences show relatively little divergence between populations. Female preferences will evolve somewhat as the costs and benefits of searching for males in different environments varies (e.g. Figure 5.2b), but the evolved differences are likely to be insufficient to generate reproductive isolation in themselves. This conclusion is apparently controverted by some artificial selection experiments (Wilkinson and Reillo 1994) and some studies of geographical variation (Houde and Endler 1990) which have demonstrated correlations between preference and trait. However, the selection experiments on stalk-eyed flies by Wilkinson and Reillo (1994) seem to include some direct selection on females, which is very difficult to eliminate in laboratory situations. In addition choice tests were apparently confounded by allowing females to choose between their own line versus stranger line males. In a re-evaluation and extension of their study on geographical variation in mate preferences of guppies, Endler and Houde (1995) found females preferred males with more orange coloration whatever their population of origin. The preference, although still directional, was less strong where the mean coloration was lowest and this could reflect natural selection on females in predator-rich areas (Endler and Houde 1995).

5.3. Sexual selection and population divergence in birds

The preceding review of the theory implies ample opportunity for sexually selected traits to diverge between populations, but less for female preferences to diverge in a way such that reproductive isolation can arise. In this section I concentrate on evidence that population divergence in male traits has indeed been driven by sexual selection, without considering the evolution of female preferences. In the following section I consider ways by which variation among populations in sexually selected traits might lead to restricted patterns of female choice, and hence pre-mating isolation. I concentrate on studies of birds. Many studies have demonstrated the operation of sexual selection within populations of birds (Andersson 1994). Sexually selected traits include song, plumage, and courtship displays. Song and plumage patterns have been the main foci of studies, and some results are considered here.

Song

Song variation has had a major role in bird speciation theory (Thielcke 1973; Baker 1996; Martens 1996) and the various competing hypotheses regarding song evolution are of interest for the differing weight they attribute to sexual selection. In many ways these hypotheses encapsulate debates about possible ways species recognition cues could diverge, incorporating arguments about small population size and founder effects, and natural and sexual selection. The debate does not center on the evolution of female preferences. Rather it revolves around the mechanisms of song divergence. If song differences between populations are generated, recognition of those differences is automatically assumed to follow.

The mere fact that songs are largely sung during the breeding season indicates their importance in mating success, and suggests a role for sexual selection in their evolution. The most consistent finding with respect to sexual selection is that females exhibit preferences for complex songs (e.g. ones with more syllable types; Searcy 1992; Hasselquist et al. 1996). Such preferences are seen even in species which sing very simple songs (Searcy 1992), again illustrating the presence of hidden female preferences, and a failure of preference and trait to co-evolve. Across species in the northern hemisphere, there is an association between song complexity and northerly breeding (Read and Weary 1992) suggesting that sexual selection is stronger, or opposing natural selection weaker, in more northern latitudes.

Sexual selection is an integral component to divergence in other aspects of song, in addition to complexity. In particular, some features of songs, such as frequency, are known to be adapted to physical characteristics of the environment, suggesting a role for habitat in driving song divergence (Morton 1975; Wiley 1991). Song divergence may also be affected by character displacement (or reinforcement) as a means of avoiding matings between closely related species, but there is very little evidence for this in birds (Thielcke 1969). Instead, song recognition rather than the song itself becomes refined in areas of sympatry (e.g. Gill and Murray 1972; Ratcliffe and Grant 1985; Lynch and Baker 1991); this may often be a result of learning.

Other hypotheses for song evolution are based on drift, and do not directly involve sexual selection. Songs have varying degrees of both learned and innate components depending on the species (Baptista 1996). The learning can be from fathers (Grant and Grant 1996), neighbours or other conspecifics (Payne and Payne 1993; Baptista 1996). Thielcke (1973) suggested that song evolution may result from a failure to learn components of songs in low density founding populations and there are several apparent examples (Thielcke 1973; Lynch and Baker 1991). It is not clear that the divergence in song is indeed driven by low population density of founders, however, and it is possible that the more rudimentary songs sung by outlying populations and species represent a response to relaxed sexual selection, assuming a cost to singing complex songs.

Most evidence for evolution by drift comes from changes in the form of the components of songs (i.e. syllables; Lynch 1996). For example, changes in syllable structure appear to play an important role in the evolution of songs in Darwin's finches, most likely as a consequence of drift (Grant and Grant 1996). While syllable structure may be neutral, the evidence that syllable number is also determined by random drift (Baker 1996; Lynch 1996) is more questionable. For example, Baker (1996) showed that island populations of an Australian honeyeater have fewer syllable types (and that mainland and island populations did not recognize each other's songs). He suggested that syllables on the island had been lost through drift, but a relaxation of sexual selection pressures on the island remains a viable alternative hypothesis.

Plumage

Most of the hypotheses discussed above for song have also been applied to the evolution of other putative sexually selected traits, such as courtship and plumage patterns. For example, geographical variation in patterns of sexual dichromatism have been explained on the basis of variation in sexual and natural selection pressures, species recognition, and drift (Peterson 1997). As in the case of song, some role for sexual selection has been established (Møller and Birkhead 1994), but a critical evaluation is still needed. Two studies have examined geographical variation in both the male trait and in female preferences from different geographical locations. Hill (1994) found that female house finches from all locations prefer red males, and Møller and de Lope (1994) found that female swallows from all locations prefer long-tailed males. The substantial geographical variation in the male trait was attributed in both cases to variation in the strength of natural selection opposing the sexual selection (Hill 1994; Møller et al. 1995). Both studies are thus in accord with the general idea that evolution of male traits need not be accompanied by evolution of female preference. As in song, there is a suggestion of increased sexual selection or reduced natural selection on plumage in more northerly latitudes, because dichromatism is more prevalent further north in Europe and in North America, both within species (Hill 1994; Møller and de Lope 1994) and between species (Curson 1994, pp. 11–12; Fitzpatrick 1994).

In summary, there is clearly a role for sexual selection in divergence of both song and plumage. In the case of song, drift also seems to play a major part. The relative roles of drift and sexual selection largely remain to be evaluated, with perhaps a larger role for sexual selection than has been assigned.

5.4 Pre-mating reproductive isolation in birds

The evidence that sexual selection is involved in the evolution of traits such as song and plumage is strong. In this section I ask how females might come to use male traits which have diverged as a result of sexual selection, as species recognition cues. Again, I restrict my evaluation to birds. In this group sexual imprinting is often of central importance in the establishment and maintenance of pre-mating isolation (Immelman 1975). Sexual imprinting has actually been defined as the process by which young birds learn species-specific cues that enable them to find a conspecific mate when adult (Clayton 1993). The learning is usually from the parents (Clayton 1993). The cues are largely auditory and visual, and both are typically employed (e.g. Baker and Baker 1990; Grant and Grant 1997a,b). The evidence for the importance of sexual imprinting in the field comes from heterospecific clutch swaps in gulls, sparrows, flycatchers and geese (Harris 1970; Fabricius 1991), which have always resulted in some adults entering into mixed species pairs. The mixed pairs are often due to just one mis-imprinted bird, and one raised elsewhere in a normal brood, thus demonstrating the remarkable power of imprinting to determine mating preferences. In captivity, similar results have been shown for many other taxa, including ducks, doves, quail and finches (Immelman 1975; ten Cate and Mug 1984; Clayton 1993). Imprinting itself may evolve to avoid mating with distantly related conspecifics (ten Cate and Bateson 1988) or to avoid mating with heterospecifics, if some post-mating isolation is present.

Sexual imprinting should stabilize mate choice towards the common type and makes it more difficult for traits to evolve under directional selection (Laland 1994). The apparent conflict between the stabilizing nature of any mechanism of mate choice involved in species recognition and the directional nature of sexual selection has long been used as an argument against the importance of sexual selection in the origin of mate recognition cues, as discussed by Ryan and Rand (1993a). Several resolutions to the conflict have been suggested, specifically with reference to imprinting. Female preferences can at times over-ride the imprinting process, which may be stronger in males than females (Immelman 1975; ten Cate and Mug 1984). ten Cate and Bateson (1988) suggest that imprinting provides a crude internal representation, and sensitivity to contrasts or bright colours also contributes to preferences. ten Cate and Bateson (1988) and Weary *et al.* (1993) argue that the imprinting itself, by producing aversions to some types, incidentally produces preferences for deviant types in the opposite direction. Such arguments imply a creative role for imprinting and indeed Immelman (1975) suggests that those groups which provide the most typical examples of imprinting (e.g. ducks, doves, estrildid finches) are also known for their extensive radiation, with several closely related species occurring in the same area. More work needs to be done to quantify Immelman's suggestion.

Sexual imprinting is clearly a powerful mechanism for maintaining pre-mating isolation, and can work on traits produced by natural selection and drift, as well as sexual selection. In Darwin's ground finches (*Geospiza*) the main species recognition cues are morphological differences (Ratcliffe and Grant 1983) produced as a result of natural selection (Schluter and Grant 1984) and song differences, produced as a result of cultural drift (Grant and Grant 1996). The employment of these cues in species recognition is attributable to the early experience of the female (Grant and Grant 1997a,b). The most extreme example of a role for imprinting comes from the parasitic viduine finches of Africa. Males mimic the songs of their foster species, and these songs are used as mate recognition cues. A group of closely related species has been produced by the invasion of several estrildid finch hosts. Payne (1973) labels these 'cultural species'. They are the best candidates for sympatric speciation in birds.

Obviously, not all species recognition arises through sexual imprinting. For example adult parasitic cuckoos, *Cuculus,* are not likely to interact with their young at all, and up to five similar congeneric species co-occur in the Himalayas. The selection pressures that cause the evolution of innate recognition mechanisms are unclear and need further study, but presumably include the disadvantages of mating with the wrong species (Mayr 1942, 1992). Habitat selection (through an imprinting-like mechanism) also seems to be involved in species recognition. It has been noted that hybridization between some closely related species occurs mainly in human-altered habitats (Chapin 1948; Sibley 1954).

5.5 The role of sexual selection in adaptive radiation

Sexually selected traits may rapidly diverge between populations, and these traits are used in species recognition. Potentially, therefore, sexual selection could elevate the rate of speciation, and there is some evidence for this in birds:

groups which are apparently subject to strong sexual selection are particularly speciose (Barraclough *et al.* 1995; Mitra *et al.* 1996; Møller and Cuervo 1998).

The influence of sexual selection on adaptive radiation is unexplored. One possibility is that sexual selection simply creates a large number of species that are ecological equivalents and allopatric replacements of one another (West-Eberhard 1983), with no influence on adaptive radiation. The production of reproductively isolated, ecologically equivalent species, could create increased opportunities for the establishment of sympatry and lead to accelerated adaptive diversification driven by ecological character displacement. But this need not be the case: ecological character displacement requires that at least some ecological differences are established in allopatry (reviewed by Taper and Case 1992). These ecological differences may be accompanied by sufficient divergence in traits that can be used in species recognition, even in taxa subject to only low levels of sexual selection. If this is so, then mildly and strongly sexually selected taxa will show no differences in rates of adaptive diversification.

In this section I consider the potential roles sexual selection may play in adaptive diversification. First, I ask if mildly and strongly sexually selected taxa differ in one measure of adaptive diversification (the number of species occurring in sympatry). Second, I investigate potential roles for sexual selection in promoting divergence in species recognition traits when an adaptive radiation is proceeding. The presence of some sexual selection is necessary for adaptive diversification, but different levels of sexual selection in different groups may have little affect on rates of adaptive diversification.

Sexual selection and ecological diversity

Sympatry is the ultimate test of good species, and it is usually thought that the establishment of sympatry requires both ecological differences and reproductive isolation (e.g. Mayr 1951). As one way to assess a role for sexual selection in adaptive radiation, I investigated the extent to which sexual selection is associated with high sympatric species' diversity. To do this, I extended the analysis of Barraclough *et al.* (1995), based on the compilation of Sibley and Monroe (1990). Sexually dichromatic species seem to be subject to generally stronger sexual selection than monochromatic species, because there is a higher variance in mating success among males (Møller and Birkhead 1994). Barraclough *et al.* (1995) demonstrated an association between the fraction of species in a clade that are sexually dichromatic and the number of species diversity in that clade. The findings of Barraclough *et al.* (1995) suggest that an elevated intensity of sexual selection translates into an increased rate of speciation.

For their main analysis Barraclough *et al.* (1995) compared 20 pairs of sister tribes (the taxon just above the genus), and estimated frequency of dichromatic species in each tribe from a sample of all the species. Fifteen pairs had differences in the estimated fraction of species which are dichromatic greater than 0.2, and an analysis restricted to these pairs showed a strong association between species number and dichromatism (in 12 of the 15 comparisons tribes which were more dichromatic were also more speciose). I restricted my analysis to these 15 pairs, although similar results were obtained when all 20 pairs are used. I divided the world distribution into approximately 40 country-sized regions (e.g. Colombia, S. Africa), and tabulated distributions of all species in the tribes according to the compilation of Sibley and Monroe (1990). I recorded the number of species occurring in each region, and the number of regions spanned by a tribe. Dichromatic tribes have wider distributions (10 wider, three smaller, two equal, sign test $P < 0.05$), but not significantly more species in their most speciose region than monochromatic tribes (however, this is due to a single reversal, nine more, four less, two equal, sign test, $P > 0.1$). Because the regions I evaluated are larger than a generally accepted notion of sympatry, I also compared the residuals from a regression of number of species in the most speciose region on number of regions occupied by the tribe. Here there was no evidence of an increased number of species in a region than expected on the basis of distribution; the more dichromatic tribes had the higher residual eight times, and the more monomorphic tribes had the higher residual seven times.

This crude first test finds no evidence for an increase in the number of sympatric species—and by extension ecological diversity—due to sexual selection. Rather, the elevated number of species in sexually dichromatic clades may be due to allopatric replacements. The test clearly could be refined, for example, by studying correlates of morphological diversity and sexual dimorphism. It also needs to be extended to specific case studies. The most speciose 'sympatric' assemblage in the dataset is that of the estrildine finches in central Africa (40 species). These are colourful and often sexually dichromatic birds. While a role for sexual selection is plausible in generating the sympatric diversity in this case, correlates between ecology and sexual dichromatism have been identified in birds. Sexual dichromatism and/or plumage coloration is associated with nest height (Martin and Badyaev 1996), foraging height (Garvin and Remsen 1997), geographical distribution (this study) and, specifically in finches, with habitat (Price 1996). Thus ecological differentiation could be occurring into those habitats which also favour sexual dichromatism.

Given the ease with which song, displays and plumage characteristics can evolve, pre-mating isolation based on such traits may evolve relatively easily, even in groups that are only mildly sexually selected. Ecological differentiation of isolated populations, and not the generation of pre-mating isolation, may be the more important rate-limiting step in adaptive radiation. The conclusion is in accord with the observation that rates of speciation slowdown as an adaptive radiation proceeds (Nee *et al.* 1992; Zink and Slowinski 1995). The slowdown is best attributed to filling of ecological niche space (Nee *et al.* 1992)

Evolutionary sequences in adaptive radiations

If sexual selection does not drive ecological differentiation, what part does it play in adaptive radiations? Traits subject to natural selection may also be used in species recognition, with a classic example being body size in Darwin's finches

(Ratcliffe and Grant 1983). In this case sexual selection need play no role in speciation. Rice and Hostert (1993) suggested that speciation arising as a correlated response to natural selection is likely to be most important during the early stages of an adaptive radiation, presumably because relatively large morphological differences are needed if they are to be adequate species recognition traits, and such differences may be generated early in adaptive radiations. The implication is that other processes, such as sexual selection, may be more important later in an adaptive radiation, when the morphological differences generated between new species may be smaller. In particular, divergence into different habitats may often promote divergence in sexually selected traits (Endler 1992; Marchetti 1993; Schluter and Price 1993), but be accompanied by only subtle changes in shape (Lack 1947; Grant 1986; Richman and Price 1992). I evaluate a possible role for sexual selection in adaptive radiation, by asking when and where ecological differences arise, and especially how habitat shifts occur.

There have been only a few attempts to address the sequence whereby niche differences are built up during adaptive radiations (Diamond 1986; Richman and Price 1992; Schluter *et al.* 1997; for a similar analysis in lizards see Losos 1992). Diagrammatic representations of some major ecological splits for Darwin's finches (Schluter *et al.* 1997) and Old World warblers (Richman and Price 1992) are shown in Figure 5.3. In both Darwin's finches and the warblers, a major early split appeared to be between ground and tree habit (Lack 1947). Within these major modes there are splits based on body size. Then, within the body size clades (of the leaf warblers) there are splits based on altitudinal (habitat) partitioning. Such partitioning has hardly evolved in the ground finches (Schluter and Grant 1982). The patterns appear to be consistent across these two quite different groups, with the adaptive radiation of the Darwin's finches having progressed no further than an earlier stage seen in the warblers. These conclusions are tentative and there are many statistical difficulties (Schluter *et al.* 1997). For

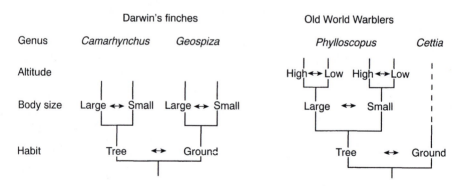

Fig. 5.3 Diagrammatic representation of the build-up of ecological differences in body size, altitude and arboreal habit, based on parsimonious reconstructions. For details of phylogeny and ecology see Grant (1986), Richman and Price (1992), Stern and Grant (1996) and Schluter *et al.* (1997). Each terminal branch represents a cluster of several species. There is little altitudinal segregation in Darwin's finches (Schluter and Grant 1982).

example, an alternative interpretation of the observation that closely related sympatric species are separated by altitude is that such segregation represents the first step in ecological differentiation whenever new species are formed, even during the early stages of the radiation. This was the interpretation given by Diamond (1986) and Richman and Price (1992) in their analyses of evolutionary sequences. It differs radically from the interpretation here, but both scenarios agree in that changes in habitat (tree/ground and/or altitude) occur both early and late in the adaptive radiations. Firm conclusions await more studies.

Taking the analysis as it stands, there is only modest support for the contention that reproductive isolation, arising solely as a side-effect of natural selection, predominates early in adaptive radiations (cf. Rice and Hostert 1993). Major morphological differences in size which are known to be involved in species recognition, appear after a tree/ground split. Both the tree/ground split and altitudinal partitioning seem to be accompanied by subtle shape changes (Lack 1947; Grant 1986; Richman and Price 1992), which may be less useful as species recognition cues than large shifts in body size. However, both tree/ground and altitude involve large changes in habitat. Habitat differences can affect the evolution of both song (Morton 1975) and colour patterns (Endler 1992; Marchetti 1993). Specifically, in both the finches and the warblers variation in song and plumage patterns is correlated with habitat (Bowman 1979; Grant 1986; Marchetti 1993; Badyaev and Leaf 1997). Thus ecological change associated with exploitation of new habitat will often be accompanied by change in mate recognition traits; traits which may be used in species recognition. Divergence in sexually selected traits, through the changing selection pressures in association with habitat may play an important role in speciation both early and late in adaptive radiations.

5.6 Discussion

Mayr's (e.g. 1942, 1951, 1958, 1963) influential views on speciation were derived mainly from studies on birds. He studied geographical variation in particular and identified many patterns of the sort illustrated in Figures 5.1 and 5.4. Mayr's 1958 paper starts with an account of the paradise kingfishers: there are three very similar subspecies on New Guinea, but four of the six taxa on offshore islands were classified as different species, and some still are (Figure 5.4). Mayr discounted selection alone as the cause of the differentiation between mainland and island, because the substantially varying conditions across mainland New Guinea were not associated with much geographical variation in plumage and morphology. This led him to emphasize the importance of gene flow in promoting homogeneity, and the necessity of its absence for differentiation. Mayr (1992) noted that one of the most surprising aspects of the evolutionary synthesis was the almost total neglect of sexual selection. In this discussion, I consider Mayr's views on speciation, in the light of sexual selection.

Mayr (e.g. 1963, p. 604) noted that behavioural changes in habitat and food

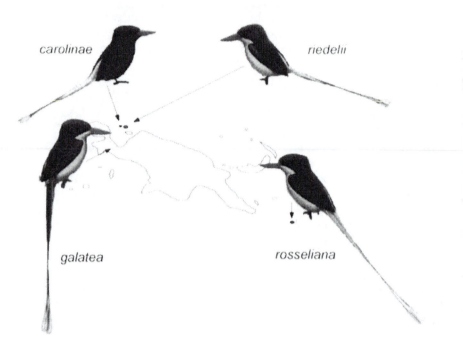

Fig. 5.4 Four forms of the paradise kingfisher, genus *Tanysiptera*. The two forms illustrated to the south are currently classified as conspecific (members of the superspecies *T. galatea*). The form *riedelii* is sometimes classified as a member of the *T. galatea* superspecies complex and sometimes classified as a separate species. *T. carolinae* is considered to be a good species (drawings and taxonomy from Fry and Fry 1992).

selection play a major role in the shift into new adaptive zones, with structural changes in morphology acquired secondarily. He also emphasized that an incipient species 'can complete the process of speciation only if it can find a previously unoccupied niche' (Mayr 1963, p. 574). The limited evidence to date suggests that while sexual selection can speed up speciation rates it has little effect on rates of adaptive diversification. However, habitat creates a link between adaptation and speciation via sexual selection: movement into new habitats is often associated with ecological diversification, and this will usually result in the evolution of communication signals used in mate identification.

Variation in habitat almost inevitably leads to variation in sexual selection pressures, by affecting the ease with which different traits are perceived (e.g. sound in a dense forest, colour in the open; Endler 1992), and also which traits are the best indicators of male quality (for example, a particular kind of parasite may occur in one habitat only). Variation in predators, parasites, or resources affects natural selection pressures on sexually selected traits. The net result is that environmental variation may often lead to substantial variation in male traits. In addition to environmentally driven variation, a potentially important mechanism of sexually selected population divergence is the spread and fixation of new male traits as a consequence of pre-existing female preferences for many kinds of novelty (Arak and Enquist 1993; Ryan and Rand 1993b). If populations are isolated from one another, fixed mutations will often differ among populations (Muller 1942; Sibley 1957).

Roles of small population size and genetic drift have been emphasized in many works on speciation, including Mayr's (1958) founder effect theory. However, the two mechanisms of population divergence discussed here are likely to be accelerated by large population size. The rate of divergence due to fixation of new mutations by genetic drift is independent of population size, and the rate of divergence due to fixation of favourable new mutations under selection is actually higher in large populations (Crow and Kimura 1970, pp. 418–430). The kind of mass-selected change induced by a changing environment is also likely to be more efficient in large than small populations (see Weber 1996 for a recent discussion).

I have argued that female preferences may rarely evolve as a genetically correlated response to sexual selection on males in a manner which generates pre-mating isolation between populations. If this is the case, the most critical need is an evaluation of the mechanisms by which sexually selected traits become involved in species recognition. Mayr (1992) stated that 'owing to change of function, sexual preference features become species recognition marks when two incipient species first come in contact with each other'. Such a change of function mechanism is implicated by the widespread occurrence of sexual imprinting, whose connection with sexual selection remains unclear. While it is apparent that the selection pressures which cause the evolution of species recognition sometimes result from contact between incipient species, as suggested by Mayr (and seen in the many cases of refined song recognition in areas of sympatry) it is not clear that this need generally be the case. Indeed there are many allopatric replacements which are classified as good species. There are many selection pressures in the environment, such as the presence of other more distantly related species, or individuals from differently adapted populations, which may directly favour the evolution of mechanisms used in species recognition, whether the recognition is by sexual imprinting or innate.

In this review I have concentrated on the evolution of pre-mating isolation via sexual selection in birds, and it is worth asking how generally applicable the conclusions are. Sexual imprinting, a major mechanism of pre-mating isolation in birds, seems to be restricted to a few groups. In other groups, the selection pressures which generate innate mechanisms of pre-mating isolation remain largely an open question. Avoidance of mating with (not necessarily closely related) heterospecifics possibly plays a major role. One of the most spectacular adaptive radiations is that of the cichlid fish in East African lakes. The species are colourful, and both sexual selection and sexual imprinting are thought to be rampant. Dominey (1984) argued an important role for sexual selection, and Immelman (1975) for sexual imprinting. As in birds, most of the evidence for speciation by sexual selection comes from studies of allopatric replacements of ecological equivalents. Key innovations affecting ecological abilities have also been suggested as the cause of the rapid diversification in the cichlids, and Meyer (1993) concludes his review on the cichlid radiation with no conclusion: 'the jury is still out on a role for sexual selection'. (See also Chapter 6 by Turner.)

In conclusion, sexual selection is strongly implicated in the production of cues which are used in species recognition. The overall importance of sexual selection in leading to reproductively isolated sympatric species remains to be determined. It will require a better understanding of the way by which sexually selected traits become species recognition cues, and on the environmental pressures generating population divergence in sexually selected traits.

Acknowledgements

I thank M. Benowitz-Fredericks, C. M. Garcia, P. Grant, D. Irwin, K. Marchetti, D. Schluter and A. Uy for advice.

References

Andersson, M. B. 1994. *Sexual selection.* Princeton University Press.
Arak, A. and Enquist, M. 1993. Hidden preferences and the evolution of signals. *Phil. Trans. R. Soc. London Ser.* B **340**, 207–213.
Badyaev, A. V. and Leaf, E. S. 1997. Habitat associations of song characteristics in *Phylloscopus* and *Hippolais* warblers. *Auk* **114**, 40–46.
Baker, M. C. 1996. Depauperate meme pool of vocal signals in an island population of singing honeyeaters. *Anim. Behav.* **51**, 853–858.
Baker, M. C. and Baker, A. 1990. Reproductive behavior of female buntings—isolating mechanisms in a hybridizing pair of species. *Evolution* **44**, 332–338.
Baptista, L. F. 1996. Nature and its nurturing in avian vocal development. In *Ecology and evolution of acoustic communication in birds* (eds. D. Kroodsma and E. Miller), pp.39–60. Ithaca: Cornell University Press.
Barraclough, T. G., Harvey, P. H. and Nee, S. 1995. Sexual selection and taxonomic diversity in passerine birds. *Proc. R. Soc. London* Ser. B **259**, 211–215.
Bowman, R. I. 1979. Adaptive morphology of song dialects in Darwin's finches. *J. Ornit.* 120, 353–389.
Burley, N., Krantzberg, G. and Radman, P. 1982. Influence of colour banding on the conspecific preferences of zebra finches. *Anim. Behav.* **30**, 444–455.
ten Cate, C. and Bateson, P. 1988. Sexual selection: the evolution of conspicuous characteristics in birds by means of imprinting. *Evolution* **42**, 1355–1358.
ten Cate, C. and Mug, G. 1984. The development of mate choice in zebra finch females. *Behaviour* **90**, 125–150.
Chapin, J. P. 1948. Variation and hybridization among the paradise flycatchers of Africa *Evolution* **2**, 111–126.
Clayton, N. S. 1993. Song, sex and sensitive phases in the behavioural development of birds. *Trends Ecol. Evol.* **4**, 82–84.
Crow, J. and Kimura, M. 1970. *An introduction to popuation genetics theory.* Minneapolis: Burgess.
Curson, J., Quinn, D. and Beadle, D. 1994. *New World warblers.* London: Christopher Helm.

Diamond, J. 1986. In *Community ecology* (eds. J. Diamond and T. Case), pp. 98–125. New York: Harper and Row.

Dominey, W. J. 1984. Effects of sexual selection and life histories on speciation: species flocks in African cichlids and Hawaiian *Drosophila*. In *Evolution of fish species flocks* (ed. A. A. Echelle and I. Kornfield), pp. 231–249. Maine: Orono Press.

Endler, J. A. 1992. Signals, signal conditions, and the direction of evolution. *Am. Nat.* **139**, s125–s153.

Endler, J. A. and Houde, A. E. 1995. Geographic variation in female preferences for male traits in *Poecilia reticulata*. *Evolution* **49**, 456–468.

Fabricius, E. 1991. Interspecific mate choice following cross-fostering in a mixed colony of Grey lag geese (*Anser anser*) and Canada geese (*Branta canadensis*). A study on development and persistence of species preferences. *Ethology* **88**, 287–296.

Fitzpatrick, S. 1994. Colourful migratory birds—evidence for a mechanism other than parasite resistance for the maintenance of good genes sexual selection. *Proc. R. Soc. London Ser. B* **257**, 155–160.

Fry, C. H. and Fry, K. 1992. *Kingfishers, bee-eaters and rollers: a handbook.* Princeton University Press.

Garvin, M. C. and Remsen, J. V. 1997. An alternative hypothesis for heavier parasite loads of brightly colored birds: exposure at the nest. *Auk* **114**, 179–191.

Gill, F. B. and Murray, B. G. 1972. Discrimination behavior and hybridization of the Blue-winged and Golden-winged warblers. *Evolution* **26**, 282–293.

Grant, B. R. and Grant, P. R. 1996. Cultural inheritance of song and its role in the evolution of Darwin's finches. *Evolution* **50**, 2471–2487.

Grant, P. R. 1986. *Ecology and evolution of Darwin's finches.* Princeton University Press.

Grant, P. R. and Grant, B. R. 1997a. Hybridization, sexual imprinting, and mate choice. *Am. Nat.* **149**, 1–28.

Grant, P. R. and Grant, B. R. 1997b. Genetics and the origin of bird species. *Proc. Natl Acad. Sci. USA* **94**, 7768–7775.

Harris, M. 1970. Abnormal migration and hybridization of *Larus argentatus* and *L. fuscus* after interspecies fostering experiments. *Ibis* **112**, 488–498.

Hasselquist, D., Bensch, S. and von Schantz, T. 1996. Correlation between male song repertoire, extra-pair paternity and offspring survival in the great reed warbler. *Nature* **381**, 229–232.

Heard, S. B. and Hauser, D. L. 1994. Key evolutionary innovations and their ecological mechanisms. *Hist. Biol.* 10, 151–173.

Hill, G. E. 1994. Geographic variation in male ornamentation and female mate preference in the house finch—a comparative test of models of sexual selection. *Behav. Ecol.* **5**, 64–73.

Houde, A. E. and Endler, J. A. 1990. Correlated evolution of female mating preferences and male color patterns in the guppy *Poecilia reticulata*. *Science* **248**, 1405–1408.

Immelman, K. 1975. Ecological significance of imprinting and early learning. *Annu. Rev. Ecol. Syst.* **6**, 15–37.

Iwasa, Y. and Pomiankowski, A. 1995. Continual change in mate preferences. *Nature* **377**, 420–422.

Kaneshiro, K. Y. 1980. Sexual isolation, speciation and the direction of evolution. *Evolution* **34**, 437–444.

Kirkpatrick, M. 1985. Evolution of female choice and male parental investment in polygynous species: the demise of the 'sexy son' *Am. Nat.* **125**, 788–810.

Kirkpatrick, M. 1996. Good genes and direct selection in the evolution of mating preferences. *Evolution* **50**, 2125–2140.

Krakauer, D. C. and Johnstone, R. A. 1995. The evolution of exploitation and honesty in animal communication: a model using artificial neural networks. *Phil. Trans. R. Soc. London Ser. B* **348**, 355–361.

Lack, D. 1947. *Darwin's finches.* Cambridge University Press.

Laland, K. N. 1994. On the evolutionary consequences of sexual imprinting. *Evolution* **48**, 477–489.

Lande, R. 1981. Models of speciation by sexual selection on polygenic traits. *Proc. Natl Acad. Sci. USA* **78**, 3721–3725.

Lande, R. 1982. Rapid origin of sexual isolation and character divergence in a cline. *Evolution* **36**, 213–223.

Losos, J. B. 1992. The evolution of convergent structure in Caribbean *Anolis* communities. *Syst. Biol.* **41**, 403–420.

Lynch, A. 1996. The population mimetics of birdsong. In *Ecology and evolution of acoustic communication in birds* (eds D. Kroodsma and E. Miller), pp. 181–197. Ithaca: Cornell University Press.

Lynch, A. and Baker, A. J. 1991. Increased vocal discrimination by learning in sympatry in 2 species of chaffinches. *Behaviour* **116**, 109–126.

Marchetti, K. 1993. Dark habitats and bright birds illustrate the role of the environment in species divergence. *Nature* **362**, 149–152.

Martens, J. 1996. Vocalizations and speciation of Palearctic birds. In *Ecology and evolution of acoustic communication in birds* (eds D. Kroodsma and E. Miller), pp. 221–240. Ithaca: Cornell University Press.

Martin, T. E. and Badyaev, A. V. 1996. Sexual dichromatism in birds: importance of nest predation and nest location for females versus males. *Evolution* **50**, 2454–2460.

Mayr, E. 1942. *Systematics and the origin of species.* New York: Columbia University Press.

Mayr, E. 1951. Speciation in birds. *Proc. Xth Int. Ornithol. Congress, Uppsala*, pp. 91–131.

Mayr, E. 1958. Change of genetic environment and evolution. In *Evolution as a process,* 2nd. edn (eds J. Huxley, A. Hardy and E. Ford.), pp. 157–180. London: Allen and Unwin Ltd.

Mayr, E. 1963. *Animal species and evolution.* Cambridge: Belknap Press.

Mayr, E. 1992. Controversies in retrospect. *Oxf. Surv. Evol. Biol.* 8, 1–34.

Meyer, A. 1993. Phylogenetic relationships and evolutionary processes in east African cichlid fishes. *Trends Ecol. Evol.* **8**, 279–284.

Mitra, S., Landel, H. and Pruett-Jones, S. 1996. Species richness covaries with mating system in birds. *Auk* **113**, 544–551.

Mitter, C., Farrell, B. and Wiegmann, B. 1988. The phlogenetic study of adaptive zones: has phytophagy promoted insect diversification? *Am. Nat.* **132**, 107–128.

Møller, A. P. and Birkhead, T. R. 1994. The evolution of plumage brightness in birds is related to extra pair paternity. *Evolution* **48**, 1089–1100.

Møller, A. P. and Cuervo, J. J. 1998. Speciation and feather ornamentation in birds. *Evolution* **52**, 859–869.

Møller, A. P. and De Lope, F. 1994. Differential costs of a secondary sexual character—an experimental test of the handicap principle. *Evolution* **48**, 1676–1683.

Møller, A. P., De Lope, F. and Caballero, J. M. 1995. Foraging costs of a tail ornament— experimental evidence from two populations of barn swallows *Hirundo rustica* with different degrees of sexual size dimorphism. *Behav. Ecol. Soc. Biol.* **37**, 289–295.

Morton, E. S. 1975. Ecological sources of selection on avian sounds. *Am. Nat.* **109**, 17–34.

Muller, H. J. 1942. Isolating mechanisms, evolution and temperature. *Biol. Symp.* **6**, 71–125.

Nee, S., Mooers, A. O. and Harvey, P. H. 1992. Tempo and mode of evolution revealed from molecular phylogenies. *Proc. Natl Acad. Sci. USA* 89, 8322–8326.

Payne, R. B. 1973. Behavior, mimetic songs and song dialects, and the relationships of the parasitic indigobirds (*Vidua*) of Africa. *Ornithol. Monogr.* **11**, 1–333.

Payne, R. B. and Payne, L. L. 1993. Song copying and cultural transmission in indigo buntings. *Anim. Behav.* **46**, 1045–1065.

Payne, R. J. H. and Krakauer, D. C. 1997. Sexual selection, space, and speciation. *Evolution* **51**, 1–9.

Peterson, A. T. 1997. Geographic variation in sexual dichromatism in birds. *Bull. Br. Ornithol. Club.* **116**, 156–172.

Price, T. 1996. An association of habitat with color dimorphism in finches. *Auk* **113**, 256–257.

Price, T., Schluter, D. and Heckman, N.E. 1993. Sexual selection when the female directly benefits. *Biol. J. Linn. Soc.* **48**, 187–211.

Ratcliffe, L. and Grant, P. R. 1983. Species recognition in Darwin's finches (*Geospiza,*-Gould). II. Geographic variaiotn in mate preference. *Anim. Behav.* **31**, 1154–1165.

Ratcliffe, L. and Grant, P. R. 1985. Species recognition in Darwin's finches (*Geospiza, Gould*). III. Male responses to playback of different song types, dialects, and hetero-specific songs. *Anim. Behav.* **33**, 290–307.

Read, A. F., and Weary, D. M. 1992. The evolution of bird song: comparative analyses. *Phil Trans. R. Soc. London Ser. B* **338**, 165–187.

Rice, W. R. and Hostert, E. E. 1993. Laboratory experiments on speciation—what have we learned in 40 years? *Evolution* **47**,1637–1653.

Richman, A. D. and Price, T. 1992. Evolution of ecological differences in the Old World leaf warblers. *Nature* **355**, 817–821.

Ricklefs, R. E. 1987. Community diversity: relative roles of local and regional processes. *Science* **235**, 167–171.

Ryan, M. J. and Keddy-Hector, A. 1992. Directional patterns of female mate choice and the role of sensory biases. *Am. Nat.* **139**, s4–s35.

Ryan, M. J. and Rand, A. S. 1993a. Species recognition and sexual selection as a unitary problem in animal communication. *Evolution* **47**, 647–657.

Ryan, M. J. and Rand, A. S. 1993b. Sexual selection and signal evolution—the ghost of biases past. *Phil Trans. R. Soc. London Ser. B* **340,** 187–195.

Schluter, D. 1998. Ecological causes of speciation, In *Endless forms: species and speciation* (ed. D. Howard and S. Berlocher), pp. 000–000. Oxford University Press.

Schluter, D., and Grant, P. 1982. The distribution of *Geospiza difficilis* in relation to *G. fuliginosa* in the Galapagos islands: tests of three hypotheses. *Evolution* **36**, 1213–1226.

Schluter, D., and Grant, P. 1984. Determinants of morphological patterns in communities of Darwin's finches. *Am. Nat.* **123**, 175–196.

Schluter, D. and Nagel, L. M. 1995. Parallel speciation by natural selection. *Am. Nat.* **146**, 292–301.

Schluter, D. and Price, T. 1993. Honesty, perception and population divergence in sexually selected traits. *Proc. R. Soc. London Ser. B* **253**, 117–122.

Schluter, D., Price, T., Mooers, A and Ludwig, D. 1997. Likelihood of ancestral states in adaptive radiation. *Evolution* **51**, 1699–1711.

Searcy, W. A. 1992. Song repertoire and mate choice in birds. *Am. Zool.* **32**, 71–80.

Sibley, C. G. 1954. Hybridization in the red-eyed towhees of Mexico. *Evolution* **8**, 252–290.

Sibley, C. G. 1957. The evolutionary and taxonomic significance of sexual dimorphism and hybridization in birds. *Condor* **59**, 166–191

Sibley, C. G. and Monroe, B. L. 1990. *Distribution and taxonomy of birds of the World.* New Haven: Yale University Press.

Simpson, G. G. 1953. *The major features of evolution.* New York: Columbia University Press.

Snow, D. W. 1976. *The web of adaptation.* New York: Quadrangle.

Snow, D. W. 1982. The cotingas: bellbirds, umbrellabirds and other species. Ithaca, NY: Comstock Press.

Stern, D. L. and Grant, P. R. 1996. A phylogenetic reanalysis of allozyme variation among populations of Galapagos finches. *Zool. J. Linn. Soc.* **118**, 119–134.

Taper, M. and Case, T. 1992. Coevolution among competitors, In *Oxford surveys in evolutionary biology,* Vol. 8 (ed. D. Futuyma and J. Antonovics), pp. 63–109. Oxford University Press.

Thielcke, G. 1969. Geographic variation in bird vocalizations. In *Bird vocalizations* (ed. R. Hinde), pp. 311–339. Cambridge University Press.

Thielcke, G. 1973. On the origin of divergence of learned signals (songs) in isolated populations. *Ibis* **115**, 511–516.

Weary, D. M., Guilford, T. and Weisman, R. G. 1993. A product of discriminative learning may lead to female preferences for elaborate males. *Evolution* **47**, 333–336.

Weber, K. E. 1996. Large scale genetic change at small fitness cost in large populations of *Drosophila melanogaster* selected for wind tunnel flight: rethinking adaptive surfaces. *Genetics* **144**, 205–213.

West-Eberhard, M. J. 1983. Sexual selection, social competition, and speciation. *Q. Rev. Biol.* **58**, 155–183.

Wiley, R. H. 1991. Associations of song properties with habitats for territorial oscine birds of eastern North America. *Am. Nat.* **138**, 973–993.

Wilkinson, G. S. and Reillo, P. R. 1994. Female choice response to artificial selection on an exaggerated male trait in a stalk-eyed fly. *Proc. R. Soc. London Ser. B* **255**, 1–6.

Zink, R. M. and Slowinski, J. B. 1995. Evidence from molecular systematics for decreased avian diversification in the Pleistocene epoch. *Proc. Natl Acad. Sci. USA* 92, 5832–5835.

6

Explosive speciation of African cichlid fishes

George F. Turner

6.1 Introduction

Lakes can be considered as inverted islands, as far as freshwater organisms are concerned, and the great lakes of East Africa, clustered in and around the African Rift valley, can be considered an archipelago where a single family of fishes, the Cichlidae, has radiated to produce more than 1500 species. Even though they were only known from a small and unrepresentative collection of preserved material, the unusual diversity of these fishes came to the attention of major figures in the modern synthesis of evolutionary biology in the first half of the twentieth century (Huxley 1942; Mayr 1942). The cichlid fishes rapidly became established as a textbook example of 'adaptive radiation' or 'explosive speciation'. Recent studies are only now revealing that the number of species has been greatly under-estimated, while the radiations in individual lakes may be much younger than has been suspected. The mechanisms underlying this extraordinary radiation are still unclear, but a combination of field studies, laboratory breeding experiments and application of molecular techniques has already overturned a number of theories.

6.2 Speciation rates

Species richness

Prior to the 1980s, most cichlid taxonomists operated with morphological species concepts, distinguishing species on the basis of anatomical differences among preserved material. In 1978, Holzberg demonstrated that a pair of sympatric 'colour morphs' of *Pseudotropheus zebra* at Nkhata Bay, Lake Malawi demonstrated subtle behavioural and ecological differences and courted assortatively in their natural habitat. Interpreting this finding in the light of Paterson's recently proposed Recognition Concept of species, Ribbink *et al.* (1983) granted specific status to any taxon exhibiting a clear difference in male breeding colours, providing preliminary identification features for 147 new putative species of cichlid fishes of the 'mbuna' lineage from the rocky shores of the western and southern portions of Lake Malawi alone. To date, molecular studies have been consistent with Ribbink *et al.*'s hypothesis for all sympatric taxa examined. Allozyme studies have demonstrated significant differences in allele frequencies among three *Petrotilapia* species (McKaye *et al.* 1982) and two species of *Pseudotropheus* (*Maylandia*) (McKaye *et al.* 1984). More recently microsatellite DNA and behavioural observations have been used to demonstrate significant assortative mating among three putative species of *Ps.* (*Maylandia*) and six of *Ps.* (*Tropheops*) (van Oppen *et al.* 1998).

Most researchers now recognize African great lake cichlid species on the basis of differences in colours, in addition to morphology. Furthermore, recent field studies of previously neglected habitats have also led to increased estimates of species richness. Turner (1996) has listed 199 species, 79 undescribed, from the deep-water and offshore habitats of Lake Malawi, while Seehausen (1996) recorded 173 from the rocky habitats of Lake Victoria, 100 of which were discovered since 1986. Both of these studies were conducted over small areas of the respective lakes, and both reported that many taxa had restricted distributions even with the areas surveyed, suggesting that many more species await discovery in other localities. A recent survey of the eastern shore of Lake Tanganyika (Snoeks *et al.* 1994) has also led to an increase in the estimate of cichlid species in that lake from 173 to more than 200.

On present estimates, it is likely that Lake Victoria contains at least 500 species, Tanganyika 200 and Malawi 700–1000 species. Some caution must still be attached to these estimates, as they depend on the assumption that specific rank should be given to allopatrically distributed taxa which differ in coloration by roughly the same extent as sympatric taxa which are known to mate assortatively.

Age of radiations

Much of the early speculation about cichlid diversity was based on the assumption that the lakes were of great age. For example, in the 1940s Lakes Malawi and Tanganyika were thought to be of early Cenozoic origin, i.e. 40–65 Myrs old (Huxley 1942; Mayr 1942). Recent studies have indicated that the lakes are much younger.

Geological surveys suggest that Lake Tanganyika arose as a shallow depression linked to the Congo River system about 20 Myrs ago (Coulter 1994), and that it probably became largely isolated about 9–12 Myrs ago (Meyer *et al.* 1994). The cichlid fishes of Lake Tanganyika are morphologically, behaviourally and genetically much more differentiated than those of Lakes Malawi and Victoria. Molecular clock estimates derived from mitochondrial DNA sequences indicate that some Tanganyikan lineages diverged at least 5 Myrs ago (Meyer *et al.* 1994).

Geologists estimate that Lake Malawi is about 1–2 Myrs old at most (Ribbink 1994). MtDNA clock estimates put the origin of the radiation at about 700 000 years ago (Meyer 1993). Although incomplete lineage sorting leads to considerable problems in the use of mtDNA for phylogenetic reconstruction for

Malawi cichlids, coalescence theory suggests that the maximum age of divergence of known mtDNA genotypes is about 1.8 Myrs, although this may pre-date the radiation (Parker and Kornfield 1997).

Until recently, Lake Victoria was thought to be about 250 000–750 000 years old (Meyer 1993). Geologists have recently discovered that the deepest part of the lake was completely dry as recently as 12 400 years ago and hydrological modelling indicates that no peripheral refugia could have existed in the lake's vicinity during this time, suggesting that the entire cichlid radiation must have occurred since then (Johnson *et al.* 1996). The Victorian cichlids are undoubtedly the least differentiated of those of the three great lakes, both anatomically and genetically. In a study of 14 species from nine endemic genera, Meyer *et al.* (1990) found no variation at all in a 363 base pair (bp) sequence of the mitochondrial cytochrome-b gene and only two to three substitution differences in a 440-bp region which includes the normally hypervariable control region. This is consistent with a very recent radiation.

Monophyly of cichlid radiations

All three lakes contain cichlid lineages which have not radiated: *Tilapia* in all three lakes; *Astatoreochromis* in Lake Victoria; *Oreochromis* in Lakes Victoria and Tanganyika; *Tylochromis* in Lake Tanganyika; *Serranochromis* in Lake Malawi. *Oreochromis* in Lake Malawi is represented by two lineages, only one of which has radiated and that into a mere three species (Sodsuk *et al.* 1996). Mitochondrial DNA sequence analysis has provided strong evidence that the remainder of the cichlids in all three lakes have radiated within the catchments of the lakes they presently inhabit. All but six of the Lake Malawi cichlid species form a monophyletic group, with the riverine *Astatotilapia calliptera* as the probable sister taxon to all the lacustrine members of the clade (Meyer 1993). The cichlids endemic to Lake Victoria and its satellite lakes, such as Kyoga, Nabugabo and Edward, also form a monophyletic group, and also appear to be derived from (a different species of) *Astatotilapia* (Meyer 1993).

The molecular evidence is less easy to interpret in the case of Lake Tanganyika. Long thought to be polyphyletic (e.g. Nishida 1991), Meyer (1993) has made the intriguing suggestion that the Tanganyikan radiation is actually monophyletic, but that several taxa have escaped from the lake and colonized other water bodies. This theory seems plausible in the case of the Lamprologine tribe, since the riverine *Lamprologus* presently found in the Congo River seem to be a monophyletic group descended, relatively recently, from one of the Tanganyikan endemics (Sturmbauer *et al.* 1994). Sturmbauer and Meyer (1993) have also found that the endemic Tanganyikan tropheini is the sister group to the haplochromine tribe and that this split also occurred relatively recently, perhaps 2–3 Myrs ago. The haplochromines comprise more than 1500 species, including not only the entire Malawian and Victorian radiations, but also riverine species found as far apart as Algeria, Syria and South Africa (Daget *et al.* 1991). By contrast, only eight haplochromine species are found in Lake Tanganyika

(Coulter 1991), and six of these are mostly found in peripheral water bodies rather than the main lake itself. Genetic analyses have not yet been performed on haplochromines from outside of the East African region.

Relative rates of cichlid and other radiations

In the light of this recent work and similar studies on other island species flocks, how do the speciation rates of cichlid radiations compare with those of other textbook examples of explosive speciation? Three parameters must be considered: the age of the radiation, the number of species at present, and the number of founding lineages (Table 6.1). These can be combined to estimate the mean interval between speciation events, assuming zero extinction, and a speciation rate which is constant with time and invariant among all lineages within each radiation. The mean speciation interval is mathematically equivalent to the mean doubling time in an exponentially growing population, and is given by:

$$L = \frac{t.\log_e(2)}{(\log_e N_t - \log_e N_0)},$$

where N_t = number of extant species, N_0 = number of founding species, and t = age of radiation.

This comparison (Table 6.1) shows that the Hawaiian *Drosophila* have a comparatively low speciation rate, as do the Lake Baikal amphipods, although in the latter molecular dating of the radiation has not yet been attempted. The speciation rate of the Tanganyikan cichlids is about average for an explosive speciation event, and comparable with or slower than those of the Baikal sculpins,

Table 6.1 Estimates for species richness and rates of speciation of some major 'island' radiations. (Mean speciation interval calculated from exponential growth model, see text)

Radiation interval	No. of species	Number of founder lineages	Age of radiation (Myrs)	Mean speciation interval (th. years)
Baikal sculpins[a]	27–29	1–2	2–2.5	400–600
Baikal amphipods[b]	254	1–5	25–30	3000–5000
Galapagos finches[c]	13	1	0.6–2.8	160–760
Hawaiian *Drosophila*[d]	800	1	23–40	2000–4000
Hawaiian crickets[e]	206–210	4	2.5	ca. 440
Tanganyikan cichlids[f]	170–200	1	5–12	650–1,600
Malawian cichlids[g]	600–1000	1	0.7–2	70–240
Victorian cichlids[h]	250–500	1	0.01–0.2	1–25

[a] Molecular estimates of age of radiation from Kirilchik *et al* (1995), other data from Martens *et al.* (1994).
[b] Martens (1997), Martens *et al.* (1994), age of radiation estimated from age of lake.
[c] Vincek *et al* (1997), age of radiation estimate from allozymes.
[d] Powell and Desalle (1995), age of radiation estimate molecular.
[e] Otte (1989), age of radiation estimated from biogeography.
[f] Age of radiation and number of founder lineages from Meyer (1993); number of species from Snoeks *et al.* (1994).
[g] Age of radiation and number of founder lineages from Meyer (1993); number of species from Turner (1996).
[h] Geological estimates of age of radiation from Meyer (1993) and Johnson *et al.* (1996); number of founder lineages from Meyer (1993); number of species from Seehausen *et al* (1997a).

Hawaiian crickets or Galapagos finches. However, the cichlids of Lakes Malawi and Victoria demonstrate outstandingly rapid speciation rates which may be approached only by the far smaller radiation of the Galapagos finches, and are clearly the premier examples of large-scale rapid radiation of any extant lineages in a limited geographical area.

6.3 Adaptive radiation and competition

Cichlid radiation not due to lack of competitors

There is no evidence that lack of ecological competition has stimulated the radiation of cichlid fishes in African lakes. Indeed, all three of the largest lakes have a rich fauna of non-cichlid fishes. Lake Tanganyika is estimated to contain 145 species of non-cichlids, 61 of which are endemic, representing 21 families (De Vos and Snoeks 1994). Lake Malawi has 45 non-cichlid species in 10 families (Ribbink 1994), and Lake Victoria 40 species in 11 families (Greenwood 1994). In comparison, Lake Baikal, which is virtually the same size and shape as Lakes Malawi and Tanganyika, but is older than any of the African lakes, contains 56 fish species, 27 of which are endemic (Martin 1994). Other large ancient lakes, such as Titicaca, Biwa and Ohrid also have high levels of fish species richness and endemicity (Martens *et al.* 1994). So, even forgetting about the cichlids, the African great lakes would be remarkable for the species-richness and endemicity of their fish faunas. In Lake Malawi, the non-cichlid fishes include specialized algal scrapers, detritus feeders, pelagic plankton feeders, benthic invertebrate eaters including specialized crab and mollusc feeders and numerous fish-eaters occupying all habitats. Among the non-cichlids are the 18 largest fish species (Jackson 1961) and probably also the two most numerous species in the lake.

Supralimital evolution?

As is well-known, cichlids in African great lakes demonstrate a remarkable diversity of feeding behaviour and structures (e.g. Fryer and Iles 1972; Liem 1991), but it may be an exaggeration to say (e.g. Fryer 1996) that they demonstrate 'supralimital' characteristics which are more complex or extreme than shown by the family throughout the rest of its range. The South American cichlids are probably more diverse in body form than those of Lakes Malawi or Victoria, and perhaps almost as diverse in feeding strategies. Some of the feeding strategies of African lake cichlids do seem to be unique within the family, for example scale-eating, fin-biting or ramming mouthbrooding females to make them disgorge their young. All of these exotic strategies occur only in a few uncommon species.

That diversification has not really been supralimital, does not mean it is unimpressive. All three lakes contain algae raspers, zooplankton feeders, visually-hunting benthic invertebrate feeders, sediment sifters, ambush piscivores, pursuit hunters, mollusc crushers and quite a few other major trophic types. In

Lake Malawi, cichlids have colonized the deep benthic and open water pelagic zones to the limits of oxygenated waters at about 200 m (Turner 1996). Similar life-styles have led to similar morphologies among cichlids in the three lakes. Comparative morphologists have frequently suggested that these anatomical similarities may indicate phyletic relationships and that cichlid lineages have frequently moved between the different lakes (e.g. Greenwood 1983; Liem 1991). This has been convincingly disproved by molecular phylogenetic reconstructions which indicate that both the Malawian and Victorian cichlid radiations are monophyletic and suggest that both are derived from very similar small generalized invertebrate feeders (Meyer 1993).

Radiation of reproductive behaviour?

Cichlids are well-known for their protracted parental care, complex visual signalling courtship displays, and rapid colour changes. However, it is not correct to add these to the list of specializations of lake cichlids (e.g. Fryer 1996), as these are widespread and almost certainly plesiomorphic traits in the cichlids.

Is reproductive behaviour as evolutionarily labile as feeding adaptations? Irrespective of their ecology and the habitat they live in, all of the endemic cichlids of Lake Malawi and Lake Victoria are maternal mouthbrooders, as are their riverine sister taxa of the genus *Astatotilapia*. The sister taxon for this haplochromine clade appears to be the Tanganyikan tropheini, also maternal mouthbrooders (Sturmbauer and Meyer 1993). Within Lake Tanganyika, most mouthbrooders belong to a single clade which encompasses five to seven tribes (Nishida 1991; Sturmbauer *et al.* 1994), while the monophyletic lamprologine clade contains all but one endemic non-mouthbrooding species (Sturmbauer *et al.* 1994). Substrate brooders in Lake Tanganyika show many ecological and anatomical parallels with mouthbrooders in Lake Malawi. So, the broad division between mouthbrooding and substrate-brooding is relatively ancient and ecological circumstances within the lakes do not seem to lead to transitions between these states.

Within each of these broad categories, the role of ecology in shaping behaviour is less clear, largely because the phylogenies are less well resolved. Tanganyikan mouthbrooders range from sexually monomorphic monogamous biparental brooders to extremely sexually dimorphic polygynous maternal brooders. The latter include *Enantiopus melanogenys* which practices an extreme form of lekking behaviour, where male territories, delimited by tiny ridges of sand, are no more than two body lengths in radius and tightly clustered in virtually perfect hexagonal close packing. It is not yet known how often maternal mouthbrooding has evolved from biparental mouthbrooding, although the reverse transition seems unlikely given the stability of the strategy among the Malawian and Victorian endemics. Among cichlids, mouthbrooding is not confined to African lake species, and has evolved independently in several lineages in Africa and South America.

Most substrate brooders appear to be monogamous and practice biparental care, but there are a number of harem polygynists, ranging from *Lamprologus brichardi,* a sexually monomorphic species where about 8% of males were observed to breed with two females (Limberger 1983), to *Lamprologus callipterus,* in which mature males weigh up to 80 times as much as females and may mate with as many as 56 females in a 4-month duration as harem-holder (Sato 1994). A mitochondrial DNA phylogeny of 25 lamprologine species, indicates polygyny has probably arisen independently as many as five times (Sturmbauer *et al.* 1994).

Pharyngeal jaws and the key innovation hypothesis

Cichlid fishes have two sets of jaws, the oral (external) set which is mainly used for prey capture and the pharyngeal (internal) set which does most of the crushing and chewing. This is not unusual in fishes, but the particular form and mode of operation of the cichlid pharyngeal apparatus is shared only with other families of the labroid assemblage, the wrasse, parrotfishes, damselfishes and surfperches. Since Liem's (1973) detailed anatomical studies of the cichlid pharyngeal apparatus, there have been numerous statements that this structure represents the 'key innovation' which has been responsible for the diversification of the cichlids (e.g. Nee and Harvey 1994). This is a difficult hypothesis to assess. Liem's study showed that the labroid pharyngeal apparatus was different from those of other fishes, so it is essentially a correlational study: labroids are diverse and species rich, and they have a particular type of pharyngeal jaws. More recently Galis and Drucker (1996) compared the cichlid pharyngeal apparatus with that of the less diverse (29 species, Moyle and Cech 1996), but phylogenetically not too distant centrarchids which are the dominant medium-sized percoid fishes in North American freshwaters. Galis and Drucker found that the upper and lower jaws of the cichlids were able to move independently, unlike those of the centrarchids, and that cichlids could transmit a greater force per unit area on to a prey item with an elliptical cross-section. However, it is not at all clear that either of these traits actually confer any great evolutionary advantage on the cichlids.

The surfperches (Embiotocidae) possess the labroid pharyngeal apparatus, but only comprise 23 rather similar species which have small mouths for picking up small benthic invertebrates (Moyle and Cech 1996). Damselfishes and anemone fishes (Pomacentridae) are species-rich (315 spp., Moyle and Cech 1996), but although numerous on coral reefs, are evolutionarily rather unadventurous feeders on plankton or algae growing on rocks. The wrasse (Labridae) are the only other family with the labroid pharyngeal apparatus to rival the cichlids in species richness (500+ spp., Moyle and Cech 1996) and in diversity of form and diet. The parrotfishes (Scaridae, 80 species) are probably best regarded as a specialized offshoot of the wrasse which use their massive dentition to feed from the surface layers of stony corals. A recent molecular phylogenetic study using single-copy nuclear DNA has demonstrated that the labroids are not a clade. Although wrasse might possibly be the sister group to the cichlids, the ecologically less varied damselfishes and surfperches seem to be closely related to each other, but not to the cichlids and wrasse (Streelman and Karl 1997).

Thus, it is clear that the cichlid-type pharyngeal apparatus is by no means a guarantee of adaptive radiation. We can consider the question from another angle. How does the adaptive radiation of cichlids compare with those of other coexisting fish families in environments other than the African great lakes? Despite the presence of cichlids, South American freshwaters are dominated by characoids and catfishes. The adaptive radiation of the South American characoids is said to rival that of the African lake cichlids (Lowe-McConnell 1975), despite the fact that most species have rather unspecialized pharyngeal teeth (Moyle and Cech 1996). In terms of species richness, both the characoids and the catfishes greatly outnumber the cichlids, the former by about 10 to 1 (Lowe-McConnell 1975). Cichlids are not much more significant in the rivers of Africa, comprising only 10–20% of the fish species (Greenwood 1991) and most are algae/sediment feeders, or rather unspecialized small- to medium-sized predators.

The case for recognizing the cichlid pharyngeal apparatus as a key innovation and an explanation for adaptive radiation remains weak.

6.4 Speciation

Adaptive radiation and explosive speciation are not the same

It is important to distinguish clearly between the processes of adaptive radiation and speciation. Classical speciation theory views species as coadapted gene complexes occupying adaptive peaks in a fitness landscape, and considers the conditions necessary to shift between peaks. Reproductive isolation is modelled as arising from the accumulation of genetic incompatibilities between populations, either in terms of their effects on developmental processes or on fitness in relation to its environment. Adaptive change is seen as essential in permitting each species to occupy a different peak and so is inextricably linked to the speciation process.

However, in terms of the biological species concept, speciation is the establishment of reproductive isolation between two populations and not primarily the scaling of a new adaptive peak. Coexistence of species need not depend on niche separation (Silvertown and Law 1987) and indeed hybrids are not necessarily unfit relative to the parental species (Arnold and Hodges 1995). Among closely related species of *Drosophila*, prezygotic isolation tends to be stronger than postzygotic isolation (Coyne and Orr 1989). If sexual selection leads directly to reproductive isolation, newly-split species may not differ at all in characteristics of their niche, so both may lie on the same adaptive peak (Turner and Burrows 1995).

While many studies have claimed to demonstrate niche partitioning, no experimental studies have unambiguously demonstrated competition, let alone competitive exclusion, between sympatric African cichlid fishes. Unquestionably, many species have narrowly specialized diets and habitat preferences, but it also noteworthy that many species seem to occur syntopically and feed on the same things (Turner 1994; Rossiter 1995). Among lamprologines and biparental mouthbrooders in Lake Tanganyika, further niche partitioning may occur through segregation of breeding territories (Rossiter 1995), but this cannot be true of most maternal mouthbrooders where territories are used only for spawning and the eggs are brooded in the mouths of the non-territorial females.

A striking feature of African lake cichlid communities is the frequent co-occurrence of several anatomically and ecologically very similar species which differ strikingly in male breeding colours (Ribbink *et al.* 1983; Turner 1994, 1996; Seehausen 1996). This suggests that speciation may occur through the establishment of reproductive isolation prior to the evolution of traits leading to niche differentiation. Under this scenario, adaptive radiation might well be important in permitting the coexistence of sibling species, but may not be important in the process of speciation *per se*.

Extralacustrine speciation

Has speciation occurred within lakes? The establishment of the monophyly of Malawian and Victorian haplochromine radiations (Meyer 1993) effectively rules out the idea that they arose from multiple colonization events. However, geological studies have demonstrated marked fluctuations in water levels of all three great lakes.

At low water levels, Lake Tanganyika was divided into two, or sometimes three, isolated lakes (Coulter 1991). The present-day distribution of mitochondrial DNA haplotypes closely follows the outline of these ancient lakes in *Tropheus* (Sturmbauer and Meyer 1992) and the tribe eretmodini (Verheyen *et al.* 1996), although not in *Simochromis* (Meyer *et al.* 1996). All are shallow water rocky-shore algal raspers, but Meyer *et al.* suggest that *Simochromis* are less strictly confined to rocky shores, permitting large-scale gene flow subsequent to the re-unification of the isolated palaeo-lakes. It is thus likely that these ancient basins must have persisted long enough to permit substantial genetic differentiation by mutation and drift, and there has clearly been the opportunity for allopatric speciation. However, there is no evidence that mitochondrial DNA variation associated with the outlines of these ancient lakes is a reflection of species status, rather than merely an indication of persistent low dispersal capabilities. Indeed, the gene trees produced by mtDNA conflict strikingly with current morphological taxonomy.

There is no evidence that either Lake Malawi (Owen *et al.* 1990) or Lake Victoria (Johnson *et al.* 1996) has even been subdivided into large basins. Proponents of extralacustrine speciation have thus proposed 'the Lake Nabugabo scenario' (Greenwood 1965). Nabugabo, with a surface area of ca. 30 km^2, is a swampy lagoon separated from Lake Victoria by a thin sand bar which formed about 3500 years ago. It contains five endemic haplochromine cichlids which differ in male colour from their relatives in Victoria. This is an exciting finding, suggesting extremely rapid evolution, perhaps by sexual selection, but it still remains to be shown that the Nabugabo forms are really biological species. Greenwood's hypothesis is that Lake Victoria's cichlid radiation is a result of many such events, but it would seem unlikely to be a general explanation for the diversification of Malawian and Victorian cichlids. Only a few species are habitually found in shallow swampy areas likely to be cut off by such barriers. If the Nabugabo scenario were generally true, it would be expected that rocky shore, pelagic and deep-water assemblages would not contain many monophyletic groups, rather most species would have arisen independently from inshore swamp-living ancestors. The only published evidence is against the Nabugabo scenario: in a broad phylogenetic study of Malawian cichlids using mtDNA RFLP Moran *et al.* (1994) found a well-supported clade comprising only rocky shore and deep-water species.

In conclusion, the evidence for extralacustrine speciation is convincing for Lake Tanganyika, but for Lakes Malawi and Victoria, the only realistic possibility is isolation in small peripheral lagoons, which seems unlikely to explain diversification of most species.

Allopatric speciation within lakes

The shorelines of these large lakes are comprised of various habitats: rocks, sand, papyrus swamps, and river mouths. All three lakes also contain offshore islands. It has long been known that species or colour forms of littoral cichlids are often confined to small geographic areas within Lakes Malawi (Fryer 1959) and Tanganyika (Poll 1950), and it has recently been demonstrated also for those of Victoria (Seehausen 1996). Rocky shore cichlids are rarely observed over open sand, lack a dispersal phase in their life history and have poor depth equilibration abilities (cichlids have closed swimbladders), implying that populations isolated by habitat barriers, such as non-rocky beaches or deep channels, were free to speciate in allopatry within each lake (Fryer and Iles 1972; Ribbink *et al.* 1983).

McKaye and Gray (1984) tested the dispersal capabilities of rocky shore cichlids in Lake Malawi by constructing an artificial reef on a sandy beach about 1 km from the nearest rocky shore. Within 6 months the reef had been colonized by representatives of three of the four most diverse species complexes of rocky shore cichlids. After 30 months, a representative of the fourth complex was also established. The reef was not colonized by any of the four low diversity complexes present. These results suggest that taxa with higher dispersal capabilities are also those with greater species diversity and that sandy shores might not be as great a barrier to gene flow as had been suggested (Turner 1994). Hert (1992) found that territorial male *Pseudotropheus aurora* on Thumbi West Island, Lake Malawi, have a strongly developed homing ability and returned to their territories within a few days even when transplanted 2.5 km along continuous rocky shore. However, none returned when transplanted to the mainland shore only 1 km away, but on the other side of a deep mud-bottomed channel. When 18 individuals were released in open water over the channel at distances of 500–800 m, eight had returned within 4 days. This suggests that, although the fishes are capable of navigating back to their home territories across open water, they will not do so voluntarily, suggesting that deep channels do represent a significant barrier to these fishes. Significant genetic differences have been demonstrated between isolated island populations using allozymes (McKaye *et al.* 1984) and mitochondrial DNA (Bowers *et al.* 1994; Moran and Kornfield 1995).

Recent investigations using microsatellite DNA has demonstrated that sandy beaches may after all be a barrier to gene flow among populations of rocky shore Malawi cichlids (van Oppen *et al.* 1997). When sampled from rocky headlands separated by sandy beaches of only 700–1,400m width, all four species showed significant differences at all six loci investigated.

Founder events and genetic bottlenecking

The subdivision of the lakes into numerous isolated habitat patches and the fact that females carry broods of fertilized eggs in their mouths suggests that extreme genetic bottlenecking or founder effects could be involved in speciation in African cichlids, as has been suggested for other island radiations (Carson and Templeton 1984). Genetic studies have not supported this. Moran and Kornfield (1995) found that one Malawian rocky shore species endemic to a single island had reduced mtDNA haplotype diversity, but that four other species did not. Other studies have generally found high levels of within-population variation in microsatellite (van Oppen *et al.* 1998) and major histocompatibility complex loci (Klein *et al.* 1993). Intra-population variation in mtDNA is often very high, and even distantly related species have been found to share several mtDNA lineages, suggesting that polymorphisms have survived numerous speciation events (Moran and Kornfield 1993; Kornfield and Parker 1997). Recently, microsatellite DNA has been used to demonstrate that females of both rocky and sandy shore fishes in Lake Malawi generally mate with several males in the course of producing a single clutch, further reducing the feasibility of extreme founder effects (Kellogg *et al.* 1995; Parker and Kornfield 1996).

Sympatric speciation

Most studies of African cichlids have concentrated on the rocky shore fishes, perhaps because they live in shallow clear waters safe for scuba studies. Far less is known about deep water and pelagic areas of the lakes, but these too support large numbers of endemic species (Turner 1996), although admittedly probably not as many as the rocky shores. Sandy shores are also home to diverse cichlid communities, and although it is often stated that rocky shores represent as much of a barrier to these species as sandy shores do to rocky shore species (e.g. Fryer 1996), there is no real evidence for this and I have observed many supposedly sandy shore species on rocky habitats in Lake Malawi. Could these non-rocky shore species have evolved by sympatric speciation? There is now good evidence for sympatric origin of fish species, via disruptive ecological selection and divergent habitat use, in recently glaciated temperate lakes (Schluter 1996). This is difficult to demonstrate in large lakes where heterogeneous habitats provide possibilities for geographic isolation, but cichlids have also radiated in several tiny crater-lakes in Cameroon. These lakes have monotonous shorelines and no possibility for habitat barriers. Previously thought to have been the result of multiple invasions, mtDNA studies have shown that the cichlid radiations of Lakes Barombi Mbo and Bermin are monophyletic, strongly suggesting sympatric speciation within each lake (Schliewen *et al.* 1994).

Sexual selection and speciation

All the endemic cichlids of Lakes Malawi and Victoria, and many of those of Lake Tanganyika are maternal mouthbrooders. Females provide all of the parental care and are generally small and drab. Males are larger, brighter, often have long fins. Sand-dwelling species often build relatively huge display structures or 'bowers' from sand. Males of some Malawian and Tanganyika cichlids aggregate on leks. These observations suggest the operation of strong sexual selection by female choice for male display traits.

Males of some sandy shore species from Lake Malawi experience strongly skewed mating success, which seems to be related to their position on the lek (McKaye 1991), the height of their sand bower (McKaye *et al.* 1990), and its symmetry which is in turn correlated with the male's level of parasite infestation (Taylor *et al.* 1998). Experiments with cichlid fishes from Lake George (Lake Victoria 'superflock') and the rocky shores of Lake Malawi have shown that females prefer males with larger numbers of yellow spots on their anal fins (Hert 1989, 1991) and that the preference was lost when these spots were eliminated by the application of dry ice (Hert 1989).

Sympatric members of the *Pseudotropheus zebra* complex from the rocky shores of Nkhata Bay, Lake Malawi, which are known to be reproductively isolated in their natural habitat (van Oppen *et al.* 1998), will mate assortatively in the laboratory where environmental cues, such as microhabitat preferences or seasonal spawning triggers, are absent (M. E. Knight *et al.*, unpubl.). Seehausen and van Alphen (1998) have shown that the red and blue forms of *Haplochromis nyererei* from Lake Victoria will court assortatively in the laboratory under white light, but females court at random under monochromatic light which eliminates the colour differences between males.

In a comparative study of Lake Malawi rock cichlids, Deutsch (1997) found no relationship between male coloration (hue or brightness) and environmental parameters, such as water depth or substrate type, suggesting that male colour was not influenced by habitat-related natural selection. There was also no indication that male colours were more divergent among sympatric species pairs than allopatric taxa, implying the absence of selection for reinforcement or reproductive character displacement. Seehausen *et al.* (1997b) found that males of Lake Victoria cichlids tend to be both brighter and more divergent in colour in areas of high water transparency, despite an increase in the abundance of visual-hunting predators. They also found that taxa which would not interbreed in clear water areas, hybridized in turbid areas, suggesting that sexual selection operates more strongly leading to reproductive isolation where permitted by environmental constraints.

6.5 Discussion

It is now clear that speciation rates in African lake cichlids have been extra-ordinarily fast. At the moment, it seems that the fastest diversification is occurring in lineages, such as the rocky shore species of Malawi and Victoria, where fine-scale population structuring resulting from habitat specificity and lack of dispersal is coupled with diversification in male courtship traits by sexual selection. However, this remains a working hypothesis and many questions remain unan-swered.

Why have tilapias not produced large species flocks? Species of the genus *Oreochromis* are present in all three lakes. Like the haplochromines, they are strongly dimorphic maternal mouthbrooders, yet the only lacustrine radiation has been three species in Lake Malawi. Why are there fewer cichlid species in Lake Tanganyika than in Lakes Malawi and Victoria? Tanganyika is much older and has a greater diversity of cichlid lineages than the other lakes. Is it possible that ecological competition has become more severe and has thinned out the number of the species? Why have cichlids radiated so dramatically in lakes, but not in rivers?

Perhaps Seehausen *et al.*'s (1997b) work may indicate the answer. If divergent sexual selection for male coloration operates weakly in turbid areas in lakes, it may not operate at all in rivers where the water is often virtually opaque during the rainy season floods.

Acknowledgements

I thank Patrick Doncaster, Martin Genner and Herbie Smith for comments on an earlier version of the manuscript.

References

Arnold, M. L. and Hodges, S. A. 1995 Are natural hybrids fit or unfit relative to their parents? *Trends Ecol. Evol.* **10**, 67–71.

Bowers, N., Stauffer, J. R. and Kocher, T. D. 1994 Intra- and interspecific mitochondrial DNA sequence variation within two species of rock-dwelling cichlids (Teleostei: Cichlidae) from Lake Malawi, Africa. *Molec. Phylogen. Evol.* **3**, 75–82.

Carson, H. L. and Templeton, A. R. 1984 Genetic revolutions in relation to speciation phenomena: the founding of new populations. *Annu. Rev. Ecol. Syst.* **15**, 97–131.

Coulter, G. 1991 *Lake Tanganyika and its life.* Oxford University Press.

Coulter, G. W. 1994 Lake Tanganyika. In *Speciation in ancient lakes* (ed. K. Martens, B. Goddeeris, and G. Coulter), pp. 13–18. Stuttgart: Schweizerbart.

Coyne, J. A. and Orr, H. A. 1989 Patterns of speciation in *Drosophila. Evolution* **43**, 362–381.

Daget, J., Gosse, J.-P., Teugels, G. G. and van den Audenaerde, T. 1991 *Check list of the freshwater fishes of Africa, Volume IV.* Brussels: ISNB.

Deutsch, J. C. 1997 Colour diversification in Malawi cichlids: evidence for adaptation, reinforcement or sexual selection? *Biol. J. Linn. Soc.* **62**, 1–14.

De Vos, L. and Snoeks, J. 1994 The non-cichlid fishes of Lake Tanganyika. In *Speciation in ancient lakes* (ed. K. Martens, B. Goddeeris, and G. Coulter), pp. 391–405. Stuttgart: Schweizerbart.

Fryer, G. 1959 The trophic interrelationships and ecology of some littoral communities of Lake Nyasa with special reference to the fishes, and a discussion of the evolution of a group of rock-frequenting Cichlidae. *Proc. Zool. Soc. London* **132**, 153–281.

Fryer, G. 1996 Endemism, speciation and adaptive radiation in great lakes. *Environ. Biol. Fishes* **45**, 109–131.

Fryer, G. and Iles, T. D. 1972 *The cichlid fishes of the Great Lakes of Africa.* Edinburgh: Oliver and Boyd.

Galis, F. and Drucker, E. G. 1996 Pharyngeal biting mechanisms in centrarchid and cichlid fishes: insights into a key evolutionary innovation. *J. Evol. Biol.* **9**, 641–670.

Greenwood, P. H. 1965 The cichlid fishes of Lake Nabugabo, Uganda. *Bull. Br. Mus. Nat. Hist. (Zool.)* **12**, 315–357.

Greenwood, P. H. 1983 The *Ophthalmotilapia* assemblage of cichlid fishes reconsidered. *Bull. Br. Mus. Nat. Hist. (Zool.)* **44**, 249–290.

Greenwood, P. H. 1991 Speciation. In *Cichlid fishes: behaviour, ecology and evolution* (ed. M. H. A. Keenleyside), pp. 86–102. London: Chapman and Hall.

Greenwood, P. H. 1994 Lake Victoria. In *Speciation in ancient lakes* (ed. K. Martens, B. Goddeeris and G. Coulter), pp. 19–26. Stuttgart: Schweizerbart.

Hert, E. 1989 The functions of egg-spots in an African mouthbrooding cichlid fish. *Anim. Behav.* **37**, 726–732.

Hert, E. 1991 Female choice based on egg-spots in *Pseudotropheus aurora* Burgess 1976, a rock-dwelling cichlid of Lake Malawi, Africa. *J. Fish Biol.* **38**, 951–953.

Hert, E. 1992 Homing and home-site fidelity in rock-dwelling cichlids (Pisces: Teleostei) of Lake Malawi, Africa. *Environ. Biol. Fishes* **33**, 229–237.

Holzberg, S. 1978 A field and laboratory study of the behaviour and ecology of *Pseudotropheus zebra* (Boulenger), and endemic cichlid of Lake Malawi (Pisces: Cichlidae). *Zool. Syst. Evol.- Forsch.* **16**, 171–187.

Huxley, J. 1942 *Evolution: the modern synthesis.* London: Allen and Unwin.

Jackson, P. B. N. 1961 *Check list of the fishes of Nyasaland.* Cambridge University Press.

Johnson, T. C., Scholz, C. A., Talbot, M. R., Kelts, K., Ricketts, R. D., Ngobi, G., Beunig, K., Ssemmanda, I. and McGill, J. W. 1996 Late Pleistocene desiccation of Lake Victoria and rapid evolution of cichlid fishes. *Science* **273**, 1091–1093.

Kellogg, K. A., Markert, J. A., Stauffer, J. R. and Kocher, T. D. 1995 Microsatellite variation demonstrates multiple paternity in lekking cichlid fishes from Lake Malawi, Africa. *Proc. R. Soc. London Ser. B* **260**, 79–84.

Kirilchik, S. V., Slobodyanuk, S. Y., Belikov, S. I. and Pavlova, M. E. 1995 Phylogenetic relatedness of 16 species of the Baikal lake cottoidei bullhead fishes deduced from partial nucleotide-sequences of mtDNA cytochrome-b genes. *Mol. Biol.* **29**, 471–476.

Klein, D. H., Ono, H., O'Huigin, C., Vincek, V., Goldschmidt, T. and Klein, J. 1993 Extensive Mhc variability in cichlid fishes of Lake Malawi. *Nature* **364**, 330–332.

Kornfield, I. and Parker, A. 1997 Molecular systematics of a rapidly evolving species flock: the mbuna of Lake Malawi and the search for phylogenetic signal. In *Molecular systematics of fishes* (ed. T. D. Kocher and C. A. Stepien) pp. 25–37. London: Academic Press.

Limberger, D. 1983 Pairs and harems in a cichlid fish, *Lamprologus brichardi. Z. Tierpsychol.* **62**, 115–144.

Liem, K. F. 1973 Evolutionary strategies and morphological innovations: cichlid pharyngeal jaws. *Syst. Zool.* **22**, 425–441.

Liem, K. F. 1991 Functional morphology. In *Cichlid fishes: behaviour, ecology and evolution* (ed. M. H. A. Keenleyside), pp. 129–150. London: Chapman and Hall.

Lowe-McConnell, R. H. 1975 *Fish communities in tropical freshwaters.* London: Longman.

Martens, K. 1997 Speciation in ancient lakes. *Trends Ecol. Evol.* **12**, 177–182.

Martens, K., Goddeeris, B. and Coulter, G. (eds) 1994 *Speciation in ancient lakes.* Stuttgart: Schweizerbart.

Martin, P. 1994 Lake Baikal. In *Speciation in ancient lakes* (ed. K. Martens, B. Goddeeris, and G. Coulter), pp. 3–11. Stuttgart: Schweizerbart.

Mayr, E. 1942 *Systematics and the origin of species.* NY: Columbia University Press.

McKaye, K. R. 1991 Sexual selection and the evolution of the cichlid fishes of Lake Malawi, Africa. In *Cichlid fishes: behaviour, ecology and evolution* (ed. M. H. A. Keenleyside), pp. 241–257. London: Chapman and Hall.

McKaye, K. R. and Gray, W. N. 1984 Extrinsic barriers to gene flow in rock-dwelling cichlids of Lake Malawi: microhabitat heterogeneity and reef colonisation. In *Evolution of fish species flocks* (ed. A. A. Echelle and I. Kornfield) pp. 169–184. Maine: University of Maine at Orono Press.

McKaye, K. R., Kocher, T. D., Reinthal, P. N. and Kornfield, I. 1982 A sympatric sibling species complex of *Petrotilapia* Trewavas from Lake Malawi analysed by enzyme electrophoresis (Pisces: Cichlidae). *Zool. J. Linn. Soc.* **76**, 91–96.

McKaye, K. R., Kocher, T. D., Reinthal, P. N., Harrison, R. and Kornfield, I. 1984 Genetic evidence for allopatric and sympatric differentiation among color morphs of a Lake Malawi cichlid fish. *Evolution* **38**, 215–219.

McKaye, K. R., Louda, S. M. and Stauffer, J. R. 1990 Bower size and male reproductive success in a cichlid fish lek. *Am. Nat.* **135**, 597–613.

Meyer, A. 1993 Phylogenetic relationships and evolutionary processes in East African cichlid fishes. *Trends Ecol. Evol.* **8**, 279–284.

Meyer, A., Kocher, T. D., Basasibwaki, P. and Wilson, A. C. 1990 Monophyletic origin of Lake Victoria cichlid fishes suggested by mitochondrial DNA sequences. *Nature* **347**, 550–553.

Meyer, A., Montero, C. and Spreinat, A. 1994 Evolutionary history of the cichlid fish species flocks of the East African great lakes inferred from molecular phylogenetic data. In *Speciation in ancient lakes* (ed. K. Martens, B. Goddeeris, and G. Coulter), pp. 407–423. Stuttgart: Schweizerbart.

Meyer, A., Knowles, L. L., Verheyen, E. 1996. Widespread geographical distribution of mitochondrial haplotypes in rock-dwelling cichlid fishes from Lake Tanganyika. *Mol. Ecol.* **5**, 341–350.

Moran, P. and Kornfield, I. 1993 Retention of an ancestral polymorphism in the mbuna species flocks (Teleostei: Cichlidae) of Lake Malawi. *Mol. Biol. Evol.* **10**, 1015–1029.

Moran, P., and Kornfield, I. 1995 Were population bottlenecks associated with the radiation of the mbuna species flock (Telecostei: Cichlidae) in Lake Malawi? *Mol. Biol. Evol.* **12**, 1085–1093.

Moran, P., Kornfield, I. and Reinthal, P. N. 1994 Molecular systematics and radiation of haplochromine cichlids (Teleostei: Perciformes) of Lake Malawi. *Copeia* **1994(2)**, 274–288.

Moyle, P. B. and Cech, J. J. 1996 *Fishes: an introduction to ichthyology*, 3rd edn. New Jersey: Prentice Hall.

Nee, S. and Harvey, P. H. 1994 Getting to the roots of flowering plant diversity. *Science* **264**, 1549–1550.

Nishida, M. 1991 Lake Tanganyika as an evolutionary reservoir of old lineages of East African cichlid fishes: inferences from allozyme data. *Experientia* **47**, 974–979.

Otte, D. 1989 Speciation in Hawaiian crickets. In *Speciation and its consequences* (ed. Otte, D. and Endler, J. A.), pp. 482–526. Sunderland, Massachusetts: Sinauer.

Owen, R. B., Crossley, R., Johnson, T. C., Tweddle, D., Davidson, S., Kornfield, I., Eccles, D. H., and Engstrom, D. E. 1990 Major low levels of Lake Malawi and implications for speciation rates in cichlid fishes. *Proc. R. Soc. London Ser. B* **240**, 519–553.

Parker, A. and Kornfield, I. 1996 Polygynandry in *Pseudotropheus zebra*, a cichlid fish from Lake Malawi. *Environ. Biol. Fishes* **47**, 345–352.

Parker, A. and Kornfield, I. 1997 Evolution of the mitochondrial DNA control region in the mbuna (Cichlidae) species flock of Lake Malawi, East Africa. *J. Mol. Evol.* **45**, 70–83.

Poll, M. 1950 Histoire du peuplement et origine des espèce la faune ichthyologique du lac Tanganyika. *Ann. Soc. R. Zool. Belg.* **81**, 111–140.

Powell, J. R. and Desalle, R. 1995 *Drosophila* molecular phylogenies and their uses. *Evol. Biol.* **28**, 87–138.

Ribbink, A. J. 1994 Lake Malawi. In *Speciation in ancient lakes* (ed. K. Martens, B. Goddeeris and G. Coulter), pp. 27–33. Stuttgart: Schweizerbart.

Ribbink, A. J., Marsh, B. A., Marsh, A. C., Ribbink, A. C. and Sharp, B. J. 1983 A preliminary survey of the cichlid fishes of rocky habitats in Lake Malawi. *S. Afr. J. Zool.* **18**, 149–310.

Rossiter, A. 1995 The cichlid fish assemblages of Lake Tanganyika: ecology, behaviour and evolution of its species flocks. *Adv. Ecol. Res.* **26**, 187–252.

Sato, T. 1994 Active accumulation of spawning substrate: a determinant of extreme polygyny in a shell-brooding cichlid fish. *Anim. Behav.* **48**, 669–678.

Schliewen, U. K., Tautz, D. and Pääbo, S. 1994 Sympatric speciation suggested by monophyly of crater lake cichlids. *Nature* **368**, 629–632.

Schluter, D. 1996 Ecological speciation in postglacial fishes. *Phil. Trans. R. Soc. London Ser. B* **351**, 807–814.

Seehausen, O. 1996 *Lake Victoria rock cichlids.* Germany: Verduijn Cichlids.

Seehausen, O. and van Alphen, J. J. M. 1998 The effect of male coloration on female mate choice in closely-related Lake Victoria cichlids (*Haplochromis nyererei* complex). *Behav. Ecol. Sociobiol.* **42**, 1–8.

Seehausen, O., Witte, F., Katunzi, E. F., Smits, J. and Bouton, N. 1997a Patterns of remnant cichlid fauna in southern Lake Victoria. *Conserv. Biol.* **11**, 890–904.

Seehausen, O., van Alphen, J. J. M. and Witte, F. 1997b Cichlid fish diversity threatened by eutrophication that curbs sexual selection. *Science* **277**, 1808–1811.

Silvertown, J. and Law, R. 1987 Do plants need niches? Some recent developments in plant community ecology. *Trends Ecol. Evol.* **2**, 24–26.

Sodsuk, P. N., McAndrew, B. A. and Turner, G. F. 1996 Evolutionary relationships of the Lake Malawi *Oreochromis* species: evidence from allozymes. *J. Fish Biol.* **47**, 321–333.

Snoeks, J., Rüber, L. and Verheyen, E. 1994. The Tanganyika problem: comments on the taxonomy and distribution patterns of its cichlid fauna. In *Speciation in ancient lakes* (ed. K. Martens, B. Goddeeris, and G. Coulter), pp. 355–372. Stuttgart: Schweizerbart.

Streelman, J. T. and Karl, S. A. 1997 Reconstructing labroid evolution with single-copy nuclear DNA. *Proc. R. Soc. London Ser. B* **264**, 1011–1020.

Sturmbauer, C. and Meyer, A. 1992 Genetic divergence, speciation and morphological stasis in a lineage of African cichlid fishes. *Nature* **358**, 578–581.

Sturmbauer, C. and Meyer, A. 1993 Mitochondrial phylogeny of the endemic mouthbrooding lineages of cichlid fishes from Lake Tanganyika in Eastern Africa. *Mol. Biol. Evol.* **10**, 751–768.

Sturmbauer, C. Verheyen, E. and Meyer, A. 1994. Mitochondrial phylogeny of the

Lamprologini, the major substrate spawning lineage of cichlid fishes from Lake Tanganyika in Eastern Africa. *Mol. Biol. Evol.* **11**, 691–703.

Taylor, M. I., Turner, G. F., Robinson, R. L. and Stauffer, J. R. 1998 Sexual selection, parasites and asymmetry of an extended phenotypic trait in a bower-building cichlid from Lake Malawi, Africa. *Anim. Behav.* **56**, 379–384.

Turner, G. F. 1994 Speciation mechanisms in Lake Malawi cichlids: a critical review. In *Speciation in ancient lakes* (ed. K. Martens, B. Goddeeris, and G. Coulter), pp. 139–160. Stuttgart: Schweizerbart.

Turner, G. F. 1996 *Offshore cichlids of Lake Malawi.* Lauenau, Germany: Cichlid Press.

Turner, G. F. and Burrows, M. T. 1995 A model of sympatric speciation by sexual selection. *Proc. R. Soc. London Ser. B* **260**, 287–292.

Van Oppen, M. J. H., Turner, G. F., Rico, C., Deutsch, J. C., Ibrahim, K. M., Robinson, R. L. and Hewitt, G. M. 1997 Unusually fine-scale genetic structuring found in rapidly speciating Malawi cichlid fishes. *Proc. R. Soc. Lond Ser. B.* **264**, 1803–1812.

Van Oppen, M. J. H., Turner, G. F., Rico, C., Deutsch, J. C., Robinson, R. L., Genner, M. J. and Hewitt, G. M. 1998 Biodiversity and evolution in rock-dwelling cichlid fishes from Lake Malawi: evidence from microsatellite markers. *Mol. Ecol.* **7**, 991–1001.

Verheyen, E., Rüber, L., Snoeks, J. and Meyer, A. 1996 Mitochondrial phylogeography of rock-dwelling cichlid fishes reveals evolutionary influence of historical lake level fluctuations of lake Tanganyika, Africa. *Phil. Trans. R. Soc. London Ser. B* **351**, 797–805.

Vincek, V., O'Huigin, C., Satta, Y., Takahata, N., Boag, P. T., Grant, P. R., Grant, B. R. and Klein, J. 1997 How large was the founding population of Darwin's finches? *Proc. R. Soc. London Ser. B* **264**, 111–118.

7

Sexual conflict and speciation

L. Partridge and G. A. Parker

7.1 Introduction

We discuss here the role of sexual conflict (or antagonism)—the different evolutionary interests of males and females—as a force regulating the potential for speciation. Although there has been awareness of sexual conflict and its ramifications for many years (e.g. Trivers 1972; Dawkins and Carlisle 1976; Dawkins and Krebs 1978, 1979; Parker 1979) there has been relatively little consideration of its possible implications for speciation (but see Parker 1974a, 1979; Wilson and Hedrick 1982). Our emphasis will be on the evolution of reproductive isolation, because of its central importance in the speciation process. We shall consider the potential role of sexual conflict in the evolution of both pre- and post-zygotic isolation.

Conflict between the sexes can occur through two genetic routes (Chapman and Partridge 1996a). Sexual conflict within gene loci occurs when allelic variation at one locus affects a trait in both males and females, there are different trait optima for the sexes, and hence selection for sexual dimorphism. There will then be a genetic correlation between trait values in the two sexes through pleiotropy— effects of the same alleles on males and on females. The difference between selection on the alleles when in females and in males can therefore result in one sex constraining the evolution of the other (Slatkin 1984; Rice 1984; Lande 1987). For instance, high levels of re-mating in females may be of no direct benefit to females themselves, and may instead occur as a result of a pleiotropic genetic correlation between the sexes for mating frequency, and a correlated response in females to selection for frequent re-mating in males (Halliday and Arnold 1987). (In addition, if males mate frequently then, given an equal sex ratio, females must also do so, but this has nothing to do with pleiotropy.) If frequent mating is not merely selectively neutral for females, but positively disadvantageous, perhaps because it incurs physiological costs (Fowler and Partridge 1989; Chapman *et al.* 1995), or increases vulnerability to predators (Rowe 1994; Arnqvist and Rowe 1995), then the resulting sexual conflict will lead to selection for sex-limitation, although this may be slow to evolve (Lande 1987). Before the evolution of sex-limitation is complete, there will be persistent conflicting selection on alleles at a single gene locus, depending if they are in females or in males.

Sexual antagonism can also arise between alleles at different gene loci, when the sexes come into behavioural and physiological conflict, potentially resulting in an arms race, with sequential selection on different male and female traits. There is then antagonistic coevolution among the (sex-limited) gene loci that determine behaviour in males and those that determine behaviour in females (Parker 1979). For instance, sexual harassment and coercion of females by males can result in the evolution of female behavioural and morphological counter-tactics such as avoidance of males (Clutton-Brock and Parker 1995a,b) and defensive abdominal spines (Arnqvist and Rowe 1995). Antagonistic selection between gene loci is likely to result in an array of behavioural, morphological, and physiological sex-limited adaptations that confer benefits on each sex against antagonistic adaptations in the other. Sexual arms races could result in either continuous rapid coevolution at the two sets of gene loci, or in long periods of stasis after attainment of equilibria.

Both types of sexual conflict could have an influence on the evolution of pre- and post-zygotic reproductive isolation. The best evidence on the evolutionary progression of these two kinds of barriers to gene flow comes from interspecific crosses in *Drosophila* (Coyne and Orr 1989, 1997), where speciation is presumed to be initiated in allopatry. In allopatric species pairs, pre- and post-zygotic reproductive isolation evolve at about the same rate, in both cases presumably as an incidental by-product of independent events in the allopatric populations. There are few data from other taxa, but at least one other study suggests that post-zygotic isolation accumulates progressively in allopatry (Tilley *et al.* 1990). Sexual conflict within loci could potentially have two effects. First, sexual dimorphism or sex-limitation that has evolved separately in two populations could break down in hybrids, giving rise to post-zygotic problems. Second, genetic correlations between the sexes for traits involved in mate choice, and separate evolution in allopatry, could affect the likelihood of inter-population matings when secondary contact is made. Between-locus sexual conflict could affect the rate of evolution of reproductive isolation in allopatry. Furthermore, it could influence the evolution of post-zygotic reproductive isolation in allopatry, if it leads to an arms-race and hence rapid divergence of reproductive traits. Between-locus sexual conflict could also play a role in the evolution of pre-mating isolation after secondary contact between populations, and during parapatric and sympatric speciation. In sympatric, but not allopatric, species pairs of *Drosophila*, pre-zygotic isolation apparently evolves much more quickly than does post-zygotic, providing evidence that partial post-zygotic isolation at the time that sympatry resumes may select for reinforcement of mating barriers (Coyne and Orr 1989, 1997). Theoretical models have shown that selection for mating barriers can also be initiated in both parapatry and sympatry (Lande 1982; Liou and Price 1994; Kelly and Noor 1996; Payne and Krakauer 1997). Differing selection on females and on males for acceptance of matings in inter-population encounters could produce a role for sexual conflict in determining whether reinforcement of mating barriers does or does not occur.

In this paper, we explore the potential role of conflict between the sexes in these various aspects of the speciation process.

7.2 Within-locus sexual conflict

In theory, sexual dimorphism could be achieved by the incorporation into the population of genetic variants that, from the outset, produce sexual dimorphism. There would then be no sexual conflict. However, most new mutations are not completely sex-limited, and it seems likely the evolution of sexual dimorphism often proceeds by the invasion of alleles that produce an advantage only in one sex, followed by the evolution of sex limitation (Rice 1984). Before the evolution of sex limitation the new alleles are sexually antagonistic. Theoretical work has shown that such alleles can invade populations (Rice 1984). For autosomal genes, the disadvantage to one sex must be outweighed by the advantage to the other. In contrast, for sex-linked genes with appropriate dominance relationships, such alleles can invade even when the cost to one sex far exceeds the gain to the other. Specifically, if sex-linked genes coding for the sexually antagonistic trait are dominant and favour the homogametic sex or if they are recessive and favour the heterogametic sex, then they can often invade. Theoretically, alleles producing sex limitation can also then invade (Rice 1984), although the process may be slow (Slatkin 1984; Lande 1987). The extent to which this type of sexual antagonism could have an evolutionary role in speciation will therefore depend in part upon the rates at which these processes occur.

Some empirical evidence suggests that sexually antagonistic alleles are present in natural populations. An artificial selection experiment with *Drosophila melanogaster* forced visible mutants on the autosomes to pass through only females in each generation. After 29 generations of passage through females, when the resulting mutant chromosomes were introduced to males, they resulted in markedly lowered fitness compared with appropriate controls passaged through both sexes (Rice 1992). However, it was not evident from the results of this experiment whether the alleles affected the same traits in males and in females. Antagonistic effects of the same allele on male and female fitness may be mediated through effects on different characters in the two sexes.

Sexual dimorphism or sex limitation that evolves separately in different populations could break down in hybrids, producing problems with inviability or infertility for either sex. This idea does not seem to have been directly investigated. Some evidence consistent with it comes from the finding that, in introgressions of chromosomal segments between *Drosophila simulans* and *D. mauritiana*, segments from the X chromosome had somewhat larger effects than those from autosomes in producing hybrid male sterility (True *et al.* 1996). This is consistent with an interpretation either of dominant female advantage genes evolving on the X, with the subsequent evolution of sex-limitation which then broke down in the hybrids, or with the presence of recessive X-linked male advantage alleles, which were then incompatible with one another in the hybrids. Assessment of the role of genetic correlations between the sexes in this aspect of the speciation process will become more feasible when the identities of more of the genes responsible for post-zygotic isolation are known.

Pleiotropic genetic correlations between the sexes could also affect the likelihood of matings in inter-population encounters. If some male and female reproductive traits are genetically correlated with one another, and therefore evolve together between populations, then the relative likelihoods of intra- and inter-population matings could be affected. There has been rather little work on this topic, and such data as exist are not strongly supportive of a role for this kind of effect in speciation. In general, it seems unlikely that male and female reproductive traits will be homologous, although characters such as body size certainly are, and can contribute to assortative mating (Arnqvist *et al.* 1996; Rowe and Arnqvist 1996). One trait that has received some attention is the willingness of the two sexes to mate (Halliday and Arnold 1987; Arnold and Halliday 1988, 1992; Gromko 1992). The few data that exist on this genetic correlation are somewhat conflicting (Stamencovic-Radak *et al.* 1992), but some data do support the existence of a genetic correlation between male and female mating speeds in *Drosophila melanogaster*. Artificial selection for mating speed conducted on the two sexes separately showed that the trait was heritable, and indicated a possible genetic correlation between male and female mating speed (Stamencovic-Radak *et al.* 1992, but see Stamencovic-Radak *et al.* 1993; Butlin 1993). However, the indications are that, between different geographic populations of this species, the correlation is the opposite; populations with more vigorous males have more resistant females (van den Berg *et al.* 1984). The reasons for this correlation are not known. Kaneshiro (1976) has suggested that such a correlation might arise through sexual selection during the founding of new populations, with derived populations in general composed of less vigorous males and less discriminating females, but this theory has found little theoretical or empirical support. In inter-population matings with *D. melanogaster*, the most probable mating was between more vigorous males and the more receptive females, while matings between resistant females and less vigorous males were the least likely (van den Berg *et al.* 1984). This kind of correlation would therefore result in asymmetrical pre-mating reproductive isolation between populations.

7.3 Between-locus sexual conflict

Evolution of reproductive isolation in allopatry

When populations become allopatric, evolution occurs independently in each of the isolates. The likelihood of the populations acting as, or becoming, separate species if there is secondary contact will depend upon the degree of pre- and post-zygotic isolation that has evolved in allopatry. Very little is known about how post-zygotic isolation evolves, since very few genes producing it have yet been identified. Between locus sexual conflict could have a bearing if it led to rapid divergence of reproductive traits. Rapid evolutionary change associated with sexual selection has been suggested to contribute to the increased vulnerability to sterility of male hybrids in *Drosophila* (Wu and Davis 1993). Rapid evolutionary divergence would in general be expected to give rise to a higher level of genetic incompatibility when populations meet and attempt to interbreed. But investigation of this interesting possibility is a task for the future.

Pre-mating isolation is likely to evolve in allopatry through a combination of sexual and natural selection, and sexual conflict seems likely to play a role mainly in the former process. Models of the Fisherian process of sexual selection have emphasised the importance of neutral equilibria, random factors and runaway sexual selection of female mating preferences and preferred male traits (Lande 1981; Kirkpatrick 1982). This type of sexual selection could therefore act as fuel for the evolution of pre-mating reproductive isolation. This type of divergence through sexual selection could also result in the frequently observed asymmetric pre-mating isolation between populations when secondary contact is resumed (Arnold et al. 1996). It has also been suggested that male reproductive structures that come into physical contact with females, including sperm, show particularly rapid evolutionary change, because female reproductive tracts themselves undergo rapid evolution so as to allow females to make post-copulatory mate choice (see Eberhard 1996 for a review). Recent evidence has shown that post-mating, pre-zygotic events can be important in reproductive isolation in insects (Gregory and Howard 1994; Price 1997). Whether female choice occurs before or after mating, the Fisherian picture is one of rapid co-evolution of male trait and female mating preference. Models of sexual selection based on detectability of male traits or on indicators of male quality also produce population divergence with a positive correlation between male trait and female mating preference for it, but in these cases the male trait shows much greater inter-population divergence than does the female preference (Schluter and Price 1993; Price Chapter 5). Both of these processes would tend, if anything, to lead to assortative mating on secondary contact, and hence to promote speciation, although the importance of this process may have been overemphasised by the Fisherian models of sexual selection (Price Chapter 5).

A rather different picture of intersexual selection has recently been suggested, which raises the possibility of the involvement of inter-locus sexual conflict. Females may be selected to resist frequent mating because they can incur physiological (Fowler and Partridge 1989; Chapman et al. 1995) or ecological (Rowe 1994; Arnqvist and Rowe 1995) costs of mating that impair their lifetime reproductive success. Males may then be counter-selected to induce females to mate at a frequency greater than their own optimum (Parker 1979). Sensory exploitation during courtship is one method that could be used by males to persuade the female to mate (Dawkins and Krebs 1978, 1979; West-Eberhard 1984; Ryan 1990; Kirkpatrick and Ryan 1991; Endler 1992, 1993; Krakauer and Johnstone 1995; Rice and Holland 1997). The male could produce secondary sexual characters with properties that females have been selected to favour in other contexts. Females could then respond by evolving resistance to, rather than preference for, the initially preferred male trait. There is some evidence from phylogenies of male secondary sexual characters and female mating preferences for them that this process may be important in practice. In some phylogenies, female preference for male traits are stronger in those species where the male trait has apparently never been present than in those where it is present (reviewed in Rice and Holland 1997). It will be important to determine why we apparently see female resistance evolving in some cases and female preference in others (e.g. Houde and Endler 1990).

This kind of sexual conflict could continue after insemination has occurred. For instance, sperm–ovum conflict may occur if polyspermy is a problem, leading to an arms race between eggs to resist penetration by sperm and sperm to penetrate as rapidly as possible (reviewed in Rice and Holland 1997). Males may also resort to molecular coercion, which is potentially costly to females. Female *Drosophila melanogaster* that re-mate more frequently suffer an increased death rate and reduced lifetime reproductive success (Fowler and Partridge 1989). This effect is entirely attributable to the effects of peptides in the seminal fluid of the male (Chapman *et al.* 1995). These molecules act in the reproductive interests of males by elevating female egg-laying rate (Chen *et al.* 1988), reducing receptivity to future matings (Chen *et al.* 1988) and removing or disabling the sperm of previous mates of the female (Harshman and Prout 1994; Chapman *et al.* 1995; Gilchrist and Partridge 1997). Female death rate is elevated as a (presumably) unselected side-effect. Similar findings have been made for the nematode *C. elegans* (Gems and Riddle 1996). *Drosophila* females can evolve resistance both to male courtship and to the cost of mating. The cost of mating is context-specific; it appears only at levels of nutrition above those that females normally encounter (Chapman and Partridge 1996b), presumably indicating that females have evolved a mating rate that is not inappropriately high for the conditions that they normally encounter. Female *Drosophila* prevented from co-evolving in response to males rapidly evolved increased susceptibility to the cost of mating with those males (Rice 1996), indicating that there must be an arms race between the sexes. Some support for this point of view also comes from the finding that at least some seminal fluid proteins (Acp26Aa and Acp26Ab) show a high evolutionary rate (Aguade *et al.* 1992; Tsaur and Wu 1997). However, it should be borne in mind that this rapid evolution could be the result of a sexual arms race between males; some indication of whether these molecules are in evolutionary conflict with the females would be useful.

Sexual conflict resulting from sensory exploitation, polyspermy and the cost of mating could have some complicated effects on populations in secondary contact. If females evolve various forms of resistance to male pre- and post-mating manipulation, then the outcome could depend upon the duration of allopatry. At least after short periods of allopatry, males from one population could be more successful with females from the other, because the females would have evolved resistance to their own, but not to the allopatric, males. Males would then at the same time introduce resistance alleles to the other population, and these would tend to spread because of their advantage to the females. These are not easy ideas to test, but would repay further study. On the longer term, further divergence could lead to pre-zygotic isolation.

Evolution of pre-mating isolation in secondary contact

When previously allopatric populations resume secondary contact, varying degrees of pre- and post-mating isolation may be apparent. Speciation may already be complete if inter-mating does not occur, or if hybrids are completely inviable or infertile. In contrast, if reproductive isolation is weak, the two populations may simply coalesce. If hybrids are at a disadvantage and there is free mixing of the populations, then extinction of the rarer population is inevitable unless the two populations exploit different resources. Extinction could be avoided if the populations do not mix freely, as a result of only partial overlap or habitat segregation. Further evolution could the lead to the formation of a stable hybrid zone. Finally, complete pre-mating isolation between the incipient species may evolve in secondary contact, in other words reinforcement may occur. The plausibility of reinforcement has been the subject of much discussion and some controversy (see Butlin 1987, 1995 for reviews). One problem is that gene flow from the parent populations into the zone of interbreeding could prevent reinforcement of pre-mating barriers. However, two recent models have shown that the process is plausible, both in parapatry and sympatry, if the degree of hybrid disadvantage is sufficiently great and the populations are sufficiently diverged in the signals and receptors used in mate recognition (Liou and Price 1994), or if the genetic architecture of the relevant traits is appropriate (Kelly and Noor 1996). Ecological differences, in traits such as habitat preference, could also play a role in reducing the frequency of hybridisation to a low enough level for reinforcement to occur. Reinforcement is driven by hybrid disadvantage; some degree of post-zygotic isolation is required. Under these circumstances, the strength of selection on females and males to avoid forming hybrids may be unequal, leading to sexual conflict over hybridisation.

In a given encounter between individuals of opposite sex, there are three possibilities (Parker 1974b, 1979, 1983):

(i) it may be in the interests of both sexes to mate;

(ii) it may be in the interests of neither sex to mate;

(iii) there may be mating conflict, i.e. mating is advantageous for one sex but disadvantageous for the other.

Males typically occupy a role in which mating is favourable, and females one in which it is unfavourable; roles may (atypically) be reversed (Parker 1974b, 1979, 1983). We are concerned with mating conflicts that will affect gene flow between incipient species. We now present mate-foraging models, which provide a description of the selective forces at work. Particularly in the context of speciation, where genetic parameters may also have an important bearing on the evolutionary outcome, the biological relevance of the models should be borne in mind. This point is considered further below.

Parker (1979) proposed a model for sexual conflict over mating decision in which male interests differ from female interests. This analysis relates to disparities between the sexes in terms of reproductive time budgets, with time out of the mating pool in general greater for females than for males, mainly because of their greater time investment in offspring. In any encounter between the sexes, the selection on an individual to mate will depend upon the expected reproductive gains from the mating and on the average expected gains from rejecting the mating and waiting for other mating opportunities. In general, because the time tied up by the consequences of each mating is greater for females, they lose less by refusing an inter-population mating and waiting for another encounter, because they are likely to encounter another potential mate sooner than are males. The details of the models derived from this approach are presented in Appendix 7.1. The simplest kind of model (1A) assumes that the two populations have not diverged in traits determining aptitude for encounter, such as population density and mobility, or in amount of time in and out of the mating pool for males and females. It is also assumed that inter-population encounters are relatively rare. Here, unless there is a high male investment time out, or a highly female-biased adult sex ratio, or very low encounter rates between the sexes, or hybrid disadvantage is very low, we would commonly expect selection on females to reject hybrid matings. Males, in contrast, would in general be selected to accept them. At the typical state of very low parental investment by males, only very high levels of hybrid disadvantage can generate selection favouring pre-mating isolation in both sexes, and mating conflict will therefore be the rule. Furthermore, even if hybrids between the populations are completely inviable, there may none the less be some sexual conflict over mating. If mate identification is error-prone, then it may sometimes be better for a male to try mating with any female rather than risk missing one of the same population. If the aptitude for encounter, or the relative time investment in offspring by the sexes, has evolved to different values in the two populations (i and j), similar conclusions will apply, although selection intensity for outcomes is unlikely ever to be symmetric in male i—female j and male j—female i meetings. Most plausibly (when male time out is low and male-female encounters are generally within the same population) there will be sexual conflict favouring male persistence and female rejection in both cases. Less commonly, the hybrid disadvantage may lie in a zone where the selection direction is opposed in the two types of meeting. For example, there may be positive selection for discrimination in say, j–i meetings, but conflict in i–j meetings. If male time out approaches female time out, it is even possible for there to be an exactly opposite direction of conflict in i–j and j–i meetings. In a more complex model where inter-population encounters are more common (7.1B, see Appendix 7.1), there is again an extensive zone of conflicting selection on female and male decisions to mate with the other population, with females in general selected to resist and males to attempt to mate.

This consideration of how selection acts on males and females to hybridise or not suggests that males are selected to act as a force for gene flow, while females are in general selected to resist it. What are the likely consequences of these selection pressures? If there is any inter-population mating, and some hybrid viability and fertility, then hybrids will enter the population and reproduce. As they increase in frequency and the range of hybrid genotypes increases, so the pattern of selection on mating decisions will change. Events under these circumstances rapidly become too complex to model However, models where the populations are already partially isolated by habitat (Liou and Price 1994) can produce reinforcement, because the potentially swamping effects of gene flow are reduced (model 7.1B, Appendix 7.1, can be seen as defining the direction of selection on the two sexes at some balance between within- and between-population matings). In addition, sexual asymmetry in hybrid sterility could be important (Kelly and Noor 1996). If hybrid males are more affected by sterility than are females, as is generally true with male heterogamety (Haldane's rule), this will tend to promote reinforcement. Furthermore, the empirical evidence from *Drosophila* supports the occurrence of reinforcement under at least some circumstances (Coyne and Orr 1989, 1997).

Even without further evolution of anti-hybridisation mechanisms, females may tend to resist inter-population matings (and males to attempt them) when secondary contact is resumed. When the direction of selection on the two sexes is the same, so that hybridisation is in both cases unfavourable, females will often be more strongly selected than are males to resist matings. The same principles will apply when potential mates vary in quality within populations. Females are likely to resist lower quality partners and males are likely to persist; there will be a zone of conflict over mating decisions until a very low mate quality is reached, at which point both sexes will resist, although females will be more strongly selected to do so than males. If evolution in allopatry has resulted in particular mating preferences, then females are likely, in the past, to have been the agents of enforcement, and such mating preferences could lead to assortment for the reasons already discussed. This could help to tip the balance towards reinforcement if females can win the mating conflict, without the need for the evolution of new anti-hybridisation mating behaviour.

The biology of the interaction between the sexes will mediate any effects of sexual conflict on the outcome, both in initial secondary contact and in deciding whether reinforcement will occur. Since the evolution of reinforcement is a transitory phenomenon, the evolution of the strategies for winning mating conflicts themselves is unlikely to play a major role, but they will have an influence on whether males or females are more likely to win where there is sexual conflict over hybridisation. The use of game theory approaches specifically to analyse possible outcomes of conflict over mating decisions began with Parker (1979), and continues to develop (Parker 1983, Clutton-Brock and Parker 1995a,b) often by reinterpreting other contest models (e.g. Enquist and Leimar 1983, 1987) in a sexual context. The models search for an evolutionarily stable strategy (ESS; Maynard Smith 1982) solution, and they are therefore not models of 'Red Queen' continuous coevolution in which the strategic possibilities (the rules of the game) are continuously changing. Rather, they assume a possible set of evolutionary options associated with given payoffs and seek for evolutionarily stable outcomes, as is appropriate in the present context. Both behaviour and morphology will play a role in the resolution of mating conflicts. Behavioural models have generally assumed that there will be asymmetries in armament which will determine the contest costs. Harassment contests approximate to asymmetric wars of attrition (Maynard Smith 1974; Parker 1979; Hammerstein and Parker 1982; Clutton-Brock and Parker 1995a). An important variable in these models is the level of information available to the contestants about relative fighting ability (Maynard Smith and Parker 1976 Parker and Rubenstein 1981 Enquist and Leimar 1983, 1987; Clutton-Brock and Parker 1995a). In contests where the contestants are known to one another, punishment and intimidation, usually of females by males, is a possible ESS (Clutton-Brock and Parker 1995a,b). In addition, the level of armament must itself evolve, which has been examined in theoretical models of an arms race (Dawkins and Krebs 1979; Parker 1979, 1983; Maynard Smith 1982). Which sex will typically occupy the winning role, i.e. the role with the higher distribution for armament, depends on the difference, for each sex, between the value of winning, and between the costs of increasing armaments.

In all these games, the principal determinant of the outcome of mating conflict appears to relate to the balance between:

1. 'Power', or the relative contest costs of the two sexes: e.g. relative cost of enforcing victory for the arms race game, the average relative rates of expenditure of contest cost for the asymmetric war of attrition or sequential assessment game, and the relative ability to inflict damage in the punishment game.
2. 'Value of winning', or the relative fitness difference between mating and not mating (relative opportunity costs), for the two sexes.

In Appendix 7.2, we build on models 7.1A and 7. IB (Appendix 7.1) of selection on males and females to avoid or accept hybrid matings to deduce which sex will have the greater value of winning the mating conflict. With rare encounters between populations, which is the most likely substrate for reinforcement of mating barriers during speciation, there is a large region of the mating conflict where the selection on males to mate is greater than that on females to resist. The reason is that, for most of the parameter space, the difference between the gain to the male from mating with the current female, on one hand, and from continuing to search on the other is positive and it is greater than the corresponding gain to the female from continuing to search for a mate of her own population as opposed to accepting the current male. The asymmetry occurs because of the (typically) much longer time a male must spend searching for an alternative mate. When encounters between populations are more frequent, this sex difference in value of winning becomes less marked, but the hybrid disadvantage is also likely to be less because of gene flow, and this will tend to increase the zone of conflict between the sexes. It is also a less plausible scenario for reinforcement. Either way, the odds seem stacked against a female win if only the value of winning is considered.

While the value of winning may generally be loaded in the male interests, power and contest costs in mating conflicts are much less easy to predict. Unless the male is much larger than the female, it may generally be less energetically expensive for a female to prevent mating than for a male to achieve it against female resistance. For example in most birds, fish, amphibia and other groups in which the males lack a penis, it is difficult to see how a male can mate unless the female cooperates, and even with an intromittent organ, forced copulation may be difficult and costly. If contest costs are greater for the male this could act to offset the effect of the asymmetry in value of winning, reducing our ability to make predictions about which sex is more likely to occupy the winning role. Thus even where relative armament costs favour the female, this does not necessarily mean that the female must occupy the winning role: persistent harassment by a male is likely to impede females in various deleterious ways. A further complication is that intra-sexual selection often drives male body size above that of the female, so that the costs of increasing armaments to males are subsidised by benefits in male–male combat.

Rice and Holland (1997) discuss gamete conflict over rapid versus slow penetration of sperm into the ovum; sperm favouring fast penetration for competitive reasons, and ova slow penetration to prevent polyspermy. Plant stigmas have long been known to exercise choice of pollen via self-incompatibility (e.g. Baker 1959) and other genotypic features, and recent studies (Bishop *et al.* 1996) indicate that sperm from other clones can be resisted in the ascidian *Diplosoma listerianum*. One may speculate about arms races in terms of gamete conflict, in which females are selected to produce ova (or reproductive tracts) with increased resistance to sperm from other populations, and males to produce sperm with increased ability to overcome this resistance. Such arms races would be primarily biochemical. As in mating conflict, the value of winning may be loaded in the male's favour. Perhaps power (contest costs) will be loaded in favour of the female for the following reason. Probably because of sperm competition (Parker 1970), the sperm has tended to become minimal (Parker 1982); any extra armament against the ovum or female tract may be paid for at a loss in sperm numbers or competitiveness. In contrast, the ovum (or female tract) may have resources divertable to the battle at less cost. Thus whereas male–male competition may have assisted males in the power side of mating conflict, it may have had the opposite effect in gamete conflict. Further, with internal fertilisation, gamete conflict occurs within the female, which is also likely to favour female interests in arms races of this type.

In the light of this analysis of outcomes, we therefore make three tentative predictions. First—all else equal—speciation will be more extensive in groups where females generally win mating conflicts than in those groups where males usually win. Some evidence supports this prediction, although it has not in general been interpreted in this way. Among passerine birds, net speciation rate (the net outcome of speciation and extinction) is greater in those clades where a higher proportion of the species are sexually dichromatic, presumably indicating a greater role for female mate choice (Barraclough *et al.* 1995). This correlation would also be predicted by various models of sexual selection that do not involve sexual conflict, and it would be good to have comparative evidence on this point from taxa where intrasexual selection is important. Evidence could consist of some demonstration that those groups where males are typically the bigger sex produce fewer species than those where females are typically bigger or where there is no size dimorphism. Second, and related, where there is evidence for reinforcement, we might predict that it will be the females that are responsible for the resulting pre-mating isolation. Data are few, but *Drosophila* provides one example consistent with this prediction. The closely related species *D. persimilis* and *D. pseudoobscura* are sympatric in some locations, and occur singly in others. Rare hybrids between them occur in the field, with the male hybrids sterile. Pre-mating isolation as assayed in behavioural tests is greater in the areas of sympatry, consistent with the occurrence of reinforcement (Noor 1995). Furthermore, it is the females that are entirely responsible for the pre-mating isolation—males do not discriminate (Noor 1996). Again, it would be nice to have comparative evidence from groups where the power balance between the sexes differs. Third, we predict that in those groups where females tend to win mating conflicts, the genetic variability within each of the many species will be less than within the fewer species in comparable groups where males tend to win. Failure of guppies (*Poecilia reticulata*) to speciate despite widespread genetic differentiation may well be related to male control (Magurran 1998). Any empirical evaluation of genetic variation along these lines would need to correct for the effect of population size (having fewer species may generally imply more individuals in each one).

7.4 Conclusions

Both within- and between-locus sexual conflict may have major roles to play in the progress of the speciation process. Both may affect population divergence in allopatry, and inter-locus conflict may be important in determining the outcome of secondary contact, and in promoting parapatric and sympatric speciation. The formulations presented in Appendices 7.1 and 7.2 are undoubtedly simplistic: they are phenotypic models in which the payoffs are constant and related to a fixed hybrid disadvantage, *d*. Despite this limitation, we believe that our general conclusion holds: that in conditions of incipient sympatric or parapatric speciation, or under mixing of populations that have previously diverged allopatrically, females will typically act as a force favouring premating isolation, and males as a force against it.

Rice and Holland (1997) envisage sexual conflict as fuelling speciation by increasing genetic divergence between populations. Sexual conflict over mate quality within populations is certainly likely to promote genetic divergence between populations with restricted gene flow. Whether it will catalyse or restrict the evolution of mating isolation is likely to depend on the resolution of mating conflict. Mating conflict could be either a hindrance to isolation if 'male-win' scenarios prevail, or a facilitator if females tend to win, with male persistence acting as a major catalyst to speciation by increasing selection on female resistance.

This conflict could, in theory, affect the evolution of any trait involved in the decision whether or not to invest in hybrid offspring. Post-mating, pre-zygotic events, including sperm usage could be involved, and so could parental investment after fertilisation. If there is selection on one sex to avoid fertilisation, then it will usually be advantageous to have the barrier as early in the interaction between male and female as possible (reviewed in Eberhard 1996).

Further work is required on more advanced models of the type included here. However, empirical studies are perhaps especially important at this stage, particularly in investigating the propensity for speciation and the extent of genetic diversity in cases where males are likely to be in the winning role in mating conflicts versus those where females are likely to occupy the winning role. Our prediction for greater speciosity and less diversity within species in the latter case might be testable in groups with similar ecologies but a wide range of sexual size dimorphism.

Acknowledgements

We are indebted to Nick Barton, Roger Butlin, Nick Colgrave, Jerry Coyne, Matthew Gage, Mike Ritchie and Tom Tregenza for their thoughtful and constructive input and comments.

References

Aguade, M., Myashita, N. and Langley, C. H. 1992. Polymorphism and divergence in the Mst26A male accessory gland gene region in *Drosophila*. Genetics 132, 755–770.

Arnold, S. D. and Halliday, T. R. 1988. Multiple mating: natural selection is not evolution. Anim. Behav. 36, 1547–1548.

Arnold, S. D. and Halliday, T. R. 1992. Multiple mating by females: the design and interpretation of selection experiments. Anim. Behav. 43, 178–179.

Arnold, S. J., Verrell, P. A. and Tilley, S. G. 1996. The evolution of asymmetry in sexual isolation: a model and test case. Evolution 50, 1024–1033.

Arnqvist, G. and Rowe, L. 1995. Sexual conflict and arms races between the sexes: a morphological adaptation for control of mating in a sexual insect. Proc. R. Soc. London Ser. B. 261, 123–127.

Arnqvist, G., Rowe, L., Krupa, J. J. and Sih, A. 1996. Assortative mating by size—a metaanalysis of mating patterns in water striders. Evol. Ecol. 10, 265–284.

Baker, H. G. Reproductive methods as factors in speciation in flowering plants. Cold Spring Harbour Symp. Quantitative Biol. 24, 177–191

Barraclough, T. G., Harvey, P. H. and Nee, S. 1995. Sexual selection and taxonomic diversity in passerine birds. Proc. R. Soc. London Ser. B. 259, 211–215.

van den Berg, M., Thomas, G., Hendriks, H. and van Delden, W. 1984. A reexamination of the negative assortative mating phenomenon and its underlying mechanism in *Drosophila melanogaster*. Behav. Genet. 14, 45–61.

Bishop, J. Jones, C. S. and Noble, L. 1996. Female control of paternity in the internally fertilizing compound ascidian *Diplosoma listerianum*. Proc. R. Soc. London Ser. B. 263, 401–407.

Butlin, R. 1987. Speciation by reinforcement. Trends Ecol. Evol. 2, 8–13.

Butlin, R. K. 1993. A comment on the evidence for a genetic correlation between the sexes in *Drosophila melanogaster*. Anim. Behav. 45, 403–404.

Butlin, R. 1995. Reinforcement: An idea evolving. Trends Ecol. Evol. 10, 432–434.

Chapman, T. and Partridge, L. 1996a. Sexual conflict as fuel for evolution. Nature 381, 189–190.

Chapman, T. and Partridge, L. 1996b. Female fitness in *Drosophila melanogaster*: an interaction between the effect of nutrition and of encounter rate with males. Proc. R. Soc. London Ser. B. 263, 755–760.

Chapman, T., Liddle, L., Kalb, J. M., Wolfner, M. F. and Partridge, L. 1995. Cost of mating in *Drosophila melanogaster* females is mediated by male accessory gland products. Nature 373, 241–244.

Chen, P. S., Stumm-Zollinger, E., Aigaki, T., Balmer, J., Bienz, M. and Bohjlen, P. 1988. A male accessory gland peptide that regulate reproductive behavior of female *D. melanogaster*. Cell 54, 291–298.

Clutton-Brock, T. H. and Parker, G. A. 1992. Potential reproductive rates and the operation of sexual selection. Q. Rev. Biol. 67, 437–456.

Clutton-Brock, T. H. and Parker, G. A. 1995a. Sexual coercion in animal societies. Anim. Behav 49, 1345–1365.

Clutton-Brock, T. H. and Parker, G. A. 1995b. Punishment in animal societies. Nature 373, 209–216.

Coyne, J. A. and Orr, H. A. 1989. Patterns of speciation in *Drosophila*. Evolution 43, 362–381.

Coyne, J. A. and Orr, H. A. 1997. 'Patterns of speciation in *Drosophila*' revisited. Evolution 51, 295–303.

Dawkins, R. and Carlisle, T. R. 1976. Parental investment, mate desertion and a fallacy. Nature 262, 131–133.

Dawkins, R. and Krebs, J. R. 1978. Animal signals: information or manipulation? In: Krebs, J. R. and Davies, N. B. *Behavioural Ecology: An Evolutionary Appraoch*. pp. 282–309. Sinauer Associates, Sunderland, MA.

Dawkins, R. and Krebs, J. R. 1979. Arms races between and within species. Proc. R. Soc. London Ser. B. 295, 489–511.

Eberhard, W. G. 1996. *Female control: Sexual Selection by Cryptic Female Choice*. pp. 501. Princeton University Press, Princeton.

Emlen, S. T. and Oring, L. W. 1997. Ecology, sexual selection and the evolution of making systems. Science 197, 215–223.

Endler, J. A. 1992. Signals, signal conditions, and the direction of evolution. Amer. Natur. 139, S125–S153.

Endler, J. A. 1993. Some general comments on the evolution and design of animal communication systems. Phil. Trans. R. Soc. B. 340, 215–225.

Enquist, M. and Leimar, O. 1983. Evolution of fighting behaviour: decision rules and assessment of relative strength. J. Theor. Biol. 102, 387–410.

Enquist, M. and Leimar, O. 1987. Evolution of fighting behaviour: the effect of variation in resource value. J. Theor. Biol. 127, 187–205.

Fowler, K. and Partridge, L. 1989. A cost of mating in female fruitflies. Nature 338, 760–761.

Gems, D. and Riddle, D. L. 1996. Longevity in *Caenorhabditis elegans* reduced by mating but not by gamete production. Nature 379, 723–725.

Gilchrist, A. S. and Partridge, L. 1997. Heritability of pre-adult viability differences can explain apparent heritability of sperm displacement ability in *Drosophila melanogaster*. Proc. R. Soc. London Ser. B 264, 1271–1275.

Gregory, P. G. and Howard, D. J. 1994. A postinsemination barrier to fertilisation isolates two closely related ground crickets. Evolution 48, 705–710.

Gromko, M. H. 1992. Genetic correlation of male and female mating frequency: evidence from *Drosophila melanogaster*. Anim. Behav. 43, 176–177.

Halliday, T. R. and Arnold, S. J. 1987. Multiple mating by females: a perspective from quantitative genetics. Anim. Behav. 35, 939–941.

Hammerstein, P. and Parker, G. A. 1982. The asymmetric war of attrition. J. Theor. Biol. 96, 647–682.

Harshman, L. G. and Prout, T. 1994. Sperm displacement without sperm transfer in *Drosophila melanogaster*. Evolution 48, 758–766.

Houde, A. and Endler, J. 1990. Correlated evolution of female mating preferences and male color pattern in the guppy *Poecilia reticulata*. Science 248, 1405–1408.

Kaneshiro, K. Y. 1976. Ethological isolation and phylogeny in the plantiba subgroup of Hawaiian *Drosophila*. Evolution 30, 740–745.

Kelly, J. K. and Noor, M. A. F. 1996. Speciation by reinforcement: a model derived from studies of *Drosophila*. Genetics 143, 1485–1497.

Kirkpatrick, M. 1982. Sexual selection and the evolution of female choice. Evolution 36, 1–12.

Kirkpatrick, M. and Ryan, M. J. 1991. The paradox of the lek and the evolution of mating preferences. Nature 350, 33–38.

Krakauer, D. C. and Johnstone, R. A. 1995. The evolution of exploitation and honesty in animal communication: a model using artifical neural netoworks. Phil. Trans. R. Soc. London Ser. B. 348, 355–361.

Lande, R. 1981. Models of speciation by sexual selection on polygenic traits. Proc. Natl Acad. Sci. USA 78, 3721–3725.

Lande, R. 1982. Rapid origin of sexual isolation and character divergence in a cline. Evolution 36, 213–223.

Lande, R. 1987. Genetic correlations between the exes in the evolution of sexual dimorphism and mating preferences. In: Bradbury, J. and Andersson, M. B. Eds. *Sexual Selection: Testing the Alternatives*. pp. 83–94. Wiley, Chichester.

Liou, L. W. and Price, T. D. 1994. Speciation by reinforcement of pre-mating isolation. Evolution 48, 1451–1459.

Magurran, A. 1998. Levels of morphological and behavioural variation. Phil. Trans. R. Soc. London Ser. B 353, 275–286.

Maynard Smith, J. 1974. The theory of games and the evolution of animal conflicts. J. Theor. Biol. 47, 209–221.

Maynard Smith, J. 1982. *Evolution and the Theory of Games.* Cambridge University Press, Cambridge.

Maynard Smith, J. and Parker, G. A. 1976. The logic of asymmetric contests. Anim. Behav. 24, 159–175.

Noor, M. A. F. 1995. Speciation driven by natural selection in *Drosophila*. Nature 375, 674–675.

Noor, M. A. F. 1996. Absence of species discrimination in *Drosophila pseudoobscura* and *Drosophila persimilis* males. Anim. Behav. 52, 1205–1210.

Parker, G. A. 1970. Sperm competition and its evolutionary consequences in the insects. Biol. Rev. 45, 525–568.

Parker, G. A. 1974a. Courtship persistence and female-guarding as male time-investment strategies. Behaviour, 48, 157–184.

Parker, G. A. 1974b. Assessment strategy and the evolution of fighting behaviour. J. Theor. Biol. 47, 223–243.

Parker, G. A. 1979. Sexual selection and sexual conflict. In: Blum, M. S. and Blum, N. B.,

Eds. *Sexual Selection and Reproductive Competition in Insects.* Pp. 123–166. Academic Press, New York.

Parker, G. A. 1982. Why are there so many tiny sperm? Sperm competition and the maintenance of two sexes. *J. Theor. Biol.* 96, 281–294.

Parker, G. A. 1983. Mate quality and mating decisions. In: Bateson, P. P. G., ed. *Mate Choice.* pp. 141–166. Cambridge University Press, Cambridge.

Parker, G. A. and Rubenstein, D. I. 1981. Role assessment, reserve strategy and acquisition of information in asymmetric animal conflicts. Anim. Behav. 29, 221–240.

Parker, G. A. and Simmons, L. W. 1996. Parental investment and the control of sexual selection—predicting the direction of sexual competition. Proc. R. Soc. London Ser. B. 263, 315–321.

Payne, R. J. H. and Krakauer, D. C. 1997. Sexual selection and speciation. Evolution 51, 1–9.

Price, C. S. C. 1997. Conspecific sperm precedence in *Drosophila.* Nature 388, 663–666.

Rice, W. R. 1984. Sex chromosomes and the evolution of sexual dimorphism. Evolution 38, 735–742.

Rice, W. R. 1992. Sexually antagonistic genes: experimental evidence. Science 256, 1436–1439.

Rice, W. R. 1996. Sexually antagnoistic male adaptation triggered by experimental arrest of female evolution. Nature 381, 232–234.

Rice, W. R. and Holland, B. 1997. The enemies within: intergenomic conflict, interlocus conflict evolution (ICE), and the intraspecific Red Queen. Behav. Ecol. Sociobiol. 41, 1–10.

Rowe, L. 1994. The costs of mating and mate choice in water striders. Anim. Behav. 48, 1049–1056.

Rowe, L. and Arnqvist, G. 1996. Analysis of the causal components of assortative mating in water striders. Behav. Ecol. Sociobiol. 38, 279–286.

Ryan, M. J. 1990. Sexual selection, sensory systems and sensory exploitation. Oxford Surv. Evol. Biol. 7, 156–165.

Schluter, D. and Price, T. 1993. Honesty, perception and population divergence in sexually selected traits. Proc. R. Soc. London Ser. B 253, 117–122.

Slatkin, M. 1984. Ecological causes of sexual dimorphism. Evolution 38, 622–630.

Stamencovic-Radak, M., Partridge, L. and Andelkovic, M. 1992. A genetic correlation between the sexes for mating speed in *Drosophila melanogaster.* Anim. Behav. 43, 389–396.

Stamencovic-Radak, M., Partridge, L. and Andelkovic, M. 1993. Genetic correlation between the sexes in *Drosophila melanogaster*: a reply to Butlin. Anim. Behav. 45, 405.

Stephenson *et al.* 1992.

Tilley, S. G., Verrell, P. A. and Arnold, S. J. 1990. Correspondence between sexual isolation and allozyme differentiation: a test in the salamander *Desmgnathus ochrophaeus.* Proc. Natl Acad. Sci. USA 87, 2715–2719.

Trivers, R. L. 1972. Parental investment and sexual selection. In: Campbell, B., ed. *Sexual Selection and the Descent of Man.* Pp. 136–179. Heinemann, London.

True, J. R., Weir, B. S. and Laurie, C. C. 1996. A genome-wide survey of hybrid incompatibility factors by the introgression of marked segments of *Drosophila mauritiana* chromosomes into *Drosophila simulans.* Genetics 142, 819–837.

Tsaur, S. C. and Wu, C-I. 1997. Positive selection and the molecular evolution of aa gene of male reproduction, Acp26Aa of *Drosophila.* Mol. Biol. Evol. 14, 544–549.

West-Eberhard, M. 1984. Sexual selection, competitive communication and species-specific signals in insects. In: Lewis, T., ed. *Insect Communication.* Pp. 283–324. Academic Press, Toronto.

Wilson, D. S. and Hedrick, A. 1982. Speciation and the economics of mate choice. Evol. Theory 6, 15–24.

Wu, C-I. and Davis, A. W. 1993. Evolution of post-mating reproductive isolation: the composite nature of Haldane's rule and its genetic bases. Am. Nat. 142, 187–212.

APPENDIX 7.1

Conflict between the sexes over hybridisation

For simplicity, we consider time spent on just two forms of reproductive activity: time during which an individual is available for mating and time during which an individual is unavailable for mating (see also Clutton-Brock and Parker 1992; Parker and Simmons 1996).

Consider a simple case in which each female mates once per reproductive cycle (there are no complexities of 'collateral investment' *sensu* Parker and Simmons 1996). During a given reproductive cycle of duration T time units, a female spends time S_f searching for a mate or otherwise available for mating ('time in' the mating pool) and time G_f as time unavailable for mate acquisition during parental care, gaining the energy for gametes, etc. ('time out' of the mating pool). Thus

$$T = S_f + G_f, \tag{7.1a}$$

and if the adult sex ratio is M males to each female, then

$$MT = S_m + G_m, \tag{7.1b}$$

where times S_m, G_m, are the equivalent 'times in' (mate searching time) and 'out' (gamete replenishment, paternal care) for males (see Figure 7.1a). The times out: G_f, G_m are likely to be very different for the two sexes for biological reasons, most notably (but not entirely, see Clutton-Brock and Parker 1992) because of the relative difference in parental investment between the sexes (Trivers 1972) which implies (rare sex role reversals excepted) that $G_f > G_m$, and that generally, $G_f >> G_m$. The average time it will take a receptive female to encounter a male will be inversely proportional to the density of searching males in the same area of the habitat. The searching male density is proportional to S_m/MT, the proportion of each adult male's time available for mating. Thus

$$S_f = B \, MT/S_m,$$

$$S_f = BM \, (S_f + G_f)/[M \, (S_f + G_f) - G_m],$$

where B is a constant related to the 'aptitude for encounter': the lower B, the more quickly the sexes meet. To define further, note that if male time out is negligible compared with time in (i.e. $S_m >> G_m$) so that the maximum number of adult males is available, $MT/S_m \approx 1.0$, and so $S_f \approx B$. Thus B is the average 'time in' for a female if all males were to be available all the time in a given population. The solution to the quadratic obtained from the above equation is

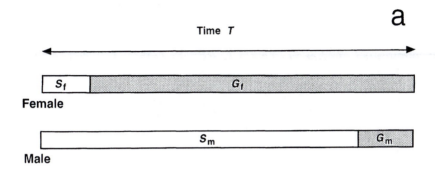

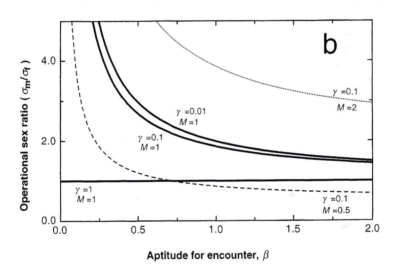

Fig. 7.1 (a) Arrangement of 'time in' (S_f for each female and S_m for each male) and 'time out' (G_f for each female and G_m for each male) of the mating pool when the adult sex ratio is unity. The 'times out' are fixed, but the 'times in' relate to each other as defined in the text. (b) Operational sex ratio, sexually-available males divided by sexually-available females (σ_m/σ_f), calculated from equations (7.2a) and (7.2b), plotted against the aptitude for encounter, β (see text) The bold curves are for an adult sex ratio (males/females) of $M = 1.0$, and show the effect of altering γ (the ratio of 'times out', G_m/G_f). The other two curves show the effect of altering adult sex ratio while keeping γ constant at 0.1.

$$S_f = \left[-(G_f - G_m/M - B) + \sqrt{(G_f - G_m/M - B)^2 + 4BG_f} \right]/2,$$

(Clutton-Brock and Parker 1992). If we set G_f to unity, so that all time periods are expressed relative to the female time out, we obtain

$$\sigma_f = \left[-(1 - \gamma/M - \beta) + \sqrt{(1 - \gamma/M - \beta)^2 + 4\beta} \right]/2, \qquad (7.2a)$$

where γ is the male time out divided by the female time out (usually $G_f > G_m$, so typically $1 > \gamma$), σ_f is the female time in scaled by female time out ($\sigma_f = S_f / G_f$), and similarly $\beta = B/G_f$. From (7.1a) and (7.1b), the relative male time out, S_m/G_f is

$$\sigma_m = M(\sigma_f + 1) - \gamma.$$

Figure 7.1b shows the operational sex ratio (OSR, Emlen and Oring 1977), i.e. the ratio of sexually available males to sexually available females, σ_m/σ_f (see also Clutton-Brock and Parker 1992) at different vales of relative male 'time out', γ, and sex ratio, M. At a sex ratio of unity (continuous bold curves), if $\gamma = 1$, the sexes are equal and the OSR is one, whatever the aptitude for encounter. Reducing γ results in highly male biased OSRs when β is low (where the sexual encounter rate is high). If the adult sex ratio is male biased (upper dotted curve), the OSR is correspondingly higher, and if the adult sex ratio is female biased (lower broken curve), the OSR is correspondingly lower, and can become female-biased at high values for β.

Model 7.1A. Conflict in rare meetings between semi-isolated populations

We now use the above formulations for σ_f, σ_m, to re-apply the logic developed by Parker (1979; which uses less realistic versions of 'times in') to analyse selection on mating decision conflicts in rare meetings between populations which occasionally meet within the same habitat; there is therefore already a consider-able degree of pre-mating isolation. We consider the simple case where there are essentially two semi-isolated populations (Figure 7.2a). Note that the isolation need not be geographic, it can relate to ecological niches within a habitat, provided that the probability of encounters between opposite sexes of the two populations is low. Imagine initially that the aptitude for encounter, β, and the ratio of times out for the two sexes, γ, are similar for the two populations, which have not diverged in these parameters. β is likely to be a function of the population densities and relative frequencies of the two populations. We first consider the case where β is a fixed property of the populations and inter-population encounters are rare. We then relax this unrealistic assumption. Since there has been some genetic divergence between the two populations, offspring that are hybrids between them have reduced Darwinian fitness relative to the fitness of pure-bred offspring.

Calling this hybrid disvavantage d, we deduce the thresholds d_m, d_f, which are the threshold levels of hybrid disadvantage for the male and female at which the outcome of mating yields the same fitness as the outcome of not mating. We

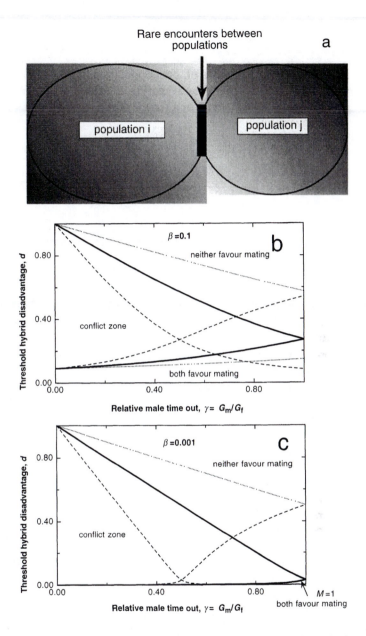

Fig. 7.2 (a) Model 7.1A—two populations meet only very infrequently at the boundary of their niches, or geographic ranges. (b) Thresholds of hybrid disadvantage in Model 7.1A (d_m for males, upper curves; and d_f for females, lower curves) at which it pays to switch decision from mating (below curve) to not mating (above curve), plotted against γ (the ratio of 'times out', G_m / G_f). Bold curves are for adult sex ratio (males/females) of $M = 1.0$; dotted curves are for $M = 2.0$; broken curves are for $M = 0.5$. Below the female threshold, both sexes favour mating and above the male threshold neither favours mating. Between the two, the sexes are in conflict (see text). The aptitude for encounter is high $\beta = 0.1$.
(c) Same as for b, but the aptitude for encounter is more typical: $\beta = 0.001$.

assume that selection acts to maximise the rate of progeny production (progeny/ time) through life. Thus if a male of one population meets a female of the alternative population, selection favours mating if the gain rate from mating exceeds the gain rate due to searching for a female of the same population, and the threshold is thus given by

$$(1 - d_m)/G_m = 1/(G_m + S_m);$$

$$d_m = \frac{S_m}{G_m + S_m}; \text{ equivalent to } \frac{\sigma_m}{\gamma + \sigma_m}; \tag{7.3a}$$

and by analogy the corresponding decision threshold for a female is

$$d_f = \frac{S_f}{G_f + S_f}; \text{ equivalent to } \frac{\sigma_f}{1 + \sigma_f}. \tag{7.3b}$$

Some solutions for the zone of conflict over mating decisions are shown in Figures 7.2b and 7.2c, in which the thresholds d_m, d_f above are plotted against relative male time out, γ. In Figure 7.2b, $\beta = 0.1$, so that the sexes have a low rate of encounter: with maximum male availability the female must wait on average one-tenth of her investment time (time out) in order to find a mate. Given the protracted time investment by most females to produce a clutch of offspring, this β must be regarded as extreme. The bold lines show the thresholds d_m (upper curve) and d_f (lower curve) for the case where the adult sex ratio is unity ($M = 1$). Between these two curves there is mating conflict, with the male under selection to mate and the female under selection to refuse. If the sexes have equal times out, as might be approached if male and female parental investment is similar, $\gamma = 1$, and the sexes have equal thresholds ($d = 0.27$). At the other extreme, $\gamma = 0$, and the male has negligible time out. Then the male should always attempt to mate even if hybrids are almost inviable ($d \to 0$); effectively if sperm cost nothing (no time out costs), any potential offspring are worthwhile. For the female, this is not so and she has a relatively low threshold ($d = 0.091$) of hybrid disadvantage above which she should refuse matings. If the disadvantage d lies above the upper bold curve, neither sex favours mating; if it lies below the lower bold curve, both favour mating. The remaining curves show the corresponding thresholds when the adult sex ratio is $M = 2$ (dotted) and $M = 0.5$ (broken) males/female. When the sex ratio is male-biased as $M = 2$, even when the sexes are equal in 'times out' ($\gamma = 1$), the male has a higher threshold because he has a much longer 'time in' searching for a mate than the female (dotted curves). In contrast, when the adult sex ratio is female biased as $M = 0.5$, equality of thresholds occurs at $\gamma = 0.5$ (broken curves), and for higher values of γ the male threshold falls below that of the female so that the conflict zone is inverted (females should persist in courtship of alternative populations and males should reject). Note that such conditions require a high male investment time out due to mating, or a highly female-biased adult sex ratio. Equality of thresholds occurs only when $\gamma = M$.

A more typical situation would be that females spend a much smaller average time in awaiting a mating. In Figure 7.2c we examine a case where $\beta = 0.001$, so that with maximum male availability each female waits on average one-thousandth of her investment time (time out) in order to find a mate. For unity adult sex ratio, the male threshold now drops almost linearly between $\gamma = 0$ and $\gamma = 1$ (upper bold curve), whereas the female threshold stays close to zero throughout the range (lower bold curve). The female time in is so short that it pays her to await a meeting with her own population rather than to mate with a male who reduces—even only marginally—the viability of her offspring. The conflict zone is now extensive, and as γ increases the zone where neither favour mating increases. The zone where both favour mating is restricted to low values of d at relatively high γ (see Figure 7.2c). When the sex ratio is male biased as $M = 2$ (dotted curves), the female threshold is even more restrictive, being close to zero across the entire range of γ from 0 to 1. The male threshold is now higher, making the conflict zone more extensive. When the sex ratio is female-biased as $M = 0.5$ (broken curves), there is again the inversion of conflict for $\gamma > 0.5$ (females should persist and males resist).

The above analysis is simplistic in that it assumes that the parameters β and γ have not diverged for the two populations (see above). A more realistic analysis would take account of such possibilities (see also Parker 1979). When these parameters differ, the thresholds d_m and d_f differ for males and females of the two populations. Calling the populations i and j, there are two types of rare meeting to consider: (i) male i female j; (ii) male j female i. A theoretical possibility would be that there is, say, conflict in case (i) but no conflict in case (ii). The implications of such cases for the probability of evolution of isolation of populations deserve further analysis (see also a parallel case in Model 7.1B, below). However, some properties of models with asymmetric β and γ can be deduced. Suppose, for example, ecotypes are equal in γ but differ in β with population i having $\beta = 0.1$ and population j having $\beta = 0.001$. When a male i meets a female j the conflict zone would lie between the appropriate d_m curve from Figure 7.2b, and the appropriate d_f curve from Figure 7.2c. At any given value for γ, the conflict zone would be increased. In contrast, when a male j meets a female i, the conflict zone would lie between the appropriate d_m curve from Figure 7.2c, and the appropriate d_f curve from Figure 7.2b. At low values for γ, the conflict zone would be reduced, and at high values for γ, the curves may cross over giving an inversion of conflict so that females persist and males reject.

Model 7.1B. Conflict over isolation between populations that meet frequently

In Model 7.1A, encounters with another population were very rare so that search times, S_m, S_f, were not significantly affected by cross-population matings. Consider now the situation where two populations commonly meet in part of their ranges or habitats (Figure 7.3a) where there is no isolation between the two populations. We consider the fate of rare mutations that cause their bearers to mate discriminately (i.e. only with an individual of their own population) rather than indiscriminately. Such mutants increase their search time ('time in') awaiting

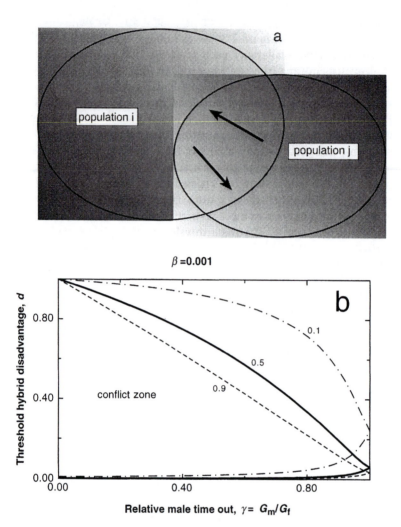

Fig. 7.3 (a) Model 7.1B—two populations meet frequently in part of the geographic ranges or habitat. (b) Thresholds of hybrid disadvantage in Model 7.1B (d_m for males, upper curves; and d_f for females, lower curves) at which it pays to switch decision from mating (below curve) to not mating (above curve), plotted against γ (the ratio of 'times out', G_m/G_f). All curves are for adult sex ratio (males/females) of $M = 1.0$. Bold curves are for p_{ii} and $q_{jj} = 0.5$; dotted curves are for p_{ii} and $q_{jj} = 0.9$; broken curves are for p_{ii} and $q_{jj} = 0.1$. Below the female threshold, both sexes favour mating and above the male threshold neither favours mating. Between the two, the sexes are in conflict (see text). The aptitude for encounter is $\beta = 0.001$.

a mating—they restrict their mating opportunities in order to obtain matings of higher value. We again seek thresholds for hybrid disadvantage at which the decision to be discriminate has equal payoffs to the decision to be indiscriminate.

The relations for the population for 'times in' and 'out' are as described in equations (7.1) and (7.2).

We adopt the following notation: males of population i meet i female with probability p_{ii} and j females with probability p_{ij} (where $p_{ii} + p_{ij} = 1$); females of population j meet i males with probability q_{ij} and j males with probability q_{jj} (where $q_{jj} + q_{ij} = 1$). Since there is only partial mixing of the two populations, it is generally likely that there is a greater probability of meeting the similar population (i.e. $p_{ii} > p_{ij}$; $q_{jj} > q_{ij}$), though we shall consider cases where is not so.

Consider a rare mutant male of type i that ignores all females of type j. It therefore takes on average time S_m/p_{ii} to encounter a female of its own population i, rather than time S_m. Its offspring have a fitness of 1.0 relative to an average of $[p_{ii} + (1 - p_{ii}) (1 - d_{mi})]$ for indiscriminate males, in which d_{mi} is the disadvantage of i–j offspring viewed against i–i offspring. The gain rates of the mutant equals that of indiscriminate males when

$$1/(G_m + S_m / p_{ii}) = [p_{ii} + (1 - p_{ii}) (1 - d_{mi})]/(G_m + S_m),$$

which occurs when the hybrid disadvantage is

$$d_{mi} = \frac{S_m}{p_{ii} G_m + S_m}; \text{ equivalent to } \frac{\sigma_m}{p_{ii}\gamma + \sigma_m}. \tag{7.4a}$$

when all times are standardised in terms of the female time out, G_f (see above). In order to evaluate conflicting selection on mating decision between the sexes, we need to consider the best decision of a female of type j, since conflict will concern i–j meetings. A female of type j has a probability of q_{jj} of meeting her own population and q_{ij} of meeting an i male. By a similar analysis we obtain the threshold for j females

$$d_{fj} = \frac{S_f}{q_{jj} G_f + S_f}; \text{ equivalent to } \frac{\sigma_f}{q_{jj}\gamma + \sigma_f}. \tag{7.4b}$$

Note that (7.4a) and (7.4b) converge to equations (7.3a) and (7.3b) for Model 7.1A when respectively p_{ii} and q_{jj} approach 1.0 (almost all encounters are within the same population).

We now examine the range of conflict when selection is opposed (mutant i males do best to mate with j females, but j females do best to reject i males). This clearly depends now not only on all the previous time parameters, but also on the relative frequency of the two populations i and j. Figure 7.3b shows the thresholds d_m (upper curves) and d_f (lower curves) for the case where $\beta = 0.001$ (potential minimum 'time in' for the female is one hundredth of her 'time out') and where the adult sex ratio is unity ($M = 1$). The bold curves are for the case where $p_{ii} = q_{jj} = 0.5$; i.e. where i males meet i and j females equally frequently, and j females meet i and j males equally frequently. The conflict zone between the two curves (where discriminate mating is favoured in females but selected against in males) is extensive and follows a rather similar pattern to that of Model 7.1A (cf. bold curves in Figure 7.2c). The broken curves refer to the thresholds when

$p_{ii} = q_{jj} = 0.9$, where i males meet i females with probability 0.9 and j females with probability 0.1, and j females meet j males with probability 0.9 and i males with probability 0.1. This reduces the conflict zone, but not very significantly, because of the strong domination that is exerted on the thresholds by the difference in times out between the sexes.

Equality in the probability of within-population encounters of the form $p_{ii} = q_{jj}$ is very unlikely. For instance, i may overlap over only a small part of its range with j, but j may overlap over a large part of its range with i, resulting in $p_{ii} > q_{jj}$. The upper dash–dot curve represents the case where $p_{ii} = 0.1$, and the lower $q_{jj} = 0.1$, i.e. (unlikely) cases where meetings of a given sex occur most commonly with the alternative population. While it is impossible to envisage this happening concurrently for both populations, it is at least theoretically possible where the two populations have highly asymmetric overlap and occur at different population densities. Then both populations would tend to have higher probability of encounter with the common population. The two sets of curves can then be used to gain some idea of what is happening simultaneously in the two forms of encounter, i–j and j–i. For example, males of the rare population meeting females of the common population would tend to have threshold d_m in the zone between the upper bold and dash–dot curves, but females of the common population would tend to have threshold d_f between the lower bold and broken curves. The conflict zone is much more extensive than in meetings between males of the common ecotype (d_m now between the upper broken and bold curves) with females of the rare ecotype (d_f now between the lower dash–dot and bold curves).

Appendix 7.2

Which sex has the greater value of winning the mating conflict?

Call the average costs of one unit of contest time, or the cost of a unit increase in armament C_f for females and C_m for males, and the value of winning V_f for females and V_m for males. A general rule of thumb (and a specific prediction for the asymmetric war of attrition, see Parker 1974a; Hammerstein and Parker 1982) is that sex i is likely to occupy (on average) the winning role defined as $V_i/C_i > V_j/C_j$.

We can analyse the models of Appendix 7.1 above to investigate how the value of winning in mating decision conflicts between populations differs for males and females in relation to relative 'times out', γ (see also Parker 1979). The fitness of hybrid progeny is $(1-d)$, where d is the hybrid disadvantage. Suppose that d lies within the conflict zone such that it pays a male to mate but a female not to mate. We calculate the value V (V_m for the male and V_f for the female) as an opportunity cost, the difference between the gain rates achieved by winning and by losing. We then calculate the hybrid disadvantage, d_x, at which selection for winning is equal on the two players, i.e. where $V_m = V_f$. By mapping this 'equal selection intensity' threshold on to the plot of conflict thresholds, we can compare the relative areas over which males will be under more intense selection to win than females, and vice-versa. Model 7.2A investigates conditions favouring the benefits of opportunistic indiscriminateness in gaining lower quality progeny at a cost only of the 'time out' which must be paid for them. Model 7.2B investigates conditions under which selection favours benefits of discriminateness in gaining better quality progeny at a cost of increased search time ('time in') this will demand.

Model 7.2A—rare encounters between populations

When a male occasionally meets a female of the alternative population, he achieves a gain rate of $(1-d)/G_m$ if he mates with that female, where G_m is the 'time out' costs he must pay due to mating (see Appendix 7.1). If he ignores the female to search for a female of his own population, his gain rate will be reduced to the value $1/(G_m + S_m)$, where S_m is the mean search time to find such a female. The female gains are analogous but we assume that not mating yields the higher gain and examine the condition under which across-population matings are to be avoided. Thus expressing both as positive opportunity costs (mating minus no mating for the male; no mating minus mating for the female):

$$V_m = (1 - d)/G_m - 1/(G_m + S_m),$$

$$V_f = 1/(G_f + S_f) - (1 - d)/G_f.$$

Calling d_x the hybrid disadvantage at which the opportunity costs are exactly equal for the two sexes so that $V_m = V_f$ (i.e. where selection on males favouring mating balances selection on females not to mate), then we can calculate that

$$d_x = 1 - \frac{G_m G_f (M^{-1} + 1)}{(G_m + G_f)(G_m + S_f)}, \tag{7.5a}$$

cf. Parker (1979, equation 12 in which the adult sex ratio $M = 1$). If times are expressed relative to the female 'time out' cost, G_f,

$$d_x = 1 - \frac{\gamma(M^{-1} + 1)}{(1 + \gamma)(1 + \sigma_f)}. \tag{7.5b}$$

By comparing (7.5b) with d_m in equation (7.3a), and d_f in equation (7.3b), it is possible to show that $d_m > d_x > d_f$ if $\gamma < M$. That is, the hybrid disadvantage below which it pays males to mate (d_m) is greater than that for equal opportunity costs (d_x), which is in turn greater than that below which it pays females to mate, provided that the ratio times out $(\gamma = G_m/G_f)$ is smaller than the ratio of adult males to females (M). If this condition is violated, there is a reversal of sex roles and the inequalities of hybrid disadvantage are reversed (see also Figures 7.2b, 7.2c).

Figure 7.4a shows d_x for Model 7.2A (rare meetings between populations) in relation to thresholds d_m and d_f when the aptitude for encounter is $\beta = 0.001$, and the adult sex ratio is unity $(M = 1.0)$. It clearly lies closer to d_m than d_f unless the relative times out approach equality $(\gamma \to 1.0)$. To quantify, the relative closeness of d_x to the male threshold d_m can be expressed as

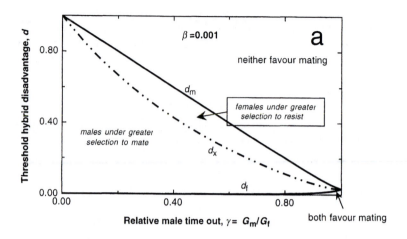

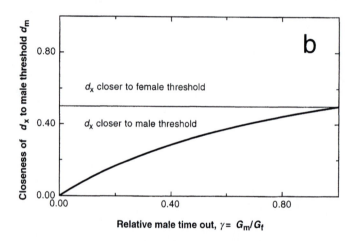

Fig. 7.4 (a) Threshold d_x (dash dot curve) at which selection on males to mate is equal to selection on females to resist, for Model 7. 2A in relation to γ (the ratio of 'times out', G_m/G_f). The upper bold curve is the male threshold, d_m, above which males should not attempt to mate and the lower bold curve the equivalent threshold, d_f, for the female (see Figure 7.2c). $\beta = 0.001$, and $M = 1.0$. Within the conflict zone, at a level of d etween d_x and d_m, the female is under greater selection to resist, and between d_f and d_x, the male is under greater selection to mate. (b) Plot of index I from equation (7.6), representing the closeness of threshold d_x to the male threshold d_m, for Model 7.2A.

$$I = \frac{d_m - d_x}{d_m - d_f}, \tag{7.6}$$

i.e. its proportionate deviation from d_m in units of the difference between d_m and d_f.

Remarkably, for Model 7.2A this index is independent of the values of β or M, and depends only on relative male time out γ. At $\gamma = 0$, $I = 0$ and $d_x = d_m$; as γ

increases, I increases monotonically towards a value of 0.5 (where d_x is equidistant between d_m and d_f) at $\gamma = 1$ (see Figure 7.4b). Provided that $\gamma < M$, there is no role reversal in sexual conflict so that $d_m > d_x > d_f$, and d_x will be closer to the male threshold. There is a larger region of the conflict zone over which the male experiences more intense selection to mate than the female to reject. In nature, the most typical case is for very small relative male time out ($\gamma \ll M$). The threshold for equal values of winning lies very close to the male threshold under such circumstances, and the conflict zone is maximal (d_m approaching 1 and d_f approaching zero). Most of the rare encounters between males and females of different populations are, therefore, not only likely to involve mating conflict, they are also likely to involve much greater selection pressure on males to win.

Model 7.2B—frequent inter-population encounters

For frequent meetings, a male of type i mating discriminately (i.e. only with type i females) gains at a rate $1/(G_m + S_m/p_{ii})$, and an i male mating indiscriminately gains at rate $[p_{ii} + (1 - p_{ii})(1 - d_{mi})]/(G_m + S_m)$. Assuming that discriminateness is selected against, we subtract the former payoff from the latter to give V_m. Similar relations apply for the female, but reverse the signs of the payoffs to find the condition under which discriminateness is favoured to give V_f. Hence

$$V_m = [p_{ii} + (1 - p_{ii})(1 - d_{mi})]/(G_m + S_m) - 1/(G_m + S_m/p_{ii}),$$

$$V_f = 1/(G_f + S_f/q_{jj}) - [q_{jj} + (1 - q_{jj})(1 - d_{fj})]/(G_f + S_f).$$

Calling d_x the hybrid disadvantage at $V_m = V_f$, and setting $m = (G_m + S_m/p_{ii})$ and $f = (G_f + S_f/q_{jj})$, we obtain

$$d_x = \frac{M + 1 - MT(m^{-1} + m^{-1})}{(1 - p_{ii}) + M(1 - q_{jj})}, \tag{7.7a}$$

and expressing all times relative to the female 'time out' cost, G_f,

$$d_x = \frac{M + 1 - M(1 + \sigma_f)[(1 + \sigma_m/p_{ii})^{-1} + (1 + \sigma_f/q_{jj})^{-1}]}{(1 - p_{ii}) + M(1 - q_{jj})}. \tag{7.7b}$$

Outcomes of Model 7.2B are illustrated in Figure 7.4c and 7.4d. Figure 7.4c shows a typical hypothetical case where the probability of encounter of like populations is predominant and equal ($p_{ii} = q_{jj} = 0.7$), and the adult sex ratio is unity. Threshold d_x for equal values of winning now lies equidistant between d_m and d_f: there is now a much less restricted zone of d over which females may be under more intense selection than in Model 7.2A. Using index I, equation (7.5), to map relative distance of d_x from the male threshold d_m, we find that d_x is independent of γ and β (Figure 7.4d). For the case where $p_{ii} = q_{jj}$, I is dependent only on M, the adult sex ratio. Where $M = 1$, d_x is equidistant between d_m and d_f; increasing M causes d_x to move closer to the female threshold, giving a bigger zone favouring females. Decreasing M has the opposite effect (Figure 7.4d).

When there is asymmetric overlap between populations ($p_{ii} \neq q_{jj}$), d_x is pushed

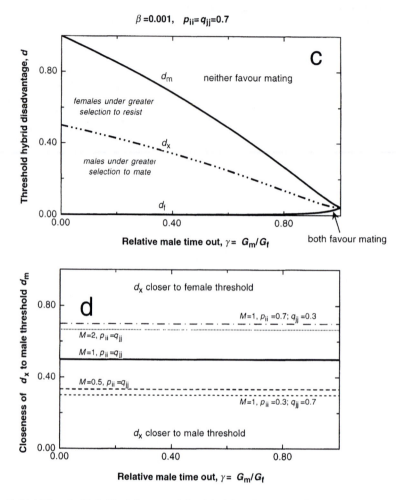

Fig. 7.4 (c) Threshold d_x (dashdot curve) for Model 7. 2B; rest as for a. (d)Plot of index I for Model 7.1B; the different values of β and M are shown against each plot (see text). (d) Plot of index I from equation (7.6), representing the closeness of threshold d_x to the male threshold d_m, for Model 7.2B; the different values of p_{ii}, q_{jj} and M are shown against each plot (see text).

closer to the male threshold in meetings of a rarer male population i with a commoner female j; and (simultaneously) towards the female threshold for the commoner males meeting rarer females (Figure 7.4d). Thus the zone over which males are favoured is greater for the rarer males than for the commoner males, generating asymmetric selection intensity (and possibly also direction) in i–j and j–i meetings.

Although the prospects for females look better under Model 7.2B, it must be remembered that this is a model essentially of selection for isolation from a prior state of mixing, in which the hybrid disadvantage d is likely to take a low value because of the extensive gene flow between populations. Such conditions of low d favour males (e.g. Figure 7.4c). When there is effective isolation, with very rare meetings, d may then be much greater, the restricted gene flow allowing much divergence of the two populations. But then the appropriate model is Model 7.2A, which allows a much greater range over which males are favoured, especially at low relative male time out.

8

Population differentiation without speciation

Anne E. Magurran

8.1 Introduction

It is conventional to view biological diversity as synonymous with species richness and/or species abundance (Magurran 1988; Rosenzweig 1995). In recent years, however, the concept of biological diversity has been extended to embrace the rich seam of diversity within species themselves (Barbault *et al.* 1995; Gaston 1996; Mallet 1996; Butlin and Tregenza 1998). Such diversity can be cryptic and only apparent once appropriate techniques are applied, an example being the use of allozymes to uncover levels of heterozygosity (Butlin and Tregenza 1998), or it may be expressed phenotypically and manifest itself as variation in behaviour, or morphology, or life histories (Mallet 1996; Foster and Endler 1999).

Current conservation practice recognises that intraspecific diversity is worth preserving in its own right. This is partly because the viability of a species may in some way be related to its variability (Bisby 1995, but see also Amos and Harwood 1998). An additional consideration is that intraspecific diversity is perceived as the raw material for evolution (Barbault and Sastrapradja 1995; Mallet 1996). It stands to reason that populations that are highly differentiated are the precursors of new species. Thus it is not simply variation as such, but also the potential to generate new species and even more biological diversity, that must be conserved (Myers 1997). However, as evolutionary biologists have been discovering, the relationship between population divergence and speciation is not straightforward. Although population differentiation can be an important step on the path to speciation, speciation is by no means the inevitable outcome of even the most marked population differentiation. This chapter illustrates the issue by reviewing recent work on a little fish that is increasingly being cited as a classic case of evolution in action.

8.2 Population differentiation in the Trinidadian guppy

Few species offer a more compelling example of population differentiation than the guppy, *Poecilia reticulata* (Endler 1995; Magurran *et al.* 1995). The guppy is endemic to northeastern South America and occurs in Venezuela, Guyana, Surinam and several of the Lesser Antilles including Trinidad and Tobago (Rosen and Bailey 1963). It is however, in Trinidad, that this species has been most intensively studied. Work by Caryl Haskins and his colleagues (Haskins and Haskins 1950, 1951; Haskins *et al.* 1961) alerted biologists to the evolutionary significance of Trinidadian populations. Half a century later interest in the Trinidadian guppy shows no signs of waning. Indeed, as the reference list of this chapter testifies, the quality and quantity of contemporary research on this little fish continues to grow. Why have guppy populations in Trinidad proved so rewarding to study? Part of the answer is provided by the geography of Trinidad itself. Trinidad was, until recently (somewhere between 10^3 and 10^4 years ago; Kenny 1989, 1995; Magurran *et al.* 1995) part of mainland South America. Sea level changes in the wake of the last ice age created the island we know today. One legacy of these continental origins is the Northern Range mountain chain running along the northern flank of Trinidad. The Northern Range is dissected by many rivers, virtually all of which support thriving guppy populations. Some of these populations are part of a rich community of fish species, including important guppy predators such as the pike cichlid, *Crenicichla alta*. In other cases, guppies co-occur with just a few species, most notably the cyprinodont *Rivulus hartii. Rivulus* is at best a minor predator of guppies. The presence of barrier waterfalls, a characteristic feature of the Northern Range, means that there can be marked changes in predation regime over very short distances. Guppies and *Rivulus* tend to be found in all but the highest reaches of a river but larger fish, including guppy predators, are often prevented from colonising upstream by barrier waterfalls. Evolutionary biologists thus have at their disposal replicate populations that vary in terms of an important selective factor: predation risk. It is not surprising therefore that Trinidadian guppy populations have proved invaluable in demonstrating the power of natural selection in the wild (Endler 1995).

Haskins and Haskins (1950) and Haskins *et al.* (1961) first noted the relationship between male coloration and predation regime and set the stage for a series of important papers by John Endler (see for example Endler (1978, 1980)). This work confirmed that guppy colour patterns in different populations result from the interaction of natural selection (for increasing inconspicuousness in more dangerous sites) and sexual selection (exerted by females for males of different hues). Ben Seghers (Seghers 1973, 1974; Liley and Seghers 1975) was in turn the first to recognise that guppy populations displayed geographical variation in behaviour and morphology that could be related to variation in predation intensity. For example, he discovered that guppies in populations that are subject to predation from a number of piscivorous species have higher schooling tendencies than those fish that exist in much less threatening environments (Seghers 1974). Seghers (1974) was also able to show that population differences in schooling behaviour were retained when fish were bred under standard conditions in the laboratory, thus confirming that the behaviour patterns had a genetic basis. As schooling behaviour is an important antipredator tactic (Magurran and Pitcher 1987; Neill and Cullen 1974; Magurran 1990; Pitcher and Parrish 1993) the adaptive nature of the geographic variation is evident. Subsequent investigations have revealed that Trinidadian guppy populations differ in virtually every feature that biologists have cared to examine. Endler (1995) lists 47 traits that covary with each other and with predation risk. Although such covariance provides strong evidence for the role of natural selection in shaping populations, these studies are essentially correlational. A direct causal link between predation risk and evolution is required. Trinidadian guppy populations have provided this too and what is particularly remarkable is the speed with which heritable changes arise.

8.3 Evolution in action

A pioneering investigation of natural selection in the wild was performed by John Endler in 1976 when he collected 200 guppies from a high-risk site in the Lower Aripo River and moved them to an upstream tributary of the same river (Endler 1980). Barrier waterfalls had prevented this tributary from being colonised by guppies, or by other fish species with the exception of *Rivulus*. The transplanted guppies thrived in their new home; indeed their descendants can be found there today. Endler monitored the population for the next 2 years. The changes in male colour pattern were as swift as they were dramatic. There was an increase in the size of colour spots, the number of colour spots, and the overall diversity of colour (Endler 1980). In other words the males became more colourful in line with the expectation for a low risk habitat. In these more benign sites females will have more opportunity to exert preferences for brighter males and brighter males will be at a much lower risk of capture by predators. A genetic correlation between male colour pattern and female choice (Houde 1994) will reinforce the trend towards more flamboyant males.

Evolution in the Aripo introduction experiment was not confined to male morphology either. Marked changes in life history traits were also observed. Reznick and Endler (1982) noted phenotypic changes in offspring size and in the proportion of body mass devoted to reproduction. A follow-up study confirmed that the life-histories of the source and transplant environments continued to diverge over the 11 years during which the experiment was monitored (Reznick *et al.* 1990) and that the changes in life-history were heritable. Once again evolution was in the direction anticipated for a low-predation community. The transplanted fish were larger and older at maturity and produced fewer but bigger offspring.

Reznick and Bryga (1987) carried out an additional transplant in the El Cedro River of Trinidad. Here they found changes in male size and age at maturity after 4 years and changes in female life-histories after 7 years. Reznick *et al.* (1997) estimate the mean generation time of guppies in low predation localities to be 210 days which is equivalent to 1.74 generations per year though Endler (1983) suggests that there can be as many as 14 generations in 23 months while Haskins *et al.* (1961) propose four generations a year.

The rate of evolution represented by these life history changes in truly remarkable. Reznick *et al.* (1997) calculated the rate of change in terms of Darwins (Stearns 1992) as (ln X2 − ln X1)/Δt, where X1 and X2 are the values of the trait in question at the beginning and end of the time period and Δt is the length of the time period in years. Table 8.1 summarises their results. They conclude that the rate of evolution they observed was on a par with that observed in artificial selection experiments, and seven orders of magnitude faster than in the fossil record. Reznick *et al.* (1997) attribute the differences in rate of evolution between the sexes to greater genetic variance for age and size at maturity in males rather than weaker selection on female traits.

Why should guppy populations evolve so rapidly? The answer lies in the fact that these transplant experiments resulted in strong directional selection which persisted for the time period over which the traits were measured. As Reznick *et al.* (1997) point out, once life histories become optimised for the prevailing conditions no more evolution is to be expected. Thus, further monitoring of the fish in the Aripo and El Cedro introductions is likely to reveal a reduction in the rate of evolution. In addition, the transplant experiments ensured that selection pressure was consistently in the same direction, unlike many natural situations where selection operates erratically as environmental conditions change.

Evolution will also occur, albeit at a slower rate, under weaker selection. The very first guppy transplant experiment in Trinidad was performed by Caryl Haskins in 1957 when he collected 200 guppies from a high predation site in the Caroni drainage and moved them to a low-risk (and prior to his experiment, guppy-free) stream in the Turure River of the Oropuche drainage. This transplant was never published and might have been overlooked were it not for serendipitous sampling of the Turure River during an investigation of the genetic diversity of guppy populations in Trinidad (Shaw *et al.* 1991). This, and a subsequent study (Shaw *et al.* 1992), revealed that guppies in the Turure River bore the genetic signature of the Caroni drainage even though they were geographically part of the Oropuche drainage. From the genetic data it was possible to conclude that the descendants of the founder fish had not only colonised the Upper Turure River but also moved downstream below the barrier waterfall and replaced the indigenous guppy population in the Lower Turure River.

The guppies that were introduced to the Upper Turure experienced a reduction in predation risk. There are costs of devoting too much time and energy to

Table 8.1 The rate of evolution in natural populations of guppies in Trinidad. The table shows the rate of change, expressed as 10^3 Darwins, in two life history traits, age at maturity and size at maturity, for both male and female guppies. Geometric mean rates of evolution in the fossil record range from 0.7 to 3.7 Darwins (Reznick *et al.* 1997). See text for details.

	Male age	Male size	Female age	Female size
Aripo River (18 generations)	16.4	10.9	8.0	13.9
El Cedro River (7 generations)	45.0	27.1	3.7	5.1
El Cedro River (13 generations)	13.9	5.3	7.9	9.3

antipredator behaviour (Magurran and Seghers 1991) but these are much less than the costs of failing to employ appropriate predator evasion tactics when required. As there is often variation in the incidence of predation risk during ecological time, for example as predator population densities change, there may be good adaptive reasons why antipredator behaviour should be retained over an evolutionary time scale. Indeed, the costs of antipredator behaviour might be so low in any case that it diminishes only as a consequence of non-adaptive decay resulting from accumulating mutations which reduce the incidence of the relevant genes in the population. For these reasons we might anticipate that antipredator behaviour is lost much more gradually than colour patterns are modified or life-histories change. None the less, evolution of antipredator behaviour is apparent over the time frame provided by the Haskins' experiment. The 34 years that elapsed between the introduction and its rediscovery were marked by a significant (heritable) decrease in the schooling tendencies of both male and female guppies (see Figure 8.1). Schooling, as was noted earlier, is an effective antipredator behaviour and varies adaptively in Trinidadian guppy populations (Seghers 1974; Magurran and Seghers 1994b). By Reznick *et al.*'s (1997) calculations, 34 years represent approximately 60 generations. An equivalent investigation of Endler's Aripo introduction 16 years after the transplant had occurred revealed only a nonsignificant trend towards a lower schooling tendency (Magurran *et al.* 1995; Carvalho *et al.* 1996; Figure 8.1)

Each of these studies leads to the same conclusion, that is that guppy populations in Trinidad show marked evolution over a time period that is not only short in evolutionary terms but also comprehensible in human ones. This occurs for traits that might be expected to be relatively resistant to evolutionary change (antipredator behaviour in the above example) as well as for ones for which rapid evolution is to be anticipated.

8.4 The relationship between reproductive isolation and population differentiation

Given the speed and magnitude of evolutionary change in Trinidadian guppy populations it seems reasonable to suppose that they are in a state of incipient speciation. After all, as was noted above, such populations differ more, in a variety of traits, than some well established species. It is therefore perplexing to discover that this is not the case. Molecular analyses (Carvalho *et al.* 1991; Fajen and Breden 1992; Magurran *et al.* 1995) reveal that guppy populations in the Caroni and Oropuche drainages diverged in the order of 500 000 years ago. This translates into a minimum of 1, and possibly 2, million generations. Despite this long separation there is no evidence that these populations are reproductively isolated. Magurran *et al.* (1996) examined female preferences in guppies from the Tacarigua River (Caroni drainage) and the Oropuche River (Oropuche drainage). These fish were as likely to solicit matings from males from the alien drainage as from their own (Figure 8.2a). It could be argued that an evolutionary history of

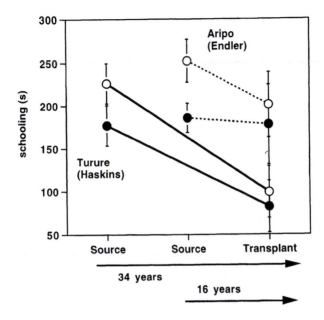

Fig. 8.1 Evolution of antipredator behaviour in Trinidadian guppies. This figure illustrates changes in schooling tendency following a relaxation of predation regime. Two introduction experiments are compared: the Turure transplant by Caryl Haskins and the Aripo transplant by John Endler. Fish were collected from the wild and then bred, raised and tested under standard conditions in the laboratory. Schooling tendency was assessed by recording the length of time (out of 5 min) that individual fish spent in the proximity of a transparent container that housed a school of guppies of the same sex. The Aripo transplant is denoted by a hatched line, the Haskins transplant by a solid line. Open symbols represent females while closed symbols signify males. The separation of the source sites on the *x*-axis symbolises the separation in time of the source and transplant populations in the two introductions. The data on which the figure is based are taken from Magurran *et al.* (1992, 1995) and the figure is redrawn from Carvalho *et al.* (1996).

predation risk tends to diminish female choice (Iwasa and Pomiankowski 1991, 1994; Pomiankowski *et al.* 1991; Pomiankowski and Iwasa 1993) and it was the case that fish in the study by Magurran *et al.* (1996) came from populations that co-occur with *Crenicichla* and other predators. However, another study (Endler and Houde 1995), which included fish from a variety of predation regimes and drainages also indicated that females do not actively discriminate against males from genetically divergent populations. Figure 8.2b illustrates the outcome for those comparisons which examined female choice across the Caroni/Oropuche divide. In none of the three cases, including the one in which both populations of fish came from low risk localities, do females treat males from the other drainage as less (or more) attractive than males from their own population. When the 11 populations in their study were taken into account, and all native and alien data pooled, there is some evidence that females are significantly more attracted to males from their own population than those from different populations

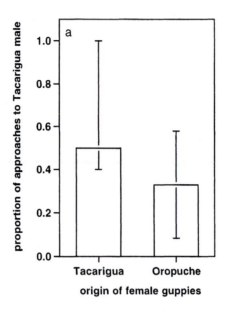

Fig. 8.2 Female guppies do not appear to discriminate against males from genetically divergent populations. (a) Summarises a study (Magurran *et al.* 1996) in which virgin female guppies from the Tacarigua River (Caroni drainage) and Oropuche River (Oropuche drainage) were given the opportunity of mating with males from both rivers. The graph shows the median proportion of approaches (plus quartiles) by both sets of females to males from the Caroni drainage. This approaching, or gliding, behaviour is a strong indicator of female preference (Liley 1966). There was no significant difference between the two sets of females in their preference for Caroni males (Mann–Whitney test, $N = 10,10$, $z = 0.24$, $P = 0.80$).

$(t_{1318} = 2.74$, $P < 0.01$; Endler and Houde 1995). However, in only five of the 11 populations is the preference for native over alien males significant and neither drainage nor predation risk affect the strength of the response. These results may be of statistical significance but their biological significance is less clear. Matings between fish from different drainages also produce viable offspring (Endler and Houde 1995; Endler 1995) so there is, to date, no evidence that post-mating isolation operates either.

8.5 Why have guppies not speciated?

How can we reconcile the evident potential for speciation in the guppy with the reality that it is not occurring? Endler (1995) lists three reasons why guppies have failed to speciate. These are ephemeral habitat gradients, the decline of female preferences in high predation areas and the scale of the selection regimes. Over part of their range in Trinidad guppies do indeed occur in ephemeral habitats.

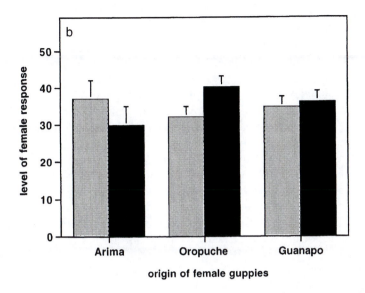

Fig. 8.2 (b) Endler and Houde (1995) also carried out an extensive analysis of Trinidadian guppies in which female preferences for males from their own population were contrasted with their preferences from males from alien populations. Some of the comparisons examined female choice between the Caroni and Oropuche drainages. These are shown here. A female's response was measured as the fraction of a male's sigmoid displays that persuaded her to cease other activities and glide unambiguously towards him. As females were virgin at the outset they remained receptive to male courtship for the duration of the trial. The mean female response (following an arcsin transformation) and standard error are illustrated for three populations: females from Arima 6 (a low predation site in the Caroni drainage) in relation to males from their own population and males from Quare 6 (a low predation population in the Oropuche drainage; females from Oropuche 1 (a high predation population in the drainage of the same name) with their own males and males from Guanapo 6 (a medium predation site in the Caroni drainage); and females from Guanapo 6 in relation to Guanapo 6 males and males from Oropuche 1. Solid bars represent males from the other drainage.

These tend to be low order streams in the upper reaches of rivers and form the classic low predation, *Rivulus*, sites that have proved indispensable to biologists. Other short-lived guppy habitats include pools, ditches and wetlands which may be subject not merely to natural perturbations but also to human ones such as pollution and drainage. It is therefore possible that guppy habitats do not persist for long enough for reproductive isolation to arise. An additional factor is that the environmental boundaries such as barrier waterfalls or riffles that help separate populations and prevent gene flow may be even more transitory than the habitats themselves. However, the impressive speed with which guppy populations evolve in response to selection (within 10^1–10^2 generations) implies that events that are ephemeral in geological terms (occurring, for example, over 10^3 or 10^4 years) are unlikely to be the sole reason why speciation has not taken place. Furthermore, as noted above, molecular analyses of populations in the Caroni and Oropuche drainages indicate that reproductive isolation has not arisen even when there has been ample time for it to have done so. This contrasts starkly with the case of cichlids in the African great lakes of Victoria, Malawi and Tanganyika (Meyer *et al.* 1990; Meyer 1993; Turner this volume). Each of these lakes is home to large numbers of endemic cichlids, some of which are very recent in origin. For example, Lake Nabugabo has been isolated from Lake Victoria for only about 4000 years yet has five endemic cichlids (Greenwood 1965). Some Lake Malawi cichlids may even have speciated within the last few hundred years (Owen *et al.* 1990; Turner this volume) while Johnson *et al.* (1996) deduce, on the basis of sediment cores, that Lake Victoria was dry some 12 500 years ago and that all its 500 cichlid species must have evolved since then.

Can female preferences provide the answer instead? It has already been noted that females from high predation regimes may be less choosy than their counterparts from safer localities. After all, it seems plausible that females that do not need to spend much time evading predators will have more opportunities to appraise the merits of potential mates. Unfortunately, data from Trinidadian guppy populations provide only partial support for this idea. Endler and Houde (1995) used the amount of variance in female responsiveness (to males) explained by male traits as a measure of the strength of preference. The average values for females (exposed to males from their own population) were 54% for low predation (*Macrobrachium*) populations versus 33% for high predation (*Crenicichla*) populations. *Macrobrachium* is a genus of freshwater prawns abundant in Trinidad, particularly in rivers that drain the northern slopes of the Northern Range. The most widespread species, which can act as a minor guppy predator, is *M. crenulatum*. *M. crenulatum* typically occurs with *Rivulus*. However, when Endler and Houde (1995) carried out the same analysis for the two other low risk (*Rivulus* only) sites in their investigation the strength of preference was only 27%. There is no doubt that low risk localities in northern flowing drainages are characterised by strong female preferences but caution is needed before concluding that low risk populations in other drainages behave in an identical fashion. In addition, low risk sites tend to be more oligotrophic than high risk ones with the result that fish, of both sexes but particularly females, need to devote a greater proportion of their time to foraging (Magurran and Seghers 1994c).

Endler's final explanation is that the distance over which gene flow occurs is large in relation to the scale at which selection takes place. (Endler 1977, 1995) has estimated the scale of gene flow as 0.75 km of river. Trinidadian rivers are relatively short with discontinuities in predation regime occurring at intervals of a few km or less. Thus, although behaviour and other traits can evolve rapidly as a consequence of selection exerted by predators, a steady influx of genes from populations under other selection regimes may undermine differentiation. Gene flow must be the key to understanding why speciation has not occurred in Trinidadian guppy populations, but what are the factors that bring it about? To answer this question it is necessary to look more closely at the guppies themselves.

8.6 Sexual dimorphism

One of the most striking features of guppies is their marked sexual dimorphism. Female guppies are larger than males. In Trinidadian populations, female standard lengths usually exceed 20 mm, often by considerable margins. The comparative figures for males are 13–18 mm (Liley and Seghers 1975). These length differences translate into much larger differences in body mass (see Figure 8.3). Sexual dimorphism in size is largely a consequence of reproductive biology. Females continue to grow after maturity and their fecundity is related to their body size (Reznick 1983), which is in turn a consequence of their foraging efficiency (Magurran and Seghers 1994a). Feeding rate is correlated with size in females but not in males (Dussault and Kramer 1981). This is undoubtedly because male growth is determinate and virtually ceases after sexual maturity. Although there is some evidence that male body size is used as a criterion of quality during female choice (Reynolds and Gross 1992), greater weight is placed upon tail length and particularly on colour pattern (Endler and Houde 1995). The trade-off between body size and reproductive success thus differs dramatically between the sexes.

Even more impressive are the colour differences. Females have a cryptic beige coloration, presumably as a defence against predation. It is the males, however, that prompt some of the local Trinidadian names for the species—'red tails' or 'seven colours'—and have led to guppies becoming one of the most popular of aquarium fish. It is probably no coincidence that, in their classification, Endler

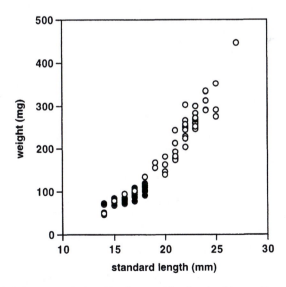

Fig. 8.3 The weight/length relationship for guppies in the Upper Tunapuna (low risk) population. Date were collected during the 1994 dry season. Females are represented by open symbols, males by closed ones. Note that females are generally larger than males.

and Houde (1995) recognised seven colour categories: fuzzy black, black, orange (including red), yellow, silver, blue and bronze-green. Haskins and Haskins (1950) first realised the significance of male colour patterns in sexual selection. Since then a series of papers have documented patterns of female preference and male coloration and the guppy is now firmly established as a model species for testing ideas about sexual selection and female choice (see Endler 1978, 1980, 1983; Kodric-Brown 1985; Breden and Stoner 1987; Houde 1987, 1988, 1997; Houde and Endler 1990; Endler and Houde 1995). Endler and Houde (1995) found that females in most of the populations they investigated had a preference for orange/ red. Caretonoid colours (orange, red and yellow) are honest indicators of foraging efficiency (Kodric-Brown 1989) and parasite load. In addition, females may use display rate as indicators of male fitness (Nicoletto 1991, 1993). Sensory drive models (Endler 1991, 1992), which predict that the environment in which the signallers live, as well as the precise nature of their sensory systems, will help shape sexual selection, can also be invoked to explain geographic variation in male coloration (Endler and Houde 1995).

Like other poeciliids, the guppy has internal fertilization and the female gives birth to live young (Wourms 1981). Wild females typically produce broods of between two and eight young once a month (Reznick and Endler 1982; Reznick and Miles 1989). This is an ovoviviparous species, in other words one in which the female provides no additional nourishment to her offspring after fertilization (Thibault and Schultz 1978). She does, however, make a major energetic investment during egg production (Reznick 1983; Reznick and Miles 1989; Reznick and Yang 1993). The resources devoted to each offspring, relative to the male's investment in his gametes, ensures that the female is the choosy sex in this species. Females can store sperm and need only a single mating to produce broods over many months (Winge 1937). They are receptive to male courtship only infrequently, that is either as virgins or in the first 24–48 h following parturition (Liley 1966; Crow and Liley 1979). Since a female's reproductive output is determined by the size and number of broods she can produce during her life at other times her priorities are predator avoidance and foraging (Magurran and Seghers 1994c).

Whereas a female's reproductive success is ensured if she has mated successfully and manages to survive, a male's fitness will depend on the number of females he inseminates. Because of female choice (Houde 1987, 1988, 1997; Houde and Endler 1990) and possibly also as a result of male–male competition (Kodric-Brown 1992) and mate copying (Dugatkin 1992; Dugatkin and Godin 1992) some males will achieve many more matings than others. Males use a modified anal fin, or gonopodium, as an intromittent organ (Constanz 1989) to transfer sperm to females and adopt two mating tactics. They can either employ a sigmoid display in which the body is arched into a characteristic S-shape and the fins extended and quivered (Liley 1966), or they may dispense with courtship and opt for sneaky mating during which the gonopodium is thrusted towards the females genital pore (Luyten and Liley 1985). Sigmoid displays test the receptivity of females and exhibit a male's colours to their best advantage. If a sigmoid display fails to elicit any response, or if two or more males are pursuing the same female, sneaky mating is used instead. All males adopt both tactics though they vary in the extent to which they use them (Magurran and Seghers 1990b) as well as in their overall level of reproductive activity (Matthews *et al.* 1997).

8.7 The battle of the sexes

Many aspects of the biology and behaviour of female guppies reinforce the population differences that natural selection has generated. For example, females, probably as a consequence of their increased investment in antipredator behaviour, have a higher schooling tendency than males. And, as a study of wild guppies in the Tacarigua River in Trinidad revealed (Griffiths and Magurran 1998), they prefer to associate with familiar schooling partners. In these choice tests, fish were given the opportunity of schooling with same sex individuals from their own school (that is the one they belonged to in the wild) or a different one. Females exerted a clear preference for fish from their natural school. Males did not. Given that it takes in the order of 12 days for female guppies to develop preferences for familiar individuals (Griffiths and Magurran 1997a), the existence of such behaviour among wild fish (Griffiths and Magurran 1997b) indicates that guppy schools are not transitory social groups. Rather it appears that the wild schools are based around female alliances and that their structure remains stable over time. Males, on the other hand, seem to move between schools as mating opportunities become available. This is reflected in their greater mobility over the short term (see Figure 8.4). Mark–recapture studies (D.N. Reznick, pers. comm.) also indicate that males are more likely to emigrate than females. It is not yet know whether guppy schools represent kin groups. A previous study using multilocus genetic fingerprinting indicated that individual members of wild guppy schools were more closely related than the average for the population (Magurran *et al.* 1995) but that the level of relatedness within schools was low, in the range of 4th order relatives (Lynch 1988). However, the schools in Magurran *et al.* (1995) were analysed without regard to gender. With hindsight it is clear that sex differences in behaviour have important implications for patterns of kinship within schools. The tendency of females to remain in cohesive groups (as well as the ability of guppies to school from birth; Magurran and Seghers 1990a) leaves open the possibility that they could be related, whereas the mobility of males means that they are probably not.

An important consequence of the tendency of females to associate with familiar schooling partners, irrespective of whether they are related or not, will be the reduction of gene flow. This ought to be reinforced by female preferences for males of certain colour patterns (Houde 1987, 1988, 1997; Houde and Endler 1990), always assuming that these preferences can be exercised in the wild, and by mate-copying (Dugatkin 1992; Dugatkin and Godin 1992), as long as it operates in nature to the extent to which it is seen in the laboratory (see also Lafleur *et al.* 1997). Lande (1981) recognised that a genetic correlation between sexually selected traits (in males) and preferences (by females) could result in rapid

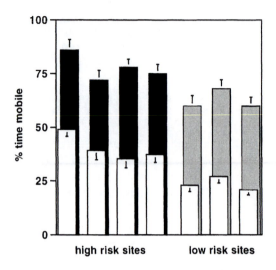

Fig. 8.4 Males are more mobile than females. This graph illustrates the proportion of time that wild fish of both sexes spend actively swimming. Mean values (and standard error) are shown. Thirty individuals were recorded per sex and population. There were four high risk (*Crenicichla*) localities (in the Aripo and Tacarigua Rivers) and three low risk (*Rivulus*) localities (in the Aripo and Tunapuna Rivers in the study. Males are represented by shaded bars, females by open ones. Data were collected as part of the investigation described in Magurran and Seghers (1994c).

speciation. Such a genetic correlation occurs in guppies (Houde 1994). In short, female partner preferences, of both the social and sexual kind, facilitate the evolution of reproductive isolation (see also: Butlin and Tregenza 1998; Coyne and Orr, this volume).

Although female guppies appear to show strong site fidelity there are infrequent occasions on which they can move considerable distances from their place of birth. This might happen, for instance, during a hurricane or tornado (there are anecdotal accounts of fish landing in people's gardens in Trinidad in the course of such events) or after a flood. Unsuccessful predation attempts may also occasionally result in fish movement. Sperm storage (Winge 1937) means that a single adult female translocated in this way has the potential to found a new population. It is already known, for example, that a single female can produce up to 41 offspring from six successive broods without re-insemination and also that guppies in more naturally isolated populations show a greater capacity to store sperm (Carvalho *et al.* 1996). It is impossible to confirm that any wild guppy populations are the product of such extreme founder effects, although the genetic evidence reveals a number of cases where there are severe population bottlenecks that probably resulted from chance colonisations of a similar kind (Carvalho *et al.* 1996). There is, however, a well documented case of a single guppy female founding a viable population. This occurred in 1981 when Professor J.S. Kenny, the then Head of the Zoology Department at the University of the

West Indies in Trinidad, collected one pregnant female from a wild population and placed her in an ornamental pond. Her descendants thrived. When the allozymic structure of the population was examined a decade later it became clear that a marked reduction in heterozygosity and significant genetic divergence had occurred (see Figure 8.5). Indeed, the differentiation between the source and transplanted populations was greater even than that between the Caroni and Oropuche drainages (Carvalho et al. 1996). Although the role of founder effects in speciation remains controversial (Barton 1989) the ability of single females to found populations, and the genetic consequences that follow, have important implications for differentiation.

Females may facilitate population differentiation and speciation but males almost certainly hinder it. There are essentially two ways in which male behaviour will increase gene flow. The first, as already noted, is greater male mobility relative to females. The second, and more important, factor is mating behaviour. One of the most striking features to the observer of a wild guppy population is the incidence of sneaky mating behaviour. A typical female is subjected to an average of one sneaky mating attempt per minute (Figure 8.6, see also Magurran and Seghers 1994c; Magurran et al. 1995). It is likely that very few of these mating attempts are successful yet the ardour with which males pursue females and the time and energy that they invest in attempting copulations implies that, in evolutionary terms, this is an important strategy. We now know that males that

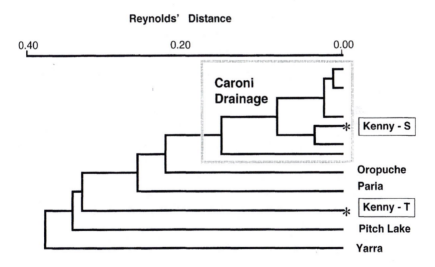

Fig. 8.5 Genetic consequences of an extreme founder effect. This figure shows a dendrogram of Reynolds' (Reynolds et al. 1983) genetic distances between samples of guppies from Trinidad. The dendrogram is a consensus tree based on bootstrapping (see Carvalho et al. 1996 for details). Calculations (UPGMA) are based on frequencies of 37 alleles across 25 gene loci. A clear separation between the Caroni and Oropuche drainages is evident. The genetic divergence between the Kenny source and Kenny transplant populations after 10 years is even more impressive. Redrawn from Carvalho et al. (1996).

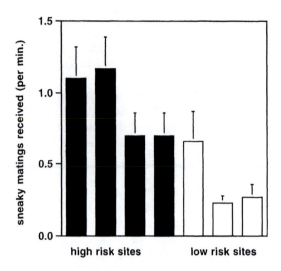

Fig. 8.6 Wild female guppies in Trinidad are subjected to a constant barrage of sneaky mating attempts. This graph illustrated the mean frequency (and standard error) of sneaky matings experienced by females in seven Trinidadian guppy populations (see Figure 8.4 for details). The figure is redrawn from Magurran *et al.* (1995).

engage most persistently in sneaky courtship have highest reserves of sperm (Matthews *et al.* 1997). We also know that some 15% of wild females, who are at a stage in their reproductive cycle during which they are not receptive to male advances, have sperm in their gonopores that could only have come from sneaky matings (I.M. Matthews and A.E. Magurran, unpublished data). Sperm competition is an inevitable outcome of the mating system of the guppy, though as yet, the fate of sperm obtained from sneaky matings is unknown. However, if even a small fraction of sneaky matings result in paternity, female choice will be undermined and gene flow facilitated. Male pursuit of reproductive success could well be the key to understanding why guppy populations have not speciated, even though many of the ingredients for rapid speciation seem to be present. Parker and Partridge (1998) note that speciation will be more rapid in circumstances where females generally win mating contests than is situations where males do. Trinidadian guppies seem to offer an excellent example of the evolutionary consequences of this battle of the sexes.

8.8 Sex and sympatric speciation

There are other ways in which the battle of the sexes may constrain differentiation. It has recently been recognised that phenotypic variation within populations may represent adaptations to particular ecological conditions (Robinson and Wilson 1994). For example, individual members of a species may develop distinct morphs associated with feeding specialisations. Most often two forms develop: a benthi-cally feeding one with a deep body and a downward pointing mouth and a pelagic form with a more fusiform shape better suited to manoeuvring and chasing mobile prey. Feeding morphs are most prevalent in low diversity systems where there are few specialised foraging competitors. Robinson and Wilson (1994) documented coexisting benthic and pelagic morphs in 38 lacustrine species of fish. These include whitefish (Coregonidae—Lindsay 1981; Todd *et al.* 1981), sunfish (Cen-trarchidea—Ehlinger and Wilson 1988), cichlids (Cichlidae—Meyer 1990a,b) and sticklebacks (Gasterosteidae—McPhail 1984, 1994). Such morphs may be a consequence of phenotypic plasticity in that the environment in which the fish develops plays a central role in shaping its feeding behaviour and body plan (Meyer 1987). For example, Wainright *et al.* (1991) found that the proportion of gastropod molluscs in the diet of pumpkin sunfish (*Lepomis gibbosus*) during ontogeny could be related to variation in the jaw morphology of the adult fish. Alternatively, feeding polymorphisms may have a genetic basis and will be expressed in offspring irrespective of the environment in which they develop (McPhail 1984; Schluter and McPhail 1992). In many cases, however, genetic and environmental factors interact during the development of such polymorphisms (Skúlason and Smith 1995). There can also be selection for phenotypic plasticity in its own right. This appears to be especially prevalent in unstable habitats (Skúlason and Smith 1995).

One of the most dramatic examples of trophic polymorphism of all is provided by the arctic charr, *Salvelinus alpinus* (Skúlason *et al.*, this volume). This species has a circumpolar distribution and occurs primarily in arctic and sub-arctic lakes and rivers which have otherwise impoverished fish faunas (Johnson 1980). Sympatric forms of arctic charr have been identified in a number of localities including Loch Rannoch in Scotland (Gardner *et al.* 1988) and Lake Hazen in Canada (Reist *et al.* 1995). It is, however, in the Icelandic lake, Thingvallavatn, that the most impressive morphological divergence occurs. This lake supports four distinctive forms of charr (Skúlason *et al.* 1989, 1992, 1993; Snorrason *et al.* 1994; Skúlason and Smith 1995; Skúlason *et al.*, this volume). These are a large and small benthivorous morph, a pelagic planktivorous morph and a piscivorous one. The morphs differ markedly in appearance and size, occupy different habitats and distinct feeding niches. This diversification, driven by intraspecific competition for limited food resources, can lead to reproductive isolation (Snorrason *et al.* 1994). Laboratory tests indicate that a number of behavioural differences between the morphs are genetically based (Skúlason *et al.* 1993). The morphs also vary in age and size at sexual maturity, spawning coloration and spawning time (Skúlason *et al.* 1989). It appears that these differences have arisen in less than 2000 generations (Snorrason *et al.* 1994). African cichlids also provide strong evidence for sympatric speciation. Schliewen *et al.* (1994) found monophyletic species flocks in two volcanic crater lakes in Cameroon. These lakes are small, 4.15 and 0.6 km², and ecologically monotonous. There are no obvious barriers to gene flow among mobile creatures such as cichlids yet the lakes are home to 11 and nine endemic species respectively. Schliewen *et al.* (1994) argue that ecological diversification may have played a key role in speciation after colonisation and highlight the fact that in one lake, Bermin, the basal lineages of the flock separate the pelagic and benthic feeders. Such ecological diversification sets the stage for assortative mating which is a prerequisite for reproductive isolation.

Guppies occur in a wide variety of habitats in Trinidad ranging from the species rich, often turbid, waters of the lowlands to the species poor communities found in clear mountain streams (Kenny 1995). The mountain populations live in oligo-trophic sites and spend a large proportion of their time foraging (Magurran and Seghers 1994c). There are few fish competitors for food and, presumably, strong selection for efficient foraging. Females have offspring to provision and males inseminations to procure. In the diverse lowland communities, on the other hand, there are many potential competitors for feeding sites, but since these localities are more productive, foraging occupies a smaller proportion of the time budget (Magurran and Seghers 1994c). Habitat use in predator rich communities is also constrained by risk (Seghers 1973). Robinson and Wilson (1995) tested the idea that guppies reared under different feeding conditions would develop appropriate morphologies. Treatments included a floating food regime and one in which fish were forced to forage benthically. Male guppies exposed to surface food developed longer bodies and skulls and took on the fusiform shape often seen in pelagic morphs, while bottom feeders became deeper bodied. Interestingly, females did not show any plasticity in body form, even in skull morphology. This led the authors to suggest that there could be sexual dimorphism in this species in their degree of phenotypic plasticity.

Robinson and Wilson (1995) used fish descended from guppies collected in 1988 from the Upper Turure, a low predation, low diversity mountain stream. These fish seem to be ideal candidates for demonstrating induced morphological diversity. However, as was noted above, Caryl Haskins transplanted guppies from a high predation site in the Caroni drainage to the Upper Turure in 1957 so the fish that Robinson and Wilson (1995) investigated were ultimately derived from a lowland, and possibly less phenotypically plastic stock. It would be interesting to test the hypothesis that the diversity of the community, as well as its productivity, is associated with the potential for morphological diversity in guppy populations.

It could also be argued that discrete feeding morphologies, in the classic sense, are unlikely to occur in guppies. This is because different morphs already exist, reflecting the separate reproductive agendas of the two sexes. Females, by virtue of their perpetual pregnancies, are deep bodied and thus pre-adapted for benthic foraging. Females are also much larger than males and thus able to deal with a wider range of food items. Males need to move rapidly through the water in search of copulations. Sneaky mating calls for agility, especially as a competing male may inseminate the female first. A streamlined, fusiform body plan is an essential requirement for a male guppy.

It might be expected therefore that, in the wild, particularly in oligotrophic low diversity sites, females will forage benthically while males will show a greater tendency towards pelagic feeding. An investigation of the behaviour of guppies in the Upper Tunapuna, a mountain stream in Trinidad's Northern Range, supported this hypothesis. The majority of females were bottom feeders, picking algae and small invertebrates off the substrate. Many males fed benthically too but a significantly greater proportion (Wilcoxon test, $z = 2.8$, $P < 0.01$; Figure 8.7) fed in open water. To do this they held station in the current and ingested small particles of food. There was no evidence of surface feeding in either sex. However, *Rivulus*, which also occurs in the Upper Tunapuna, is well-adapted for surface feeding and probably out-competes guppies in this foraging mode. In the light of these results it is perhaps not surprising that Robinson and Wilson (1995) could detect no phenotypic plasticity in the female guppies in their investigation. They were also, it seems, correct in concluding that guppies are sexually dimorphic in this trait.

8.9 Conclusions

Sexual selection has long been identified as a significant evolutionary factor. Some of the earliest papers on Trinidadian guppies (Haskins and Haskins 1949, 1950) focused on sexual selection as a potential reproductive isolating mechanism. Yet, it is only recently that the evolutionary consequences of sexual conflict, which is an

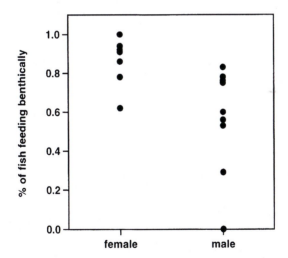

Fig. 8.7 Male and female guppies differ in their foraging behaviour. This graph shows the proportion of fish in 10 isolated pools of the Upper Tunapuna River (low risk) feeding benthically (on the bottom) as opposed to pelagically (in mid-water). Each pool was carefully scanned and the feeding location of all foraging fish noted. Clear shallow water and good visibility meant that fish could be observed without difficulty. No fish were seen to feed from the surface during the study though this behaviour is apparent on occasions at other localities. A total of 233 females and 123 males were recorded. These data were collected during June 1997.

inevitable outcome of differential investment in offspring by the two sexes, have become apparent. In some circumstances sexual conflict may fuel evolution (Rice 1996; Rice and Holland 1997), in others it seems to constrain it (Magurran 1996). The empirical insights gleaned from the Trinidadian guppy system dovetail with the theoretical ones gained from the modelling approach of Partridge and Parker (this volume). Females will, it seems, typically act as a force in favour of reproductive isolation, males as a force against it. This paper has highlighted two ways in which the battle of the sexes can inhibit speciation. First, the pursuit of copulations by males seems to be a potent force in maintaining gene flow between populations that might otherwise become reproductively isolated as a result of natural selection and female choice. Second, sexual dimorphism, which is itself a consequence of sexual conflict, reduces the opportunity for the development of feeding polymorphisms that could open the door to sympatric speciation. Many details of the relationship between sexual conflict and speciation remain to be resolved and the topic seems guaranteed to offer profitable lines of research for future investigators. What is clear is that sexual conflict plays a major role in determining whether population differentiation does translate into speciation.

Acknowledgements

I would like to thank Constantino Macías Garcia for many helpful comments on the paper, and Ben Seghers, Siân Griffiths and Iain Matthews for their insights into guppy behaviour and evolution. The work upon which much of this paper is based was supported by funding from The Royal Society, The Fisheries Society of the British Isles, the University of St Andrews and the Natural Environment Research Council (UK).

References

Amos, W. and Harwood, J. 1998 Factors affecting levels of genetic diversity in natural populations. *Phil. Trans. R. Soc. London Ser. B.* **353** 177–186.
Barbault, R. and Sastrapradja, S. (ed.). 1995 Generation, maintenance and loss of biodiversity. In *Global biodiversity assessment* (ed. V. H. Heywood), pp. 197–274. Cambridge: Cambridge University Press.
Barton, N. H. 1989 Founder effect speciation. In *Speciation and its consequences* (ed. D. Otte and J. A. Endler), pp. 229–256. Sunderland, MA: Sinauer.
Bisby, F. A. and Heywood, V. H. 1995 Characterization of biodiversity. In *Global biodiversity assessment* (ed. V. H. Heywood), pp. 25–106. Cambridge: Cambridge University Press.
Breden, F. and Stoner, G. 1987 Male predation risk determines female preference in the Trinidad guppy. *Nature* **329**, 831–833.
Butlin, R. K. and Tregenza, T. 1998 Levels of genetic polymorphism: marker loci versus quantitative traits. *Phil. Trans. R. Soc. London Ser. B.* **353**, 187–198.
Carvalho, G. R., Shaw, P. W., Magurran, A. E. and Seghers, B. H. 1991 Marked genetic

divergence revealed by allozymes among populations of the guppy *Poecilia reticulata* (Poeciliidae), in Trinidad. *Biol. J. Linn. Soc.* **42**, 389–405.

Carvalho, G. R., Shaw, P. W., Hauser, L., Seghers, B. H. and Magurran, A. E. 1996 Artificial introductions, evolutionary change and population differentiation in Trinidadian guppies (*Poecilia reticulata*: Poeciliidae). *Biol. J. Linn. Soc.* **57**, 219–234.

Constanz, G. D. 1989 Reproductive biology of Poeciliid fishes. In *Ecology and evolution of livebearing fishes (Poeciliidae)* (ed. G. K. Meffe and F. F. Snelson), pp. 33–50. New Jersey: Prentice Hall.

Crow, R. T. and Liley, N. R. 1979 A sexual pheromone in the guppy, *Poecilia reticulata* (Peters). *Can. J. Zool.* **57**, 184–188.

Dugatkin, L. A. 1992 Sexual selection and imitation: females copy the mate choice of others. *Am. Nat.* **139**, 1384–1389.

Dugatkin, L. A. and Godin, J.-G. 1992 Reversal of female mate choice by copying in the guppy (*Poecilia reticulata*). *Proc. R. Soc. London Ser. B* **249**, 179–184.

Dussault, G. V. and Kramer, D. L. 1981 Food and feeding behaviour of the guppy, *Poecilia reticulata* (Pisces: Poeciliidae). *Can. J. Zool.* **59**, 684–701.

Ehlinger, T. J. and Wilson, D. S. 1988 Complex foraging polymorphism in bluegill sunfish. *Proc. Natl Acad. Sci. USA* **85**, 1878–1882.

Endler, J. A. 1977 *Geographic variation, speciation and clines.* Princeton, NJ: Princeton University Press.

Endler, J. A. 1978 A predator's view of animal color patterns. *Evol. Biol.* **11**, 319–364.

Endler, J. A. 1980 Natural selection on color patterns in *Poecilia reticulata. Evolution* **34**, 76–91.

Endler, J. A. 1983 Natural and sexual selection on color patterns in poeciliid fishes. *Env. Biol. Fishes.* **9**, 173–190.

Endler, J. A. 1991 Variation in the appearance of guppy color pattrens to guppies and their predators under different visual conditions. *Vision Res.* **31**, 587–608.

Endler, J. A. 1992 Signals, signal conditions and the direction of evolution. *Am. Nat.* **139 suppl.**, 125–153.

Endler, J. A. 1995 Multiple-trait coevolution and environmental gradients in guppies. *Trends Ecol. Evol.* **10**, 22–29.

Endler, J. A. and Houde, A. E. 1995 Geographic variation in female preferences for male traits in *Poecilia reticulata. Evolution* **49**, 456–468.

Fajen, A. and Breden, F. 1992 Mitochondrial DNA sequence variation among natural populations of the Trinidad guppy, *Poecilia reticulata. Evolution* **46**, 1457–1465.

Foster, S. A. and Endler, J. A. 1999 *The evolution of geographic variation in behaviour.* Oxford: Oxford University Press.

Gardner, A. S., Walker, A. F. and Greer, R. B. 1988 Morphemetric analysis of two ecologically distinct forms of arctic charr, *Salvelinus alpinus* (L.), in Loch Rannoch, Scotland. *J. Fish. Biol.* **32**, 901–910.

Gaston, K. J. 1996 What is biodiversity? In *Biodiversity: a biology of numbers and difference* (ed. K. J. Gaston), pp. 1–9. Oxford: Oxford University Press.

Greenwood, P. H. 1965 The cichlid fishes of Lake Nabugabo, Uganda. *Bull. Br. Mus. Nat. Hist. (Zoology)* **12**, 315–357.

Griffiths, S. W. and Magurran, A. E. 1997a Familiarity in schooling fish: how long does it take to acquire? *Anim. Behav.* **53**, 945–949.

Griffiths, S. W. and Magurran, A. E. 1997b Schooling preferences for familiar fish vary with group size in a wild guppy population. *Proc. R. Soc. London Ser. B* **264**, 547–551.

Griffiths, S. W. and Magurran, A. E. 1998 Sex and schooling behaviour in the Trinidadian guppy. *Anim. Behav.* **56**, 689–693.

Haskins, C. P. and Haskins, E. F. 1949 The role of sexual selection as an isolating mechanism in three species of poeciliid fishes. *Evolution* **3**, 160–169

Haskins, C. P. and Haskins, E. F. 1950 Factors affecting sexual selection as an isolating mechanism in poeciliid fish. *Proc. Natl Acad. Sci. USA* **36**, 464–476.

Haskins, C. P. and Haskins, E. F. 1951 The inheritance of certain color patterns in wild populations of *Lebistes reticulatus* in Trinidad. *Evolution* **5**, 216–225.

Haskins, C. P., Haskins, E. F., McLaughlin, J. J. A. and Hewitt, R. E. 1961 Polymorphism and population structure in *Lebistes reticulatus*, an ecological study. In *Vertebrate Speciation* (ed. W. F. Blair), pp. 320–295. Austin: University of Texas Press.

Houde, A. E. 1987 Mate choice based on naturally occurring colour pattern variation in a guppy population. *Evolution* **41**, 1–10.

Houde, A. E. 1988 Genetic difference in female choice between two guppy populations. *Anim. Behav.* **36**, 510–516.

Houde, A. E. 1994 Effect of artificial selection on male colour patterns on mating preference of female guppies. *Proc. R. Soc. London Ser. B* **256**, 125–130.

Houde, A. E. 1997 *Sexual selection and mate choice in guppies.* Princeton, NJ: Princeton University Press.

Houde, A. E. and Endler, J. A. 1990 Correlated evolution of female mating preferences and male color patterns in the guppy, *Poecilia reticulata. Science* **248**, 1405–1408.

Iwasa, Y. and Pomiankowski, A. 1991 The evolution of costly mate preferences 2. The handicap principle. *Evolution* **45**, 1431–1442.

Iwasa, Y. and Pomiankowski, A. 1994 The evolution of mate preferences for multiple sexual oraments. *Evolution* **48**, 853–867.

Johnson, L. 1980 The arctic charr, *Salvelinus alpinus.* In *Charrs: salmonid fishes of the genus Salvelinus* (ed. E. K. Balon), pp. 15–98. The Hague: Dr W. Junk.

Johnson, T.C., Scholz, C.A., Talbot, M. R., Kelts, K., Ricketts, R.D., Ngobi, G., Beuning, K., Ssemmanda, I. and McGill, J.W. 1996 Late Pleistocene desiccation of Lake Victoria and rapid evolution of cichlid fishes. *Science.* **273**, 1091–1093

Kenny, J. S. 1989 Hermatypic scleractinian corals of Trinidad. *Stud. Fauna Curaçao Caribbean Islands* **123**, 83–100.

Kenny, J. S. 1995 *Views from the bridge: a memoir on the freshwater fishes of Trinidad.* St Joseph, Trinidad and Tobago: J.S. Kenny.

Kodric-Brown, A. 1985 Female preference and sexual selection for male coloration the the guppy (*Poecilia reticulata*). *Behav. Ecol. Sociobiol.* **17**, 199–206.

Kodric-Brown, A. 1989 Dietary carotenoids and male mating success in the guppy: an environmental component to female choice. *Behav. Ecol. Sociobiol.* **25**, 393–401.

Kodric-Brown, A. 1992 Male dominance can enhance mating success in guppies. *Anim. Behav.* **44**, 165–167.

Lafleur, D. L., Lozano, G. A. and Sclafani, M. 1997 Female mate-copying in guppies, *Poecilia reticulata*: a re-evaluation. *Anim. Behav.* **54**: 579–586

Lande, R. 1981 Models of speciation by sexual selection on polygenic characters. *Proc. Natl Acad. Sci. USA* **78**, 3721–3725.

Liley, N. R. 1966 Ethological isolating mechanisms in four sympatric species of Poeciliid fishes. *Behaviour Suppl.* **13**, 1–197.

Liley, N. R. and Seghers, B. H. 1975 Factors affecting the morphology and behaviour of guppies in Trinidad. In *Function and evolution in behaviour* (ed. G. P. Baaerends, C. Beer and A. Manning), pp. 92–118. Oxford: Clarendon Press.

Lindsay, C. C. 1981 Stocks are chameleons: plasticity in gill rakers of Coregonid fishes. *Can. J. Fish.Aquat. Sci.* **38**, 1497–1506.

Luyten, P. H. and Liley, N. R. 1985 Geographic variation in the sexual behaviour of the guppy, *Poecilia reticulata* (Peters). *Behaviour* **95**, 164–179.

Lynch, M. 1988 Estimation of relatedness by DNA fingerprinting. *Mol. Biol. Evol.* **5**, 584–599.

Magurran, A. E. 1988 *Ecological diversity and its measurement.* Princeton, NJ: Princeton University Press.

Magurran, A. E. 1990 The adaptive significance of schooling as an antipredator defence in fish. *Ann. Zool. Fennici* **27**, 51–66.

Magurran, A. E. 1996 Battle of the sexes. *Nature* **383**, 307.

Magurran, A. E. and Pitcher, T. J. 1987 Provenance, shoal size and the sociobiology of predator evasion behaviour in minnow shoals. *Proc. R. Soc. London Ser. B* **229**, 439–465.

Magurran, A. E. and Seghers, B. H. 1990a Population differences in the schooling behaviour of newborn guppies, *Poecilia reticulata. Ethology* **84**, 334–342.

Magurran, A. E. and Seghers, B. H. 1990b Risk sensitive courtship in the guppy *Poecilia reticulata. Behaviour* **112**, 194–201.

Magurran, A. E. and Seghers, B. H. 1991 Variation in schooling and aggression amongst guppy, *Poecilia reticulata* populations in Trinidad. *Behaviour* **118**, 214–234.

Magurran, A. E. and Seghers, B. H. 1994a A cost of sexual harassment in the guppy, *Poecilia reticulata. Proc. R. Soc. London Ser. B* **258**, 89–92.

Magurran, A. E. and Seghers, B. H. 1994b Predator inspection behaviour covaries with schooling tendency amongst wild guppy, *Poecilia reticulata*, populations in Trinidad. *Behaviour* **128**, 121–134.

Magurran, A. E. and Seghers, B. H. 1994c Sexual conflict as a consequence of ecology: evidence from guppy, *Poecilia reticulata*, populations in Trinidad. *Proc. R. Soc. London Ser. B* **255**, 31–36.

Magurran, A. E., Seghers, B. H., Carvalho, G. R. and Shaw, P. W. 1992 Behavioral consequences of an artificial introduction of guppies, *Poecilia reticulata*, in N. Trinidad: evidence for the evolution of antipredator behaviour in the wild. *Proc. R. Soc. London Ser. B* **248**, 117–122.

Magurran, A. E., Seghers, B. H., Shaw, P. W. and Carvalho, G. R. 1995 The behavioral diversity and evolution of guppy, *Poecilia reticulata*, populations In Trinidad. *Adv. Study Behav.* **24**, 155–202.

Magurran, A. E., Paxton, C. G. M., Seghers, B. H., Shaw, P. W. and Carvalho, G. R. 1996 Genetic divergence, female choice and male mating success in Trinidadian guppies. *Behaviour* **133**, 503–517.

Mallet, J. 1996 The genetics of biological diversity: from varieties to species. In *Biodiversity: a biology of numbers and difference* (ed. K. J. Gaston), pp. 13–53. Oxford: Oxford University Press.

Matthews, I. M., Evans, J. P. and Magurran, A. E. 1997 Male display rate reveals ejaculate characteristics in the Trinidadian guppy *Poecilia reticulata. Proc. R. Soc. London Ser. B.* **264**, 695–700.

McPhail, J. D. 1984 Ecology and evolution of sympatric sticklebacks (*Gasterosteus*): morphological and genetic evidence for a species pair in Enos Lake, British Columbia. *Can. J. Zool* **62**, 1402–1408.

McPhail, J. D. 1994 Speciation and the evolution of reproductive isolation in the sticklebacks (*Gasterosteus*) of south-western British Columbia. In *The evolutionary biology of the threespine stickleback* (ed. M. A. Bell and S. A. Foster), pp. 399–437. Oxford: Oxford University Press.

Meyer, A. 1987 Phenotypic plasticity and heterochrony in *Cichlasoma managuense* (Pisces, Cichlidae) and the implications for speciation in cichlid fishes. *Evolution* **41**, 1357–1369.

Meyer, A. 1990a Ecological and evolutionary consequences of trophic polymorphism in *Cichlasoma citrinellum* (Pisces, Cichlidae). *Biol. J. Linn. Soc.* **39**, 279–299.

Meyer, A. 1990b Morphometrics and allometry in the trophically polymorphic cichlid fish

Cichlasoma citrinellum: alternative adaptations and ontogenetic changes in shape. *J. Zool.* **221**, 237–260.

Meyer, A. 1993 Phylogenetic relationships and evolutionary processes in East African cichlid fishes. *Trends Ecol. Evol.* **8**, 279–284.

Meyer, A., Kocher, T. D., Basasibwaki, P. and Wilson, A. C. 1990 Monophyletic origin of Lake Victoria cichlid fishes suggested by mitochondrial DNA sequences. *Nature* **347**, 550–553.

Myers, N. 1997 The rich diversity of biodiversity issues. In *Biodiversity II* (ed. M. L. Reaka-Kudla, D. E. Wilson and E. O. Wilson), pp. 125–138. Washington, DC: Joseph Henry Press.

Neill, S. R. S. and Cullen, J. M. 1974 Experiments on whether schooling by their prey affects the hunting behaviour of cephalopod and fish predators. *J. Zool.* **172**, 549–569.

Nicoletto, P. F. 1991 The relationship between male ornamentation and swimming performance in the guppy, *Poecilia reticulata*. *Behav. Ecol. Sociobiol.* **28**, 365–370.

Nicoletto, P. F. 1993 Female sexual response to condition-dependent ornaments in the guppy, *Poecilia reticulata*. *Anim. Behav.* **46**, 441–450.

Owen, R. B., Crossley, R., Johnson, T. C., Tweddle, D., Kornfield, I., Davison, S., Eccles, D. D. and Engstrom, D. E. 1990 Major low lake levels in Lake Malawi and their implications for speciation rates in cichlid fishes. *Proc. R. Soc. London Ser. B.* **240**, 519–553.

Parker, G. A. and Partridge, L. 1998 Sexual conflict and speciation. *Phil. Trans. R. Soc. London Ser. B.* **353**, 261–274.

Pitcher, T. J. and Parrish, J. K. 1993 Functions of shoaling behaviour in teleosts. In *Behaviour of teleost fishes* (ed. T. J. Pitcher), pp. 363–439. London: Chapman and Hall.

Pomiankowski, A. and Iwasa, Y. 1993 Evolution of multiple sexual preferences by Fisher runaway process of sexual selection. *Proc. R. Soc. London Ser. B.* **253**, 173–181.

Pomiankowski, A., Iwasa, Y. and Nee, S. 1991 The evolution of costly mate preferences 1. Fisher and biased mutation. *Evolution* **45**, 1422–1430.

Reist, J. D., Gyselman, E., Babaluk, J. A., Johnson, J. D. and Wissink, R. 1995 Evidence for two morphotypes of arctic char (*Salvelinus alpinus* (L.)) from Lake Hazen, Ellesmere Island, Northwest Territories, Canada. *Nordic J. Freshw. Res.* **71**, 396–401.

Reynolds, J. B., Weir, B. S. and Cockerham, C. C. 1983 Estimation of the coancestry coefficient: basis for a short term genetic distance. *Genetics* **105**, 767–779.

Reynolds, J. D. and Gross, M. R. 1992 Female mate preference enhances offspring growth and reproduction in a fish, *Poecilia reticulata*. *Proc. R. Soc. London Ser. B* **250**, 57–62.

Reznick, D. N. 1983 The structure of guppy life histories: the tradeoff between growth and reproduction. *Ecology* **64**, 862–873.

Reznick, D. N. and Bryga, H. 1987 Life-history evolution in guppies (*Poecilia reticulata*): I. Phenotypic and genetic changes in an introduction experiment. *Evolution* **41**, 1370–1385.

Reznick, D. N. and Endler, J. A. 1982 The impact of predation on life history evolution in Trinidadian guppies (*Poecilia reticulata*). *Evolution* **36**, 125–148.

Reznick, D. N. and Miles, D. B. 1989 A review of life history patterns in poeciliid fishes. In *Ecology and evolution of livebearing fishes* (ed. G. K. Meffe and F. F. Snelson), pp. 125–148. Englewood Cliffs, NJ: Prentice Hall.

Reznick, D. N. and Yang, P. 1993 The influence of fluctuating resources on life history: patterns of allocation and plasticity in female guppies. *Ecology* **74**, 2011–2019.

Reznick, D. N., Bryga, H. and Endler, J. A. 1990 Experimentally induced life-history evolution in a natural population. *Nature* **346**, 357–359.

Reznick, D. N., Shaw, F. H., Rodd, F. H. and Shaw, R. G. 1997 Evaluation of the rate of evolution in natural populations of guppies (*Poecilia reticulata*). *Science* **275**, 1934–1937.

Rice, W. R. 1996 Sexually antagonistic male adaptation triggered by experimental arrest of female evolution. *Nature* **381**, 232–234.

Rice, W.R. and Holland, B. 1997 The enemies within: intergenomic conflict, interlocus evolution (ICE), and the intraspecific Red Queen. *Behav. Ecol. Sociobiol.* **41**, 1–10

Robinson, B. W. and Wilson, D. S. 1994 Character release and displacement in fishes: a neglected literature. *Am. Nat.* **144**, 596–627.

Robinson, B. W. and Wilson, D. S. 1995 Experimentally induced morphological diversity in Trinidadian guppies (*Poecilia reticulata*). *Copeia* **1995**, 294–305.

Rosen, D. E. and Bailey, R. M. 1963 The poeciliid fishes (Cyprinodontiformes), their structure, zoogeography and stystematics. *Bull. Am. Mus. Nat. Hist.* **126**, 1–176.

Rosenzweig, M. L. 1995 *Species diversity in space and time*. Cambridge: Cambridge University Press.

Schliewen, U. K., Tautz, D. and Pääbo, S. 1994 Sympatric speciation suggested by monophyly of crater lake cichlids. *Nature* **368**, 629–632.

Schluter, D. and McPhail, J. D. 1992 Ecological character displacement and speciation in sticklebacks. *Am. Nat.* **140**, 85–108.

Seghers, B. H. 1973 An analysis of geographic variation in the antipredator adaptations of the guppy, *Poecilia reticulata*. PhD thesis, University of British Columbia.

Seghers, B. H. 1974 Schooling behavior in the guppy (*Poecilia reticulata*): an evolutionary response to predation. *Evolution* **28**, 486–489.

Shaw, P. W., Carvalho, G. R., Magurran, A. E. and Seghers, B. H. 1991 Population differentiation in Trinidadian guppies, *Poecilia reticulata*—patterns and problems. *J. Fish Biol.* **39**, 203–209.

Shaw, P. W., Carvalho, G. R., Seghers, B. H. and Magurran, A. E. 1992 Genetic consequences of an artificial introduction of guppies, *Poecilia reticulata*, in N. Trinidad. *Proc. R. Soc. London Ser. B.* **248**, 111–116.

Skúlason, S. and Smith, T. B. 1995 Resource polymorphism in vertebrates. *Trends Ecol. Evol.* **10**, 366–370.

Skúlason, S., Noakes, D. L. G. and Snorrason, S. S. 1989 Ontogeny of trophic morphology in four sympatric morphs of arctic charr Salvelinus alpinus in Thingvallavatn, Iceland. *Biol. J. Linn. Soc.* **38**, 281–301.

Skúlason, S., Antonsson, T., Gudbergsson, G., Malmquist, H. J. and Snorrason, S. S. 1992 Variability in Icelandic arctic charr. *Icelandic Agric. Sci.* **6**, 143–153.

Skúlason, S., Snorrason, S. S., Ota, D. and Noakes, D. L. G. 1993 Genetically based differences in foraging behaviour among sympatric morphs of arctic charr (Pisces: Salmonidae). *Anim. Behav.* **45**, 1179–1192.

Snorrason, S. S., Skúlason, S., Jonsson, B., Malmquist, H. J., Jónasson, P. M., Sandlund, O. T. and Lindem, T. 1994 Trophic specialization in Arctic charr Salvelinus alpinus (Pisces: Salmonidae): morphological divergence and ontogenetic niche shifts. *Biol. J. Linn. Soc.* **51**, 1–18.

Stearns, S. C. 1992 *The evolution of life histories*. Oxford: Oxford University Press.

Thibault, R. E. and Schultz, R. J. 1978 Reproductive adaptations among viviparous fishes (Cyprinodontiformes: Poeciliidae). *Evolution* **32**, 320–333.

Todd, T. N., Smith, G. R. and Cable, L. 1981 Environmental and genetic contributions to morphological differentiation in ciscoes (*Coregonidae*) of the Great Lakes. *Can. J. Fish. Aquat. Sci.* **38**, 59–67.

Wainright, P. C., Osenberg, C. W. and Mittelbach, G. G. 1991 Trophic polymorphism in the pumpkinseed sunfish (*Lepomis gibbosus* Linnaeus): effects of environment on ontogeny. *Funct. Ecol.* **5**, 40–55.

Winge, O. 1937 Succession of broods in *Lebistes*. *Nature* **140**, 467.

Wourms, J. P. 1981 Viviparity: the maternal foetal relationship in fishes. *Am. Zool.* **21**, 473–575.

9

From genes to individuals: developmental genes and the generation of the phenotype

Diethard Tautz and Karl Schmid

9.1 Introduction

The assumption that the understanding of ontogeny should also be the basis for understanding phylogeny can be traced back into Darwin's times and has become commonplace today. However, developmental and evolutionary disciplines have followed very different routes during the past decades and have generated their own paradigms. Often, these do not provide much room for overlap. This situation is currently changing. The progress in molecular genetic methods is starting to unite the two fields. We are beginning to understand how an organism is built at the molecular level and many of the genes involved in these processes appear to be highly conserved in evolution. This provides an immediate access to comparative studies in diverse organisms and is starting to create a new discipline: molecular comparative embryology. The early results from this discipline do already shed some light on the course of the major cladogenic events. Yet, since most of the developmental genetic approaches were not designed to solve evolutionary questions, their outcomes provide only a patchy picture of the possible routes of evolution of the organisms. The genes that are most heavily studied are those involved in early embryonic decisions, rather than those required for the differentiation of the adult morphology. Evidently, evolutionary biologists would be particularly interested in the latter class of genes, as these are the ones that should be most relevant for ecological adaptations and speciation. Characterizing such genes will therefore be a major task of the future for biologists working at the interface of development and evolution.

9.2 Developmental genetics

One of the major breakthroughs in the field of developmental biology was the concept of using mutants with specific developmental defects to analyse ontogenetic processes (Lewis 1978; Nüsslein-Volhard and Wieschaus 1980). The organism of choice for such a genetic approach was *Drosophila*, since this allowed the most sophisticated types of genetic experiments. The developmental biology of *Drosophila*, on the other hand, was only poorly understood at that time, since the small size of the embryo as well as its specialized development presented major obstacles for experimental manipulation. In fact, some classic experimental manipulations on the *Drosophila* embryo were done only long after it was clear that the genetic approach was a success.

The systematic genetic screens have focused on two types of mutants, those that disrupt early embryonic pattern formation processes and those that lead to transformations of body regions. The first group of genes are required to build the basic structure of the body—the bauplan. If one of them is mutant, specific regions of the body are missing. There is no generic name for this group of genes, but the term 'bauplan genes' might be fitting. The second group of genes is required to determine the identity of body regions. Their generic name is 'homeotic genes'. If they are mutant, one gets a transformation of certain body regions into other regions, but not loss of regions as it is the case for mutants in bauplan genes.

The bauplan genes include also those genes that are necessary to generate the initial maternally determined asymmetries in the early embryo, which specify the anterior–posterior and dorso-ventral axis (St. Johnston and Nüsslein-Volhard 1992). These asymmetries are then interpreted by the genes that are expressed in the developing embryo (Pankratz and Jäckle 1993) (Fig. 9.1). They generate the segmental subdivision, as well as the regionalization of the dorso-ventral axis (Chasan and Anderson 1993). Most of the interactions between the gene products occurring at these early stages are well understood in *Drosophila*. Both the initial dorso-ventral and the anterior–posterior asymmetries are ultimately caused by the asymmetry of the cytoskeleton in the early oocyte (Grünert and St Johnston 1996). A cascade of gene functions is then required to transform this information into several morphogenetic gradients of transcription factors. These factors regulate the expression of the zygotic segmentation genes, which act again via a cascade of regulatory molecules to achieve the functional subdivision of the embryo. Most of these early gene functions are short and transient, since they are only required for setting up a set of stably expressed genes. Among these stably expressed genes are the homeotic genes that provide the different body regions with an identity, in particular in the anterior–posterior axis. If the homeotic genes were missing, the whole body would be transformed into a series of identical segments. One of the intriguing findings in this context was that the homeotic genes are clustered and that the anterior–posterior order in which they act reflects their linear position on the chromosome (Lewis 1978; McGinnis and Krumlauf 1992). The reason for this is still not understood, but the fact that this arrangement is highly conserved in evolution suggests that there must be some underlying regulatory principle that is required for this type of function.

Most of the early acting genes code for regulatory molecules which are either involved in transcriptional or translational regulation or in signalling processes.

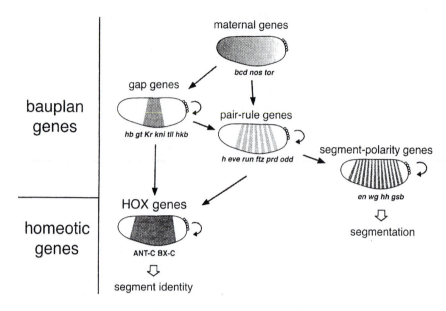

Fig. 9.1 Schematic representation of the genetic hierarchy organizing the anterior–posterior axis in *Drosophila*. Some essential genes at each step of the hierarchy are represented by their respective abbreviations (see Pankratz and Jäckle 1993 for more details). The maternal genes are those that are required for providing positional information in the egg. Their products become localized during embryogenesis and act as long range gradients to regulate the positioning of the gap genes. These are then expressed in broad domains (the example shown is the expression of *Krüppel*), which generate a series of overlapping short range gradients. The combinatorial interaction of these gradients then generates the transient expression of the pair-rule genes. These are expressed in seven stripes, whereby each stripe is regulated independently by a certain combination of the gap gene and maternal gene products. The combinatorial interaction of the pair-rule genes is required to regulate the segment-polarity genes, which eventually determine the segment boundaries. The gap genes and the pair-rule genes are also required to regulate the genes of the two homeotic complexes which eventually specify segment identity. A similar cascade of gene interactions is required to specify the subdivision of the dorso-ventral axis into different subregions (Chasan and Anderson 1993).

Many belong to well known gene families, which include protein domains that may be conserved even between prokaryotes and eukaryotes. Generally, one can say that almost all types of regulatory molecules that are known in cell biology are also involved in some aspect of early development. Still, mainly for historical reasons, one type of regulator is frequently singled out, namely the homeobox containing transcription factors. They were first identified in the homeotic gene clusters (hence the name 'homeobox'—Gehring 1994), but occur also outside the cluster and may have very different types of functions in the developmental gene hierarchy. Also, not all homeotic genes code for homeobox proteins, as for example those that are required for providing the identity to the terminal structures (Jürgens and Hartenstein 1993). To differentiate between the different types of genes, it has become customary to call genes that contain a homeobox and that are considered to function in the homeotic gene cluster 'HOX genes'.

It now seems clear that all animals contain at least one HOX gene cluster and that these are largely expressed in an anterior–posterior direction according to their position on the chromosome (McGinnis and Krumlauf 1992). This finding has been considered as so significant that it was even suggested as a defining character for all animals (the 'zootype'—Slack *et al.* 1993). However, one should note that though the HOX genes play an important role, they still constitute only a minor part of the whole developmental gene hierarchy. Moreover, it is only their clustered organization that makes them special, not their degree of conservation, as other developmental genes may be equally or even more conserved.

In contrast to *Drosophila*, the developmental analysis of vertebrate embryos has proceeded in a more traditional direction, using embryo manipulation and biochemical methods to understand the formation of the early embryonic structures. A systematic genetic approach has only recently been undertaken for the zebra fish (Granato and Nüsslein-Volhard 1996), but the genes identified in these screens still have to be characterized. However, the traditional approach has by now also provided a very detailed understanding of the basic principles of early pattern formation in a vertebrate embryo. For *Xenopus* in particular, it was possible to identify genes that are involved in axis specification and tissue determination through the early organizer regions (Harger and Gurdon 1996). One of the fascinating outcomes of these studies is that the dorso-ventral patterning cascade uses two antagonistic genes (BMP4 and *chordin*) that have complementary homologues in the *Drosophila* dorso-ventral patterning cascade (*decapentaplegic* and *short gastrulation*). However, while BMP4 specifies a ven-tralizing signal, its *Drosophila* homologue *decapentaplegic* is required for a dorsalizing signal (Ferguson 1996). This finding corroborates at the molecular level the long held view that chordates and arthropods have a reversed dorso-ventral axis with respect to each other (Arendt and Nübler-Jung 1997).

Another emerging system for evolutionary comparisons in developmental biology are the nematodes. *Caenorhabditis elegans* has become one of the best studied organisms, because many of its developmental cell fate decisions can be traced to their molecular origins. This profound knowledge now serves as a basis for comparative studies in different nematode species. The structures of particular interest are currently the generation of the vulva (Sommer *et al.* 1994) and the tail (Fitch 1997). In the case of the vulva in particular, it has been shown that genetic screens for appropriate phenotypes can easily be done also for other nematode species (Sommer *et al.* 1994). In the long run this should provide a very strong basis for inferences on the evolution of cell–cell communication pathways.

In contrast to animals, comparative developmental biology in plants is only at its beginnings. Again, it is mainly the genetic screens for specific phenotypes in either floral development or embryonic development that are about to provide a very detailed picture of the regulatory decisions that are at the basis of pattern formation in plants. The genes involved in developmental decision are also currently being molecularly characterized and it seems again that conserved gene families of regulatory genes play a role (Coen and Nugent 1994; Theissen and Saedler 1995).

9.3 The phylotypic stage

All animals (and to a certain degree also plants) seem to go through a particularly stereotypic phase during their ontogenetic development, in which the basic outline of the body pattern is generated before it is further modified to form the adult individual. This phase looks morphologically very similar, even among very distantly related taxa and is nowadays called the 'phylotypic stage' (Sander 1983). Interestingly, the morphological diversity of embryos before they reach this stage can be considerable. In particular, the first cleavage events and early embryonic development can differ markedly, even between closely related taxa. This observation of early and late morphological diversity with an intermediate stereotypic stage has become known as the hourglass model of development (Raff 1996). Moreover, because of the observation that the expression of the HOX-genes is at its peak during the phylotypic stage, it was suggested that there may be an underlying molecular principle which necessitates this stage. It was suggested that the HOX complex acts as a timing device which links anterior–posterior patterning with growth control (Duboule 1994). Such an idea seems attractive if one considers the possibility that such a timing device may have already existed in the unicellular precursors of metazoans where it might have acted to control different stages of a life cycle. This could then have been co-opted for an anterior–posterior patterning device in the first multicellular organisms. Still, the spatial deployment of the HOX genes in *Drosophila* depends heavily of the regulation by the bauplan genes. Whether this is only the case for *Drosophila* is as yet unclear, since so little is known about these genes outside of insects. Another argument for explaining the constraints on the evolvability of the phylotypic stage might be the assumed multitude of modular developmental interactions that occur at this stage (Raff 1996). The large diversity before and after this stage would be explained by a lower number of modular interactions and by a requirement for ecological adaptations to the respective environments (Sander 1983; Raff 1996). However, these models do not shed light on the question how the phylotypic stage should have evolved in the first place.

The hourglass model assumes implicitly that ontogeny has a defined start and end point. However, ontogeny is clearly a cyclical process and the asymmetric wheel model depicted in Fig. 9.2 would therefore better represent the situation. As in the hourglass model, it takes account of the fact that the adult morphology diversifies due to ecological adaptations. The adults then go on to produce eggs which themselves may be subject to adaptations. They may, for example, be designed for external or internal development or may be rich or poor in yolk content or may be tuned for fast or slow development. However, the options

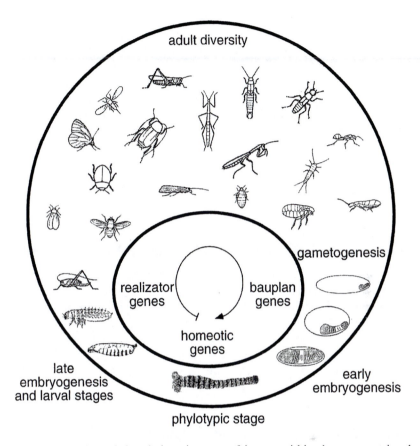

Fig. 9.2 Representation of the phylotypic stage of insects within the asymmetric wheel model. The phylotypic stage is represented by an embryo from a beetle (the flour beetle *Tribolium castaneum*), which has been stained with an antibody against the segment-polarity gene *engrailed* to visualize the segment boundaries that have been establshed at this stage. This stage looks very similar for all the different insects shown. Further development proceeds via larval stages (depicted as the larvae from *Drosophila* and *Tribolium*) which are already more diverse, but still show more similarities between the different species than the respective adults. The full diversity of the species is only realized at the adult level. The adults go on to produce sperms and eggs (gametogenesis), which are different between the species, although the range of possibilities is somewhat reduced. There are essentially three categories of eggs and early embryos in insects, the short germ, the intermediate germ and the long germ embryos which differ with respect to their yolk content and the size of their primary germ rudiment. Short germ embryos (for example the grasshopper *Schistocerca*) develop at blastoderm stage only a very small germ rudiment which consists essentially of headlobes and a growth zone, intermediate germ embryos (for example *Tribolium*) include the head lobes, thoracic segments and a growth zone, while long germ embryos (for example *Drosophila*) generate all their segments at blastoderm stage (see Tautz *et al.* 1994 for a discussion of these types). Although the examples shown are restricted to insects, all arthropods, i.e. also the crustaceans, millipeds and spiders show easily comparable phylotypic stages with those of insects.

available at that stage will already be fewer than for the adult individuals and the wheels are therefore merging again (Fig. 9.2). During early ontogeny, the germ-line cells will have to redeploy their genetic bauplan information to produce a three-dimensional embryo. This has to be done in different egg environments and will therefore go along somewhat different routes, depending on this environment. However, we assume that a more or less conserved set of bauplan genes should act at these stages which eventually lead up to the correct deployment of the HOX genes. From this stage onwards, a new set of genes may become more prevalent to regulate the details of the formation of the adult morphology. These genes may be called the 'realizator genes' (Garcia-Bellido 1975) and we suspect that they are the genes that are required for ecological adaptations.

In the asymmetric wheel model, it would seem sensible to use the phylotypic stage as the starting point of the cycle and not the point of fertilization, as it is usually done. In *Drosophila*, for example, pattern formation starts well before fertilization, namely during oogenesis. The major axes of the future embryo become specified by cell–cell signalling interactions between the oocyte and the surrounding nurse cells. After fertilization and egg laying, this prelocalized positional information is merely interpreted by the zygotic genome (St Johnston and Nüsslein-Volhard 1992). In contrast, the axis specification in chordates are apparently not achieved during oogenesis, but only after fertilization (Eyal-Giladi 1997). These examples suggest that the initiation of pattern formation does not need to be triggered by fertilization or egg laying. This has to be taken into account when making inferences about the possible conservation of developmental pathways between taxa.

9.4 Developmental pathways and modules

Early development in *Drosophila* proceeds along genetically and molecularly defined pathways. It is as yet unclear to what extent these pathways are evolutionarily conserved. They might also reflect special adaptations to the long germ mode of embryogenesis found in *Drosophila,* though we have argued that the system may be more or less conserved even in less derived insects which show a somewhat different form of embryogenesis (Tautz and Sommer 1995). Furthermore, the fact that the correct regulation of the conserved HOX gene expression domains depends on these early pathways in *Drosophila* would suggest that at least some essential components of the pathways should also be conserved. There are indeed examples of early acting genes whose homologues show persuasively similar expression patterns in chordates in particular genes involved in head development (Bally-Cuif and Boncinelli 1997). Interestingly, even some genes found to be primarily involved in segmentation in *Drosophila* were found to be similarly expressed during somite formation in chordates even though these structures cannot as yet be considered to be homologous (Müller *et al.* 1996; Holland *et al.* 1997). However, it is still largely unclear how much of the regulatory interactions among them are also conserved. Furthermore, there is an alternative explanation, namely that these genes are parts of versatile developmental modules which can be redeployed in different contexts.

One good example for such a redeployment, which is even involved in the generation of an adaptive adult character, is the involvement of the *distal-less* (*dll*) gene in the generation of butterfly eyespots (Carroll *et al*. 1994). *dll* is an important gene required for the generation of the proximo-distal axis in *Drosophila* appendages and qualifies therefore as a bauplan gene. Moreover, its expression in appendages throughout the animal kingdom (Panganiban *et al*. 1997) suggests that this function is highly conserved. However, as vertebrate and invertebrate legs are clearly not homologous structures, one has to conclude that the proximo-distal specification pathway in which *dll* is involved has frequently been recruited in different phyla for patterning body wall outgrowths (Panganiban *et al*. 1997). It therefore seems that this pathway acts as a developmental subroutine, which can be put into different contexts. The expression of *dll* in the eye spots in butterfly wings may reflect this versatility. Although it is not an appendage which is generated in this case, it is still an axis that is specified, namely an outer to inner axis of a planar field.

There is also another type of redeployment of modules during limb formation of vertebrates. It appears that some genes of the HOX cluster have become duplicated and are then specifically used to regulate limb development in a temporal–spatial progression (Sordino and Duboule 1996). Also, signalling molecules from the *wingless* gene family, which was first identified in the segmentation gene pathway in *Drosophila*, turn out to be involved in many developmental decisions at various stages of development (Nusse 1997), including leg formation in *Drosophila* (Lecuit and Cohen 1997). In fact, most of the segmentation genes in *Drosophila* are also re-expressed at later stages in multiple organs and may be functionally involved in specifying these.

Such considerations make it difficult to predict the role of bauplan genes in the generation of adult characters. Although it seems unlikely that changes in the early pathways leading to the phylotypic stage are required for late adaptive characters, it still seems possible and likely that regulatory modules are re-used for the generation of specific structures at later stages. On the other hand, the re-use of genes in different pathways would make them bad candidates for driving adaptive evolution, since mutations in them would always be expected to have multiple, probably maladaptive effects. A more specialized set of genes would be less problematic in this respect.

9.5 Screens for adaptive trait genes

Morphological traits usually show a certain degree of quantitative variation. Part of this may be environmentally induced, but it is also clear that there is a genetic basis for this variation. This is best shown by the fact that one can take almost any quantitative morphological character and subject it to artificial selection for high and low values. One of the most intensively studied traits in this context are sensory bristle numbers on adult *Drosophila* flies, which have served as a model system to understand the genetics of quantitative variation (Mackay 1996).

There are basically two opposing models of how the genetics of quantitative traits can be understood. The first assumes a large number of genes with each having only small effects on the trait, while the second assumes the existence of a few loci with major effects, but possibly complemented by other loci with minor effects (Barton and Turelli 1989; Orr and Coyne 1992; Mitchell-Olds 1995). The first model is mainly supported by population genetical reasoning. It assumes that spontaneous mutations with small effects should be more frequent than those with large effect. Also, mutations with large effects on phenotypic traits would be considered to have pleiotropic consequences on other characters. Moreover, the fixation of mutations with small effects should be easier than those with large effects. These would be expected to require strong selection over many generations to become fixed, because the associated pleiotropic and usually maladaptive effects must be overcome. However, at least the latter condition is met in artificial selection experiments and genetic analysis of the genes involved in bristle number variation in *Drosophila* has turned up several major effect loci. As predicted, some of these have pleiotropic effects on other fitness components, i.e. effect viability. In addition, both additive and epistatic effects were found for the different loci (Long *et al.* 1995). Similar conclusions were reached in another extensive study that has looked for morphological shape differences in male genitalia of *Drosophila* (Liu *et al.* 1996). In this case, it was not artificial selection that was employed, but hybridization between very closely related species that showed evolutionary fixed differences for this morphological trait. In this respect, the experiment can be considered as more 'natural' since the differences were not generated by artificial selection. Still, the analysis suggests again that there could be a small number of large effect loci which act mainly additively. On the other hand, the alternative, namely a clustering of multiple small effect loci in small chromosome regions could not be ruled out, since the mapping of the chromosomal intervals bearing the loci was relatively crude (Liu *et al.* 1996). A similar study has been done for two monkeyflower (*Mimulus*) species, which occur sympatrically, but are reproductively isolated. Still, they are fully fertile when artificially mated and it was thus possible to map eight floral traits that distinguish them. For each of the traits a chromosomal region could be identified that accounted for more than 25% of the variance (Bradshaw *et al.* 1995). Thus, major genes have likely played a role in the diversification of these plants, but there is still plenty of room for the influence of minor genes as well.

What are the genes that cause these quantitative trait differences? Unfortunately, at least for *Drosophila*, this is not clear yet, as the mapping methods will have to become much more refined (the main obstacle being the limited number of analyses that can be done for a single recombinant fly). However, at least candidate loci could be identified for the bristle number variation experiment (Long *et al.* 1995). Several of the known neurogenic genes turned out to be located in the intervals mapped for large effects. This seems sensible, since sensory bristles are products of the neurogenic pathway. Moreover, the genes involved are known to act at multiple stages and in different regulatory contexts during development, which would also explain their pleiotropic effects. Unfortunately, this would make them less likely candidates for genes that effect bristle number traits under natural conditions and it therefore remains open whether there are multiple small effect loci after all that could be the basis for natural selection.

In maize it has been possible to identify major loci that have caused a particular morphological change. The domestication of maize from its wild ancestor teosinte has involved a strong selection for apical dominance, i.e. the concentration of the resources in the main stem of the plant instead of the axillary branches. QTL mapping has allowed the identification of two epistatically acting major genes involved in this, one of which is called the *teosinte branched1* locus (Doebley *et al.* 1995). The latter gene could subsequently be cloned and the molecular analysis suggests that it might be a conserved regulatory gene (Doebley *et al.* 1997). Interestingly, it appears that the differences between maize and teosinte are due to different levels of expression rather than primary amino acid changes (Doebley *et al.* 1997). However, the transformation from teosinte into maize was again an artificial selection experiment. In this sense it corresponds to the examples discussed for *Drosophila*. Thus, the fact that only few major effect loci were identified does not need to imply that minor effect loci might not play a role under more natural evolutionary conditions as well.

9.6 A screen for naturally selected loci

Ideally, one would like to identify genes that have played a role in adaptations under natural conditions to better understand the functions of such genes. This makes it necessary to study wild-type populations and to avoid laboratory experiments, at least until a locus is identified. A pilot study in *Drosophila* shows an example of how this might be done (Schlötterer *et al.* 1997). *Drosophila melanogaster* has only recently colonized multiple habitats around the world. It seems to have originated in tropical Africa, but has adapted to very different climates and ecological conditions within only a few thousand years. Thus, the comparison of different wild-type populations from different climatic regions should allow the detection of genes that have been under positive selection to achieve the adaptations. Positive selection would result in a loss of polymorphism in the chromosomal region where the gene resides (Taylor *et al.* 1995). This hitchhiking effect should allow identification of the respective chromosomal regions. The practical approach is to survey highly polymorphic anonymous markers in different populations to see whether population and locus specific loss of heterozygosity can be detected. Indeed, in an analysis of 10 hypervariable microsatellite loci in *D. melanogaster* populations, we could identify at least one locus with a significant loss of heterozygosity, suggesting that it might be linked to a gene that has undergone a selective sweep (Schlötterer *et al.* 1997). Again, it will be necessary to conduct a much more refined study to identify a candidate gene for this effect, but our pilot study at least suggests that the initial approach might be feasible and that the chromosomal regions affected by adaptations to local environments might be tractable.

9.7 Fast evolving genes

Another, as yet rather indirect approach to find genes involved in adaptations is to look for genes that evolve very fast. The bauplan genes discussed above usually belong to well known protein families whose sequence and structure are highly conserved over large evolutionary distances. Usually this correlates with their functional conservation in diverse taxa. Thus, if one would look specifically for genes which are not conserved, one might be able to identify some that are responsible for novel or adaptive phenotypes in specific taxa. In fact, Rice and Holland (1997) have argued that genes might evolve particularly fast under conditions where they are involved in an 'interlocus contest', which could occur under a number of scenarios dealing with ecological and sexual adaptations.

Currently, the origin of evolutionary novelties is generally thought to be caused by two different processes, either evolutionary changes in regulatory networks or gene duplications with subsequent diversification of the genes. The latter process in particular would suggest that there is only a limited number of ancestral protein domains that have become duplicated and reshuffled during evolution to make up the genes that we find today in complex organisms (Dorit et al. 1990; Orengo et al. 1994). The data from the current genome sequencing projects suggest that this expectation may indeed be true. It was estimated that the number of naturally occurring protein domains which make up the universe of proteins is only between 1000 and 7000, most of which are considered to be already represented in the sequence databases (Orengo et al. 1994; Chotia 1994). On the other hand, in genome projects of higher eukaryotes one finds a relatively high number of proteins (up to about 40%) that appear to have no similarity to known sequences and whose identity and function remains unknown (Wilson et al. 1994; Goffeau 1994; Dujon 1996). One could assume that these 'orphans' are the first known members of as yet undiscovered protein families. However, as an alternative explanation we propose that the sequence of these proteins evolves so fast that their homologues cannot be identified in distant species by molecular methods. If such a class of genes exists, it might also be a source of evolutionary novelties.

To address this question, we have devised a screen to estimate the proportion of rapidly evolving genes in *Drosophila* (Schmid and Tautz 1997). A set of about 100 different cDNAs were randomly isolated from an embryonic *D. melanogaster* cDNA library and their sequence evolution was examined on three different time scales: between distantly related species (60–250 Myr), closely related species (15 Myr) and different populations of sibling species (0–2 Myr).

In the first step, the conservation of these clones was analysed by filter hybridization against genomic DNA from three insect species with increasing evolutionary distance. The hybridization conditions were chosen such that genes that evolve neutrally or close to a neutral rate would not hybridize to the genomic DNA of any of these species. Surprisingly, we found that more than one-third of the cDNAs did not even cross-hybridize with the genomic DNA from a distantly related *Drosophila* species under these conditions. To yield a signal in this assay, the probe would have had to include at least one stretch of about 100 bp with more than 65% sequence identity. Thus, genes including one or more of the known conserved protein domains (see above) should have lighted up. Partial sequencing of the clones and searching of sequence data bases showed that a similar proportion did not result in any matches with previously known genes or sequence motifs. Most interestingly, among the non-matching genes, there was only one that was already known from *D. melanogaster*, while 19 already known *D. melanogaster* genes were found in the more conserved class of clones (Table 9.1). This suggests that there is a substantial proportion of fast evolving genes in the *Drosophila* genome which are significantly under-represented among the genes that have been studied so far.

To study the nature of the fast evolving sequences in more detail, their homologues were isolated from the closely related species *D. yakuba*, which diverged about 15 Myrs ago from the ancestor of *D. melanogaster*. This is an evolutionary distance where one would expect that even neutrally evolving sequences should be recovered. For 10 fast evolving and one conserved clone, the homologous sequences were obtained and the cDNAs were completely sequenced. Nine of the 11 gene pairs contained an homologous open reading frame, indicating that they code for functional proteins. The calculation of the substitution rates between these pairs showed that they are indeed among the fastest evolving genes known from this species pair. Four of them showed amino acid replacement rates that were only half as fast as the corresponding replacement rates at third codon positions, which are considered to evolve close to the neutral rate (Table 9.2).

The third step of our analysis addressed the question of whether the high evolutionary rates are caused by neutral evolution due to low functional constraints on the protein sequence or by continuous adaptive selection for new variants as it would be predicted in the Rice and Holland (1997) scenarios. McDonald and Kreitman (1991) have suggested that a comparison of polymorphisms within populations with fixed replacements between species should allow one to assess

Table 9.1 Numbers of randomly drawn cDNA clones in the conserved and fast evolving classes and comparison with previously known *Drosophila* genes. The exact numbers provided are corrected for clones that turned out to be composed of non-coding 3'-ends only

	All cDNAs	Exact *Drosophila* matches
Conserved clones	49	19
Fast evolving clones	46	1

G-test: $G = 15.6$, $P < 0.001$.

Table 9.2 Comparison of non-synonymous (Ka) and synonymous (Ks) substitutions per site of different genes and cDNAs from *D. melanogaster* and *D. yakuba* (CI is confidence interval). The four fastest evolving cDNAs from the randomly drawn pool are at the top of the list. The other comparisons are based on previously published data (see Schmid and Tautz 1997 for details). Clone 2A12 was identified as a conserved clone in the genomic Southern blot screening experiment and codes for a kinesin-like protein

Gene	Codons	Ka	95% CI	Ks	95% CI
1G5	347	0.168	0.252–0.441	0.346	0.134–0199
1E9	393	0.116	0.092–0.139	0.279	0.205–0.356
2D9	280	0.108	0.070–0.143	0.233	0.157–0.303
1A3	227	0.092	0.063–0.120	0.196	0.119–0.279
anon-3B1.2	260	0.079	0.056–0.104	0.284	0.191–0.367
Period	1233	0.032	0.026–0.040	0.309	0.257–0.340
Amylase	494	0.020	0.012–0.028	0.110	0.072–0.142
G-S-T	208	0.017	0.002–0.030	0.114	0.058–0.168
Adh	256	0.015	0.005–0.026	0.156	0.094–0.211
G-6–P	558	0.011	0.005–0.017	0.198	0.151–0.240
Hunchback	829	0.011	0.005–0.015	0.183	0.143–0.220
CO I	498	0.007	0.002–0.0012	0.315	0.251–0.396
2A12	498	0.006	0.001–0.011	0.325	0.260–0.401

whether fixation of amino acids has mainly occurred by drift or by selection. We have analysed the fastest evolving genes in this test and have found no evidence for selection so far. In other words, the sequences showing the high replacement rates show also a high within-species polymorphism rate. However, the test is rather conservative in several respects and a more detailed analysis of the polymorphisms detected has still to be done. Another test which could point to selection is to check whether there are more positions that result in amino acid replacements between species than there are replacements at non-coding positions (compare Tsaur and Wu 1997). Again, we have not found evidence for this in our sequences if one uses them as a whole. However, there are some particularly fast evolving subregions which show this effect. Thus, it will be necessary to understand better the details of their structure and function before solid statements can be made.

It is as yet unclear what the function of the fast evolving genes may be. We find that they are expressed in similar ways as the slow evolving ones (Schmid and Tautz 1997), indicating that they may be involved in both general cellular functions as well as specific developmental ones. However, we suspect that mutants in them would not cause obvious phenotypes, as the cDNAs that we have recovered are significantly underrepresented among those *Drosophila* genes that have been studied because they have a defined genetic effect (Table 9.1). Interestingly, there is one well studied example of a *Drosophila* gene which has no apparent genetic effect. The gene was cloned because it was considered to be the homeotic gene *spalt*. It codes for a rather short protein with a functional signal sequence and it shows a very high evolutionary rate between closely related species (Reuter *et al.* 1989). However, it turned out that the real *spalt* gene is located about 20 kb away and codes for a highly conserved zinc finger protein. The former gene was therefore called *spalt-adjacent* (Reuter *et al.* 1996). Genetic analysis of *spalt-adjacent* showed that it does not affect viability if it carries a premature stop codon. In fact, some standard laboratory strains are homozygous for this mutation. Moreover, even an artificial overexpression construct did not yield any recognizable phenotype (Reuter *et al.* 1996). Thus it seems likely that *spalt-adjacent* has only a minor function, which may only be relevant for flies under natural conditions.

There are other examples of fast evolving genes in *Drosophila* as well. Among them are the *period* gene, which is involved the species-specific song rhythms of the fly (Thackeray and Kyriacou 1990), the *transformer* gene, which is involved in the sex-determination cascade (O'Neil and Belote 1992) and the male ejaculatory protein gene *Acp-26A* (Tsaur and Wu 1997). It seems significant that all three of these genes are positioned in pathways where one would expect adaptive evolution to take place. Rice and Holland (1997) list more such examples from other species. There is also evidence for a substantial number of possibly fast evolving genes from the yeast (Dujon 1996) and the nematode (Wilson *et al.* 1994) genome projects which are called 'orphans' in these cases. The evolutionary origins of these orphan genes remain as yet unknown, because the majority of sequences from eukaryotes that are available in databases come from a few model organisms that are separated by large evolutionary distances. Studying the evolution of such genes over close evolutionary distances should help to understand them better in the future.

9.8 Conclusions

The genetics of the evolution of phenotypic adaptive traits still remain a puzzle. However, at least the framework within which it occurs is becoming clearer now. There is on the one hand a huge and increasing knowledge on the genes involved in embryogenesis and there are on the other hand feasible approaches to study quantitative trait loci involved in generating adaptive characters of adult individuals. Intuitively, one would expect that the former group of genes is mainly involved in the generation of major evolutionary novelties or changes in the bauplan, while the quantitative trait loci might be more important for ecological adaptations and speciation events. In other words, there might be a group of genes involved in the generation of macro-evolutionary novelties and a second group of genes required for micro-evolutionary adaptations (see Orr and Coyne 1992 for a relevant discussion of these terms). Accordingly, one would expect that changes in the former group of genes are relatively infrequent and that one should not normally find polymorphisms in them which have phenotypic consequences. However, even this expectation is not born out. Gibson and Hogness (1996) showed that there is a natural polymorphism in the *Ultrabithorax* locus of wild-type *Drosophila melanogaster* populations that results in the transformation of a whole body segment under environmental stress conditions.

In fact this polymorphism behaves like a typical major QTL locus, as it responds to artificial selection and the magnitude of its effect is influenced by other non-linked minor loci. Evidently, since segment transformations are not normally seen among wild-type flies, one would have to suspect that this polymorphism has a different role for late expressed characters under natural conditions. Still, this result together with the results from the bristle number experiments, suggests that the major quantitative trait loci might indeed uncover genes that are otherwise primarily involved in early developmental decisions. On the other hand, it is also evident that there are many minor loci influencing any particular quantitative trait. Interestingly, one possible reason for this might be that developmental decisions need to be safeguarded by redundant pathways (Tautz 1992; Nowak *et al.* 1997). Modelling of such safeguarding effects shows that more and more genes would become recruited during evolution for any developmental process which has a certain chance of making errors (Nowak *et al.* 1997). It would then be expected that each of the genes might have a small effect on the process as a whole. Evidently, it will be difficult to identify such genes with small or redundant effects unequivocally, but candidates for them might be found among the fast evolving genes discussed above. Although the evidence for this is rather indirect as yet, these genes would at least provide a large source of polymorphisms that could be exploited by natural selection on phenotypic traits and we believe therefore that they should be studied more closely in the future.

References

Arendt, D. and Nübler-Jung, K. 1997 Dorsal or ventral: similarities in the fate maps and gastrulation patterns in annelids, arthropods and chordates. *Mech. Dev.* **61**, 7–21.

Bally-Cuif, L. and Boncinelli, E. 1997 Transcription factors and head formation in vertebrates. *BioEssays* **19**, 127–135.

Barton, N. H. and Turelli, M. 1989 Evolutionary quantitative genetics—how little do we know? *Annu. Rev. Genet.* **23**, 337–370.

Bradshaw, H.D., Wilbert, S.M., Otto, K.G. and Schemske, D.W. 1995 Genetic mapping of floral traits associated with reproductive isolation in monkeyflowers (*Mimulus*). *Nature* **376**, 762–765.

Carroll, S.B., Gates, J., Keys, D.N., Paddock, S.W., Panganiban, G.E., Selegue, J.E. and Williams, J.A. 1994 Pattern formation and eyespot determination in butterfly wings. *Science* **265**, 109–114.

Chasan, R. and Anderson, K.V. 1993. Maternal control of dorsal-ventral polarity and pattern in the embryo. In *Development of* Drosophila malanogaster, (ed. M. Bate and A. Martinez-Arias). pp. 387–424. New York: Cold Spring Harbor Laboratory Press.

Chotia, C. 1994 Protein families in the metazoan genome. *Development 1994 Suppl.* 27–33.

Coen, E.S. and Nugent, J.M. 1994 Evolution of flowers and inflorescences. *Development 1994 Suppl.*, 107–116.

Doebley, J., Srec, A. and Gustus, C. 1995 *Teosinte branchedl* and the origin of maize: evidence for epistasis and the evolution of domonance. *Genetics* **141**, 333–346.

Doebley, J., Stec, A. and Hubbard, L. 1997 The evolution of apical dominance in maize. *Nature* **386**, 485–488.

Dorit, R.L., Schoenbach, L. and Gilbert, W. 1990 How big is the universe of exons? *Science* **250**, 1377–1382.

Duboule, D. 1994 Temporal colinearity and the phylotypic progression: a basis for the stability of the vertebrate Bauplan and the evolution of morphologies through heterochrony. *Development 1994 Suppl.* 135–142.

Dujon, B. 1996 The yeast genome project: what did we learn? *Trends Genet.* **12**, 263–270.

Eyal-Giladi, E. 1997 Establishment of the axis in chordates: facts and speculations. *Development* **124**, 2285–2296.

Ferguson, E. L. 1996 Conservation of dorsal-ventral patterning in arthropods and chordates. *Curr. Opin. Gene Dev.* **6**, 424–431.

Fitch, D. H. A. 1997 Evolution of male tail development in rhabditid nematodes related to *Caenorhabditis elegans*. *Syst. Biol.* **46**, 145–179.

Garcia-Bellido, A. 1975 Genetic control of wing disc development in *Drosophila*. In *Cell Patterning* (ed.R. Porter and K. Elliott). pp. 161–178. Elsevier, Amsterdam.

Gehring, W. 1994 A history of the homeobox. In *Guidebook to the homeobox genes*, (ed. D. Duboule), pp. 1–10. Oxford University Press.

Gibson, G. and Hogness, D. S. 1996 Effect of polymorphism in the *Drosophila* regulatory gene *Ultrabithorax* on homeotic stability. *Science* **271**, 200–203.

Goffeau, A. 1994 Genes in search of functions. *Nature* **369**, 101–102.

Granato, M. and Nüsslein-Volhard, C. 1996 Fishing for genes controlling development. *Curr. Opin. Gene Dev.* **6**, 461–466.

Grünert, S. and St Johnston, D. 1996 RNA localization and the development of asymmetry during *Drosophila* embryogenesis. *Curr. Opin. Gene Dev.* **6**, 395–402.

Harger, P.L. and Gurdon, J.B. 1996 Mesoderm induction and morphogen gradients. *Semin. Cell Dev. Biol.* **7**. 87–93.

Holland, L.Z., Kene, M., Williams, N.A. and Holland, N.D. 1997 Sequence and embryonic expression of the Amphioxus *engrailed* gene (*AmphiEn*): the metameric pattern of transcription resembles that of its segmentpolarity homolog in *Drosophila*. *Development* **124**, 1723–1732.

Jürgens, G. and Hartenstein, V. 1993. The terminal regions of the body pattern. In *Development of* Drosophila melanogaster, (ed. M. Bate and A. Martinez-Arias). pp. 687–746. New York: Cold Spring Harbor Laboratory Press.

Lecuit, T. and Cohen, S.M. 1997 Proximal-distal axis formation in the *Drosophila* leg. *Nature* **388**, 139–145.

Lewis, E.B. 1978 A gene complex controlling segmentation in *Drosophila*. *Nature* **276**, 761–769.

Liu, J., Mercer, J.M., Stam, L.F., Gibson, G.C., Zeng, Z.-B. and Laurie, C. 1996 Genetic analysis of a morphological shape difference in the male genitalia of *Drosophila simulans* and *D. mauritiana*. *Genetics* **142**, 1129–1145.

Long, A.D., Mullaney, S.L., Reid, L.A., Fry, J.D., Langley, C.H. and Mackay, T.F.C. 1995 High resolution mapping of genetic factors affecting abdominal bristle number in *Drosophila melanogaster*. *Genetics* **139**, 1273–1291.

Mackay ,T.F.C. 1996 The nature of quantitative variation revisited: lessons from *Drosophila* bristles. *BioEssays* **18**, 113–121.

McDonald, J.H. and Kreitman, M. 1991 Adaptive protein evolution at the ADH locus in *Drosophila*. *Nature* **351**, 652–654.

McGinnis, W. and Krumlauf, R. 1992 Homeobox genes and axial patterning. *Cell* **68**, 283–302.

Mitchell-Olds, T. 1995 The molecular basis of quantitative genetic variation in natural populations. *Trends Ecol. Evol.* **10**, 324–328.

Müller, M., v.Weizäcker, E. and Campos-Ortega, J.A. 1996. Expression domains of a zebrafish homologue of the *Drosophila* pair-rule gene *hairy* correspond to primordia of alternating somites. *Development* **122**, 2071–2078.

Nowak, M. A., Boerlijst, M. C., Cooke, J. and Smith, J. M. 1997 Evolution of genetic redundancy. *Nature* **388**, 167–171.

Nusse, R. 1997 A versatile transcriptional effector of *wingless* signaling. *Cell* **89**, 321–323.

Nüsslein-Volhard, C. and Wieschaus, E. 1980. Mutations affecting segment number and polarity in *Drosophila*. *Nature* **287**, 795–801.

Orengo, C.A., Jones, D.T. and Thornton, J.M. 1994 Protein superfamilies and domain superfolds. *Nature* **372**, 631–634.

Orr, H.A. and Coyne, J.A. 1992 The genetics of adaptations—a reassessment. *Am. Nat.* **140**, 725–742.

O'Neil, M-T. and Belote, J.M. 1992 Interspecific comparison of the *transformer* gene of *Drosophila* reveals an unusually high degree of evolutionary divergence. *Genetics* **131**, 113–128.

Panganiban, G., Irvine, S.M., Lowe, C., Roehl, H., Corley, L.S., Sherbon, B., Grenier, J.K., Fallon, J.F., Kimble, J., Walker, M., Wray, G.A., Swalla, B.J., Martindale, M.Q. and Carroll, S.B. 1997 The origin and evolution of animal appendages. *Proc. Natl Acad. Sci. USA* **94**, 5162–5166.

Pankratz, M. and Jäckle, H. 1993. Blastoderm Segmentation. In *Development of* Drosophila melanogaster, (ed. M. Bate and A. Martinez-Arias). pp. 467–516. New York: Cold Spring Harbor Laboratory Press.

Raff, R. A. 1996 *The shape of life: genes, development, and the evolution of animal form.* pp. 173 ff. Chicago: The University of Chicago Press.

Reuter, D., Schuh, R. and Jäckle, H. 1989 The homeotic gene *spalt* (*sal*) evolved during *Drosophila* speciation. *Proc. Natl Acad. Sci. USA* **86**, 5483–5486.

Reuter, D., Kühnlein, R.P., Frommer, G., Barrio, R., Kafatos, F. C., Jäckle, H. and Schuh, R. 1996. Regulation, function and potential origin of the *Drosophila* gene *spalt adjacent*, which encodes a secreted protein expressed in the early embryo. *Chromosoma* **104**, 445–454.

Rice, W.R. and Holland, B. 1997 The enemies within: intergenomic conflict, interlocus contest evolution (ICE), and the intraspecific Red Queen. *Behav. Ecol. Sociobiol.* **41**, 1–10.

Sander, K. 1983. The evolution of patterning mechanisms: gleanings from insect embryogenesis and spermatogenesis. In *Development and evolution: The sixth symposium of the British Society for Developmental Biology*, (ed. B.C. Goodwin, N. Holder and C.C. Wylie), pp. 137–159. Cambridge: Cambridge University Press.

Schlötterer, C., Vogl, C. and Tautz, D. 1997 Polymorphism and locus-specific effects on polymorphism at microsatellite loci in natural *Drosophila melanogaster* populations. *Genetics* **146**, 309–320.

Schmid, K.J. and Tautz, D. 1997 A screen for fast evolving genes from *Drosophila*. *Proc. Natl Acad. Sci. USA* **94**, 9746–9750.

Slack, J.M.W., Holland, P.W.H. and Graham, C.F. 1993 The zootype and the phylotypic stage. *Nature* **361**, 490–492.

Sommer, R.J., Carta, L.K. and Sternberg, P.W. 1994 The evolution of cell lineage in nematodes. *Development Suppl. 1994*, 85–94.

Sordino, P. and Duboule, D. 1996 A molecular approach to the evolution of vertebrate paired appendages. *Trends Ecol. Evol.* **11**, 114–119.

St Johnston, D. and Nüsslein-Volhard, C. 1992 The origin of pattern and polarity in the *Drosophila* embryo. *Cell* **68**, 201–219.

Tautz, D. 1992 Redundancies, development and the flow of information. *BioEssays* **14**, 263–266.

Tautz, D. and Sommer, R.J. 1995. Evolution of segmentation genes in insects. *Trends Genet.* **11**, 23–27

Tautz, D., Friedrich, M. and Schröder, R. 1994 Insect embryogenesis—what is ancestral and what is derived? *Developement 1994 Suppl.* 193–199.

Taylor, M.F.J., She, Y. and Kreitman, M. E. 1995 A population genetic test of selection at the molecular level. *Science* **270**, 1497–1499.

Thackeray, J.R. and Kyriacou, C.P. 1990 Molecular evolution in the *Drosophila yakuba period* locus. *J. Mol. Evol.* **31**, 389–401.

Theissen, G. and Saedler, H. 1995. MADS-box genes in plant ontogeny and phylogeny: Haeckel's biogenetic law revisited. *Curr. Opin. Gene Dev.* **5**, 628–639.

Tsaur, S. C. and Wu, C.-I. 1997 Positive selection and the molecular evolution of a gene of male reproduction, Acp26Aa of *Drosophila. Mol. Biol. Evol.* **14**, 544–549.

Wilson, R. *et al.* 1994 2.2 Mb of contiguous nucleotide sequence from chromosome III *of C. elegans. Nature* **368**, 32–38.

10

Revealing the factors that promote speciation

Timothy G. Barraclough, Alfried P. Vogler and Paul H. Harvey

10.1 Introduction

Life is made up of species, and so understanding the evolution of species richness is fundamental to our understanding of the natural world. Traditionally, however, this area is plagued by the difficulties inherent in studying a phenomenon operating over timescales many orders of magnitude greater than our own lifespans (Panchen 1992). Recent accumulation of phylogenetic information through the use of molecular techniques has provided novel possibilities for statistical tests of hypotheses concerning the evolution of species richness in extant taxa (Mooers and Heard 1996; Nee *et al.* 1996a; Purvis 1996; Sanderson and Donoghue 1996). We illustrate this approach by concentrating on two issues. First, can biological attributes of lineages influence the tendency for those lineages to accumulate species richness? Second, can we use species-level phylogenies to identify modes of speciation and patterns of subsequent phenotypic change?

10.2 Biological factors promoting species richness

Lineages vary in the number of species they contain. Possible causes for this variation include differences in the environment experienced by those lineages (Cracraft 1985; Ricklefs and Schluter 1993; Rosenzweig 1995; Kerr and Packer 1997), or mere chance variation in probabilities of speciation and extinction (Raup *et al.* 1973; Raup 1985). An additional possibility is that lineages may vary with respect to biological attributes that influence the net rate of cladogenesis, either through an effect on speciation rate, or extinction rate, or the equilibrium number of species a lineage can realise. For example, many authors have suggested that strong sexual selection by female choice may promote speciation, and ultimately species richness (Darwin 1871; West-Eberhard 1983). Similarly, species range size has been postulated as an attribute with a strong effect on extinction probability (Jackson 1974; Jablonski 1987), and small body size has been proposed as an attribute permitting high equilibrium species numbers within lineages (Hutchinson and MacArthur 1959; Morse *et al.* 1985).

To test ideas of this kind, we need to look for replicate evidence for an association between the trait of interest and species richness. A recent approach is to compare sister-groups that differ in their expression of the trait in question. There are four reasons why sister-group comparison is currently the best approach for identifying evolutionary correlates of species richness (Mitter *et al.* 1988; Zeh *et al.* 1989; Barraclough *et al.* 1998a).

1. *Replication.* By including several sister-group comparisons in our tests, we gain replicated statistical evidence for any trend we observe. This replication increases our ability to make general conclusions about an effect, reducing the possibility of detecting accidental, historical associations between traits and species richness (Cracraft 1990).
2. *Rates of diversification.* Our interest is in testing whether lineages with a particular attribute accumulate more species than those without it. However, different taxa have different ages, and so we might expect older taxa to have more species than younger taxa, if they are growing, simply because they have had time to accumulate more species. Sister taxa are, by definition, the same age, and so by comparing the number of species between sister-groups we obtain direct estimates of relative net rates.
3. *Non-independence of taxa.* Several taxa may share a high species accumulation rate and the same value of a biological attribute simply because they inherited both from a common ancestor. In this case, these taxa do not represent independent data points supporting an evolutionary correlation between species diversity and the biological variable since both traits have evolved only once. Sister-group comparisons overcome this problem because differences between sister taxa have evolved subsequent to their divergence and so necessarily represent independent evolutionary events (Møller and Birkhead 1992).
4. *Confounding variables and noise.* Taxa differ in many attributes apart from the trait of primary interest, X, and diversification rate. Some of these attributes may influence diversification rate, causing error variation or noise in the data that may obscure any pattern arising from an effect of trait X on diversification rates. In addition, trait X may have no direct influence on diversification rates, but we may observe a correlation through the influence of unaccounted for confounding variables. The shared common ancestry of sister taxa means they will tend to be similar to each other in many respects. By comparing sister taxa we are comparing like with like, thereby controlling for much potential noise and confounding variation which might otherwise afflict our analysis (Read and Nee 1995; Harvey *et al.* 1995; Nee *et al.* 1996b).

Two additional features of sister-group comparison should be mentioned with respect to drawing conclusions. First, these tests detect a correlation, and so ultimately it is not possible to determine the direction of causation between species richness and the study variable, nor to rule out entirely the possibility of confounding variables. Second, the approach compares the net rate of cladogenesis between sister taxa, and so cannot determine whether effects on speciation or extinction are the cause of any observed patterns. Nonetheless, sister-group analysis has provided new evidence for the role of several biological traits in promoting species richness. Three examples are provided.

1. *Plant-feeding (phytophagy) in insects.* The first use of sister-group analysis was by Mitter et al. (1988) in their test of the widely suggested hypothesis that plant-feeding (phytophagous) insects have enhanced species richness (Erhlich and Raven 1964; Southwood 1973). They identified all monophyletic groups of exclusively phytophagous insects with a putative non-phytophagous sister taxon, and compared the numbers of species between sister taxa. Under the null model of no association, we predict the phytophagous taxon should contain more species than its non-phytophagous sister taxon in roughly half the comparisons. In fact, 11 out of 13 comparisons display greater species richness in the phytophagous taxon, which is significant under a sign test ($P = 0.011$, one-tailed sign test). This result is consistent with the view that plant-feeding is a factor which promotes high diversity in insect groups, perhaps through the increased capacity for trophic niche specialisation entailed by this lifestyle. This work was followed by similar analyses on the effects of resin canals in plants (positive relationship: Farrell *et al.* 1991), and carnivorous parasitism in insects (no relationship: Wiegmann *et al.* 1993). A general summary of these results is found in Farrell and Mitter (1993).

2. *Sexual selection by female choice in birds.* The second example is the use of sister-group analysis to perform a statistical test of the age-old hypothesis that sexual selection by female choice can increase the rate of origin of reproductive isolation, and thereby increase speciation rates and ultimately species richness (Darwin 1874; Fisher 1958; Lande 1981). Barraclough *et al.* (1995) tested the hypothesis in passerine birds, using sexual dichromatism as an index of the strength of intersexual selection, and the DNA–DNA hybridisation phylogeny of Sibley and Ahlquist (1990) to identify sister taxa. They found a significant relationship between the proportion of sexually dichromatic species within clades, and the species richness of those clades, with about 75% of nodes following the pattern expected under the Darwin hypothesis. The result is robust to likely errors in the phylogeny, and so is consistent with the Darwin hypothesis. A full discussion of the assumptions underlying this test is presented elsewhere (Barraclough *et al.* 1995). Subsequent work has confirmed a similar pattern with an alternative measure in birds (mating system: Mitra *et al.* 1996), and found evidence for a related effect in flowering plants (nectar spur length: Hodges and Arnold 1995).

3. *Rate of molecular evolution in birds and plants.* The final example is the use of sister-group comparisons to investigate the relationship between the rate of species accumulation and rates of genetic change in flowering plants (Barraclough *et al.* 1996). Chase *et al.* (1993) present a phylogeny of the majority of plant families based on sequence data for the chloroplast gene rbcL. Their

phylogeny includes information on the number of nucleotide substitutions occurring on branches, and so we can compare branch lengths between sister taxa to compare relative rates of nucleotide substitution (Bromham *et al.* 1996; Mindell and Thacker 1996). Sister-taxon comparisons reveal a significant, positive relationship between branch length (i.e. rate of nucleotide substitution) and the number of species in each family, which is apparently robust with respect to likely errors in the phylogeny, and possible artefacts arising from parsimony reconstrucion (Sanderson 1990, see original paper for details). A similar positive relationship is found in passerine birds, using DNA–DNA hybridisation data from Sibley and Ahlquist (1990) (unpublished results). The finding of a similar relationship for different molecular measures in very different organisms points intriguingly towards a potentially general association between rates of cladogenesis and rates of genetic change. If indeed general, this pattern may reflect the effect of circumstances associated with, or necessary for, speciation on the rate of genetic change (such as population structure: Mayr 1954; Coyne 1992; Templeton 1996; Slatkin 1996), an influence of rates of genetic change on the probability of speciation (Orr 1995; Orr and Orr 1996), or a reduction of extinction risk in lineages with high rates of genetic change. Further publication of large phylogenies will allow systematic tests for a general effect, and may allow tests of possible mechanisms underlying these patterns.

In conclusion, sister-group analysis is a simple, but powerful tool for testing hypotheses concerning biological factors that promote species richness, providing the first statistical evidence for several long standing hypotheses (see list in Table 10.1). Identification of sister-groups relies on the availability of large-scale phylogenies, and so inevitably this approach will increase in utility as more phylogenies are published.

Table 10.1 Sister-group analyses of correlates of species richness

(a) Insects and insect/plant relations		
Mitter *et al.* (1988)	phytophagy in insects	positive
Zeh *et al.* (1989)	novel oviposition sites	positive
Farrell *et al.* (1991)	resin/latex canals in plants	positive
Wiegmann *et al.* (1993)	parasitic carnivory in insects	none
Connor and Taverner (1997)	leaf-mining in insects	negative
(b) Sexual selection, breeding systems		
Barraclough *et al.* (1995)	(a) Insects and insect/plant relations	positive
Hodges and Arnold (1995)	nectar spurs in plants	positive
Mitra *et al.* (1996)	promiscuity in birds	positive
Mooers and Møller (1996)	colonial breeding in birds	none
(c) Miscellaneous		
Barraclough *et al.* (1996)	rate of molecular evolution in plants	positive
Gittleman and Purvis (1998)	body size in carnivores/primates	negative/none
McCall *et al.* (submitted)	disperal ability and island endemism	trade-off

10.3 Identifying modes of speciation and subsequent phenotypic change

The previous section dealt with identifying biological factors associated with the number of species within lineages of organisms. At a finer scale, phylogenies at the species level provide an estimate of the pattern of splits leading to present day species, and so may potentially provide information into processes operating during the radiation of a group (Harvey *et al.* 1994; Nee *et al.* 1994). In this section we describe some ways we might extract that information.

In general terms, we can recognise four dynamic events leading to present-day patterns of species richness. First, new species are formed by speciation, usually through splitting of an ancestral species into two, or potentially more, daughter species (see Grant 1981; Bullini 1994 for exceptions). Second, the geographic ranges of species may change, either by contraction/expansion or by range movements. Third, there are changes in the genotype and phenotype of species. Fourth, there is extinction. Our ultimate goal then is to understand factors influencing each event and any interactions among them. In combination with geographic and ecological data, species-level phylogenies may provide a trace of this process. However, two features must be kept in mind if we are to use this information. First, ranges and phenotypes change over time, as species are buffeted by changing circumstances, and so information of conditions at past events may not be accurately preserved among present-day species. Second, there are a large number of possible interactions between the events outlined above and so it may not always be possible to discriminate precisely between all possible scenarios.

Here, we outline ways in which species-level phylogenies may be used by focusing on two issues (i) the relationship between geography and species splitting (modes of speciation), and (ii) the interaction between geography and phenotypic evolution, using two lineages of tiger beetles (family: Cicindelidae) from North America as examples.

The role of geography in species splits

Traditionally, the most important factor causing the splitting of a single species into two is believed to be geography, either through the environmental origin of a barrier to gene flow, gradual divergence of geographically structured populations, or establishment of isolated populations through the colonisation of a new area (Mayr 1963; Allmon 1992). The empirical basis for this viewpoint is the observation that closely related sister species tend to be geographically separated with either discontinuous (allopatric) or adjacent (parapatric) geographic ranges (Lynch 1989; Chesser and Zink 1994).

With species-level phylogenies we can look at the pattern of geographic ranges with reference to the entire phylogeny (Lynch 1989). Our approach is to plot the degree of sympatry at each node in the phylogeny against the 'height' of that node, where degree of sympatry is calculated as the mean of the proportion of each clade's area overlapped by the other, and the area of a clade is the range occupied by any of its member species. Note that sympatry in this context refers to the large-scale overlap of species ranges, rather than implying necessary co-occurrence at the fine-scale. Node height is simply the relative distance of a node from the tips of the phylogeny. This plot can provide some insight into the geographic pattern of speciation.

Under a null model that the geographic ranges of species are random with respect to phylogeny, we expect an increase in sympatry with node height, as shown in Figure 10.1. The increase is expected since the ranges of sister clades become larger at higher levels, simply because they include more species, and so the chance of overlap between them is greater, given a finite area. This scenario corresponds to the situation in which ranges have changed so much since species splits that they no longer contain information about ranges at the time of splits.

Different qualitative patterns are expected under alternative geographic models of speciation, shown in Figure 10.2. The expectations depend on the geographic mode of speciation, and whether range changes have occurred subsequent to speciation. The geographic mode of speciation sets a limit for the expected intercept. If speciation is allopatric, we expect an intercept of zero since very recently split species are expected to display no overlap in ranges. If speciation is sympatric, we expect an intercept greater than 0.5, since one of the species must be entirely contained within the other's geographic range, immediately following the split. The subsequent pattern of sympatry with node height depends on whether ranges change over time, either through expansion/contraction or drift. If ranges do not change, then there will be no change in sympatry with node height. Otherwise, sympatry will tend to a rough level at which further range expansions/drifts do not tend to increase or decrease the degree of sympatry. In the case of allopatric speciation, this necessarily entails an increase in sympatry with node height. This is based simply on the idea that ranges are separated initially, due to the allopatric nature of speciation, but are then 'mixed' over time by random influences. In the case of sympatric speciation, there may be a slight increase, decrease or no change

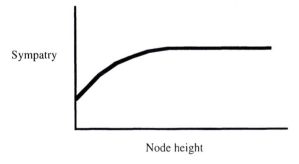

Node height

Fig. 10.1 Expected relationship between degree of sympatry and node height under null model that geographic ranges are random with respect to phylogeny.

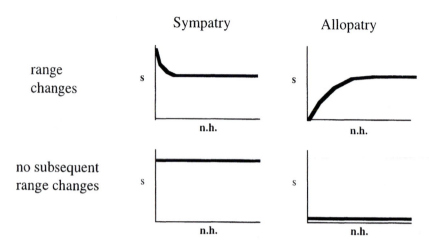

Fig. 10.2 Expected patterns between degree of sympatry (s) and node height (n.h.)under alternative geographical models of speciation.

depending on the exact relationship between the expected intercept and the expected rough equilibrium level of sympatry between older sister clades.

Note that the scenario for allopatric speciation and subsequent range changes is qualitatively similar to the null expectation of random geographic ranges. However, we expect a zero intercept in the former case, whereas under the null model we are likely to observe some recent splits with considerable geographic overlap, simply by chance. Hence, these scenarios may be distinguishable in some circumstances. Note also that the expectations are based on the assumption of uniform process for all nodes, which of course need not be the case. If there is one sympatric speciation event in a group for which allopatric speciation is the norm, this may show up as an outlier with unusually large sympatry for its node height. Similarly, ancient vicariances may show up as nodes with unusually low sympatry for their height. Examples of each case are shown in Figure 10.3.

Can we use this approach to distinguish broad geographical patterns of speciation? We present the example of two lineages of North American tiger beetles from the genus *Cicindela*. The 'repanda clade' is a lineage of 10 species from the subgenus *Cicindela*, the North American radiation of a group found widely distributed across the Holoarctic and more northerly latitudes (Figure 10.4). The subgenus *Ellipsoptera* is a group of 13 species restricted to more central and southern areas of the USA, although the two lineages overlap widely in their distributions (Figure 10.5). Both lineages comprise species living predominantly in sandy habitats, mostly associated with streams and riverbanks. However, both groups also include species living on sand dunes well away from water, and species living on ocean beaches (Freitag 1965; Graves and Pearson 1973). Phylogenies of these species were reconstructed using DNA sequence data, described in detail elsewhere (Vogler *et al.* 1998, Barraclough *et al.*, in prep.).

The plots of range sympatry against node height are shown in Figures 10.6 and

(a) North American *Rhagoletis* fruit flies

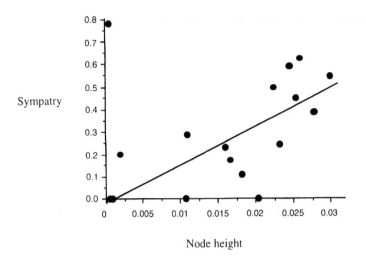

Node height

Fig. 10.3 Two examples of plots between sympatry and node height. (a) Phylogeny taken from Smith and Bush (1997), range data from Foote *et al.* (1993). The overall trend is for increasing sympatry with node height, consistent with a model of allopatric speciation, with subsequent range changes. However, a recent node with unusually high sympatry represents an outlier suggestive of a sympatric speciation event (regression line drawn excluding this point). Note that the genus *Rhagoletis* provides classic examples of sympatric speciation by host shifts (Bush 1969), and yet overall trend appears to be allopatric.

10.7. In the 'repanda clade' we observe a gradual increase in sympatry with node height, although with marked scatter about the line such that some quite recent splits are characterised by fifty-percent range overlap. This pattern is consistent with expectations under the model of allopatric speciation followed by the range changes outlined above, although we cannot rule out the possibility that ranges may have changed so much as to be effectively random with respect to phylogeny (the null model outlined above). Therefore, all we can say with certainty is that we observe strong evidence for post-speciational range changes in this group.

In the *Ellipsoptera,* we observe a similar pattern, with some recent nodes displaying relatively high sympatry, and a weak positive slope. Hence, in both groups we find evidence for post-speciational range changes: intercepts are midway between expected values for allopatric or sympatric speciation without range changes, and many sister-species display sympatries between 0 and 0.5. These analyses rely on accurate reconstruction of species-level phylogenies for the study groups. Using a preliminary phylogeny, we found evidence for vicariance with low levels of range changes in *Ellipsoptera.* Additional sequencing has since improved our estimate of phylogeny, leading to slight changes to recent relationships, and the changes in results we present here (Barraclough *et al.* 1998b, in prep).

(b) World primates

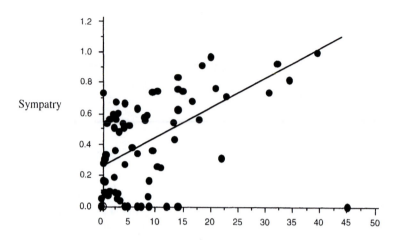

Sympatry

Fig. 10.3 Two examples of plots between sympatry and node height. (b) Phylogeny from Purvis (1995), range data from Purvis (unpublished). Only nodes with quoted date estimates in Purvis (1995) are included. Although noisy, the general trend is for increasing sympatry with node height, again consistent with allopatric speciation and range changes. However, the oldest dated split, between Madagascan lemurs and African bushbabies/ lorises is characterised by zero sympatry, and so is an outlier suggestive of an ancient vicariance event.

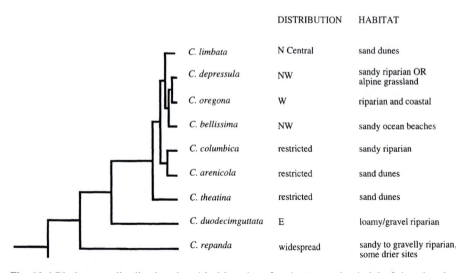

	DISTRIBUTION	HABITAT
C. limbata	N Central	sand dunes
C. depressula	NW	sandy riparian OR alpine grassland
C. oregona	W	riparian and coastal
C. bellissima	NW	sandy ocean beaches
C. columbica	restricted	sandy riparian
C. arenicola	restricted	sand dunes
C. theatina	restricted	sand dunes
C. duodecimguttata	E	loamy/gravel riparian
C. repanda	widespread	sandy to gravelly riparian, some drier sites

Fig. 10.4 Phylogeny, distributional and habitat data for the 'repanda clade' of tiger beetles (genus: *Cicindela*).

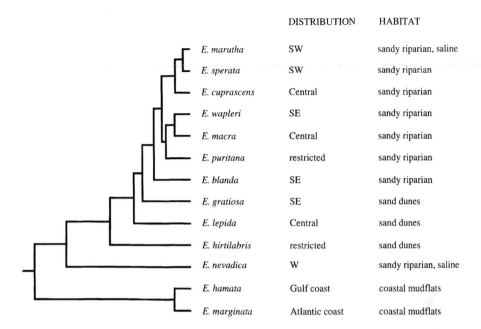

	DISTRIBUTION	HABITAT
E. marutha	SW	sandy riparian, saline
E. sperata	SW	sandy riparian
E. cuprascens	Central	sandy riparian
E. wapleri	SE	sandy riparian
E. macra	Central	sandy riparian
E. puritana	restricted	sandy riparian
E. blanda	SE	sandy riparian
E. gratiosa	SE	sand dunes
E. lepida	Central	sand dunes
E. hirtilabris	restricted	sand dunes
E. nevadica	W	sandy riparian, saline
E. hamata	Gulf coast	coastal mudflats
E. marginata	Atlantic coast	coastal mudflats

Fig. 10.5 Phylogeny, distributional and habitat data for the *Ellipsoptera* clade of tiger beetles (genus: *Cicindela*). The tree differs slightly from the tree presented in a previous version of this paper, due to inclusion of additional sequence data (Baraclough *et al.*, 1998; in prep.).

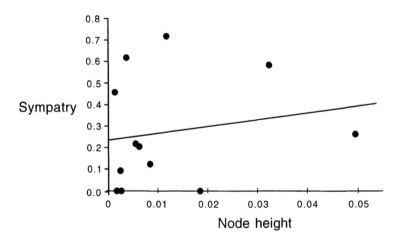

Fig. 10.6 Relationship between the degree of sympatry and node height in the *Ellipsoptera* clade.

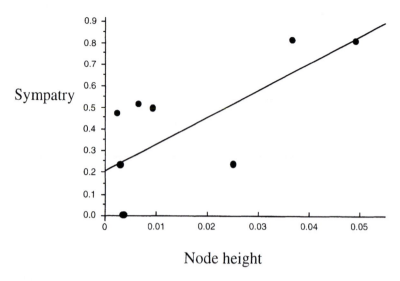

Node height

Fig. 10.7 Relationship between the degree of sympatry and node height in the 'repanda clade'.

The relationship between phenotypic variation and species ranges

Species differ phenotypically in many respects including morphology, ecology, behaviour, life history, and sexual characteristics. These differences are the focus of the ecological perspective of species, and there is a large literature on the possible role of differences among species in promoting and maintaining species richness (Brown and Wilson 1956; Hutchinson 1959; Strong *et al.* 1984; Martin 1996; Saetre *et al.* 1997). But do these phenotypic differences evolve in association with any of the stages of species accumulation, or do they arise arbitrarily with respect to the build-up of species richness? One approach to this question is to partition phenotypic variation among nodes within the phylogeny, in the form of contrasts (Felsenstein 1985; Harvey and Pagel 1991). From this, we can identify where phenotypic variation arose during the radiation of the group.

We use this to investigate the relationship between phenotypic changes and geography. As before, we consider the degree of sympatry between sister clades. There are three possible relationships between phenotypic contrasts and the degree of sympatry. First, there may be a positive relationship, which may be expected if phenotypic differences are necessary for sympatric species to coexist (Taper and Case 1992; Butlin 1995). Second, there may be a negative relationship, which may be expected if phenotypic variation evolves as a consequence of geographical variation in environmental conditions. Third, there may be no relationship if phenotype evolves at random with respect to sympatry.

We look at the origin of habitat differences among species in the two lineages of tiger beetles described above. Habitat contrasts were calculated from habitat

(a)   (b)

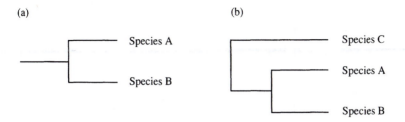

Fig. 10.8 Calculation of habitat contrasts. (a) Sister species: if A and B have the same habitat type, contrast = 0; if A and B have different habitat types, contrast = 1. (b) Higher nodes: if C has same habitat as A and B, contrast = 0; if C has same habitat as either A or B, but not both, contrast = 0.5; if C has different habitat to both A and B, contrast = 1.

data as shown in Figure 10.8. Note that we plot rates of habitat change (i.e. scaled with respect to branch length), but that the same general patterns are observed with unscaled habitat contrasts. The plots of habitat contrasts against the degree of sympatry are shown in Figures 10.9 and 10.10. In the *Ellipsoptera*, there is a positive relationship, such that the origin of sympatry is associated with habitat differences between lineages, suggesting that habitat differences may play a role in species coexistence in this clade. In contrast, in the 'repanda clade' there is a negative relationship, such that sympatric sister clades are more likely to occupy similar habitats. This observation suggests these species have less fixed habitat requirements, and that observed habitat type may be simply a consequence of the opportunities available to species in the area they are found.

In summary, our analyses of species-level phylogenies in two lineages of tiger beetle provide the following results. First, both clades appear to have experienced post-speciational range changes, with the consequence that the geographic modes of

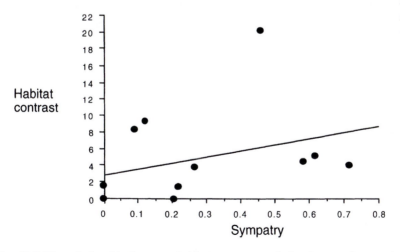

Fig. 10.9 The relationship between habitat contrast and the degree of sympatry in the *Ellipsoptera*.

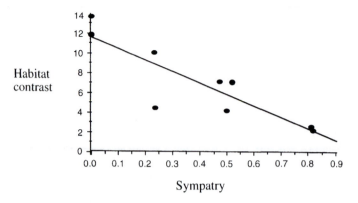

Fig. 10.10 The relationship between habitat contrast and the degree of sympatry in the 'repanda clade'.

speciation may have been obscured in these groups. Second, habitat differentiation is associated with sympatry between lineages in *Ellipsoptera*, whereas in the 'repanda clade' it appears to be associated with differences in local opportunities between different areas. Having discovered these patterns, future work is necessary to explain the observed similarities and differences between the two lineages. If we observe significantly different patterns between clades, possible explanations include differences in the environment they experience, intrinsic differences in the rate of range changes over time, or intrinsic differences in the strength of processes maintaining geographical or ecological segregation, such as interspecific competition. Detailed analysis and discussion of the radiation of these tiger beetles will be presented elsewhere (Vogler *et al.* 1998, Barraclough *et al.* in prep.).

In conclusion, species-level phylogenies can provide information for testing ideas about the evolution of species richness within groups. Future general application of this approach requires detailed simulation analysis of expected patterns under alternative scenarios, and publication of a sufficient number of suitable phylogenies to test these ideas.

10.4 Conclusions

Phylogenies provide an estimate of the sequence of events leading to present-day patterns of species richness, and so provide indispensable information towards understanding the processes operating during the evolution of diversity (Cracraft 1981; Doyle and Donoghue 1993; Mooers and Heard 1996; Purvis 1996). We have outlined some ways in which this information may be used, and in this final section suggest some possible future developments in this area.

In section 10.2 we described how sister-group comparisons provide replicate evidence for correlates of species richness, allowing tests of whether intrinsic, biological factors may influence net rates of cladogenesis. This approach has already provided evidence supporting several long-standing hypotheses, but several future developments can be expected. First, similar methods may be used to perform multivariate analyses to investigate what combination of factors explains variation in species richness between clades, perhaps assessing the relative roles of biological and environmental factors in determining species richness. Second, most currently available large phylogenies are complete to a particular higher taxonomic level, such as tribe or family, and so sister taxa are chosen at that level to maximise sample size. With the appearance of large phylogenies complete to the species-level (for example Purvis 1995; Bininda-Emonds *et al.* submitted), it will become possible to localise changes in trait values and/or diversification rates to specific nodes in the phylogeny, thereby to compare sister groups at the point where changes relevant to the hypothesis occurred. Methods for localising changes in diversification rate already exist (Sanderson and Donoghue 1994; Nee and Harvey 1994; Sanderson and Donoghue 1996), but have not currently been used on a replicate basis to identify correlates of diversity, rather to detect shifts in diversification rate during the evolution of a single clade (Sanderson and Wojciechowski 1996; Vogler and Barraclough 1998).

In section 10.3 we discussed possible uses of species-level phylogenies to investigate processes operating during the radiation of clades. To date, the relative paucity of suitable phylogenies for implementing these methods means that many studies of this kind take a specific, historical approach, for example relating vicariant events in a clade's history to a particular historical environmental event (for example Linder and Crisp 1995; Brumfield and Capparella 1996; Collins *et al.* 1996). But the current explosion of sequence data is providing the first opportunities to investigate general patterns of diversification across a range of taxa. Careful simulation studies will be necessary to interpret these patterns. In particular, we need to know: (i) what processes leave a unique, detectable trace among extant species; (ii) how subsequent evolution and range changes among species influence our ability to detect past processes; and (iii) how sensitive are the methods to errors in phylogeny.

In conclusion, there will always be a limit to the certainty with which we can reveal the factors promoting speciation and species richness, but with careful consideration of expected patterns under alternative scenarios, guided by phylogeny, we can gain some genuine empirical understanding of how species richness and diversity evolves.

Acknowledgements

We acknowledge grant support from the Welcome Trust (PHH), the BBSRC (PHH and TGB) and the NERC (APV and TGB: Grant no. GR3/10632), and thank Sean Nee, Andy Purvis and Andrew Rambaut.

References

Allmon, W. D. 1992. A causal analysis of stages in allopatric speciation. In *Oxford Surveys in Evolutionary Biology*. Eds. D. Futuyma and J. Antonovics. pp. 219-257. Oxford University Press, Oxford.

Barraclough, T. G., Harvey, P. H., and Nee, S. 1995. Sexual selection and taxonomic diversity in passerine birds. *Proc. R. Soc. London Ser. B* **259**, 211–215.

Barraclough, T. G., Harvey, P. H. and Nee, S. 1996. Rate of *rbc*L gene sequence evolution and species diversification in flowering plants (angiosperms). *Proc. R. Soc. London Ser. B* **263**, 589–591.

Barraclough, T. G., Nee, S. and Harvey, P. H. 1998a. Sister-group analysis in identifying correlates of diversification. *Evol. Ecol.* **12**, 751–754.

Baraclough, T. G., Vogler, A. P. and Harvey, P. H. 1998b. Revealing the factors that promote speciation. *Phil. Trans. R. Soc. Lond. B* **353**, 241–249.

Barraclough, T. G., Hogan, J. and Vogler, A. P. (in prep.) Testing whether ecological factors promote cladogenesis in a group of tiger beetles (Coleoptera: Cicindelidae).

Bininda-Emonds, O. R. P., Gittleman, J. L. and Purvis, A. Subm. A complete phylogeny of extant Carnivora (Mammalia).

Bromham, L., Rambaut, A. and Harvey, P. H. 1996. Determinants of rate variation in mammalian DNA sequence evolution. *J. Mol. Evol.* **43**, 610–631.

Brown Jr., W. L. and Wilson, E. O. 1956. Character displacement. *Syst. Zool.* **5**, 49-64.

Brumfield, R. T. and Capparella, A. P. 1996. Historical diversification of birds in North-western South America—a molecular perspective on the role of vicariant events. *Evolution* **50**, 1607–1624.

Bullini, L. 1994. Origin and evolution of animal hybrid species. *Trends. Ecol. Evol.* **2**, 8–13.

Bush, G. L. 1969. Sympatric host race formation and speciation in frugivorous flies of the genus *Rhagoletis* (Diptera: Tephritidae). *Evolution* **23**, 237–251.

Butlin, R. K. 1995. Reinforcement—an idea evolving. *Trends Ecol. Evol.* **10**, 432–434.

Chase, M. W., Soltis, D. E. and Olmstead, R. G. *et al.* 1993. Phylogenetics of seed plants: an analysis of nucleotide sequences from the plastid gene rbcL. *Ann. Missouri. Bot. Gard.* **80**, 528–580.

Chesser, R. T. and Zink, R. M. 1994. Modes of speciation in birds: a test of Lynch's method. *Evolution* **48**, 490–497.

Collins, T. M., Frazer, K., Palmer, A. R., Vermeij, G. J. and Brown, W. M. 1996. Evolutionary history of northern hemisphere Nucella (Gastropoda, Muricidae): Molecular, morphological, ecological, and paleontological evidence. *Evolution.* **50**, 2287–2304.

Connor, E. F. and Taverner, M. P. 1997. The evolution and adaptive significance of the leaf-mining habit. *Oikos.* **79**, 6–25.

Coyne, J. A. 1992. Genetics and speciation. *Nature* **355**, 511–515.

Cracraft, J. 1981. Patterns and process in paleobiology: the role of cladistic analysis in systematic paleontology. *Paleobiology* **7**, 456–458.

Cracraft, J. 1985. Biological diversification and its causes. *Ann. Miss. Bot. Gard* **72**, 794–822.

Cracraft, J. 1990. The origin of evolutionary novelties: pattern and process at different hierachical levels. In *Evolutionary innovations.* (Ed. M. H. Nitecki). pp. 21–44. University Chicago Press, Chicago.

Darwin, C. 1871. *The descent of man and selection in relation to sex*. John Murray, London.

Doyle, J. A. and Donoghue, M. J. 1993. Phylogenies and angiosperm diversification. *Paleobiology* **19**, 141–286.

Ehrlich, P. R. and Raven, P. H. 1964. Butterflies and plants: a study in coevolution. *Evolution.* **18**, 586–608.

Farrell, B. D. and Mitter, C. 1993. Phylogenetic determinants of insect/plant community diversity. In *Species diversity in ecological communities: historical and geographical perspectives.* (Ed. R. E. Ricklefs and D. Schluter). pp. 52–65. University of Chicago Press, Chicago.

Farrell, B. D., Dussourd, D. E. and Mitter, C. 1991. Escalation of plant defense: do latex and resin canals spur plant diversification? *Am. Nat.* **138**, 881–900.

Felsenstein, J. 1985. Phylogenies and the comparative method. *Am. Nat.* **125**, 1–15.

Fisher, R. A. 1958. *The genetical theory of natural selection.* Dover, New York

Foote, R. H., Blanc, F. L. and Norrbom, A. L. 1993. *Handbook of the fruit flies (Diptera: Tephritidae) of America north of Mexico.* Comstock, Ithaca, NY.

Freitag, R. 1965. A revision of the North American species of the *Cicindela maritima* group with a study of hybridization between *Cicindela duodecimguttata* and oregona. *Quest. Entomol.* **1**, 87–170.

Gittleman, J. L. and Purvis, A. 1998. Body size and species-richness in carnivores and primates. *Proc. R. Soc. Lond. B* **265**, 113–119.

Grant, V. 1981. *Plant speciation.* Columbia University Press, New York.

Graves, R. C. and Pearson, D. L. 1973. The tiger beetles of Arkansas, Louisiana, and Mississippi (Coleoptera: Cicindelidae). *Trans. Am. Entomol. Soc.* **99**, 157–203.

Harvey, P. H. and Pagel, M. D. 1991. *The comparative method in evolutionary biology.* Oxford University Press, Oxford.

Harvey, P. H., May, R. M. and Nee, S. 1994. Phylogenies without fossils. *Evolution.* **48**, 523–239.

Harvey, P. H., Read, A. F. and Nee, S. 1995. Why ecologists need to be phylogenetically challenged. *J. Ecology* **83**, 535–536.

Hodges, S. A. and Arnold, M. L. 1995. Spurring plant diversification—are floral nectar spurs a key innovation? *Proc. R. Soc. LondonSer. B* **262**, 343–348.

Hutchinson, G. E. 1959. Homage to Santa Rosalia, or why are there so many kinds of animals? *Am. Nat.* **93**, 145–159.

Hutchinson, G. E. and MacArthur, R. H. 1959. A theoretical ecological model of size distributions among species of animals. *Am. Nat.* **93**, 117–125.

Jablonski, D. 1987. Heritability at the species level: analysis of geographic ranges of Cretaceous mollusks. *Science* **238**, 360–363.

Jackson, J. B. C. 1974. Biogeographic consequences of eurytopy and stenotopy among marine bivalves and their evolutionary signficance. *Am. Nat.* **108**, 541–560.

Kerr, J. T. and Packer, L. 1997. Habitat heterogeneity as a determinant of mammal species richness in high-energy regions. *Nature.* **385**, 252–254.

Lande, R. 1981. Models of speciation by sexual selection on polygenic characters. *Proc. Natl Acad. Sci. USA* **78**, 3721–3725.

Linder, H. P. and Crisp, M. D. 1995. Nothofagus and Pacific biogeography. *Cladistics* **11**, 5–32.

Lynch, J. D. 1989. The gauge of speciation: on the frequency of modes of speciation. In *Speciation and its consequences.* (Ed. D. Otte and J. A. Endler). pp. 527–553. Sinauer Associates, Sunderland, MA.

Martin, T. E. 1996. Fitness costs of resource overlap among coexisting bird species. *Nature.* **380**, 338–340.

Mayr, E. 1954. Change of genetic environment and evolution. In *Evolution as a process.* (Ed. J. Huxley, A. C. Hardy and E. B. Ford). pp. 157–180. George Allen and Unwin, London.

Mayr, E. 1963. *Animal species and evolution.* Harvard University Press, Cambridge, MA.

McCall, R. A., Harvey, P. H. and Nee, S. subm. Avian dispersal ability and the tendency to form island-endemic species on continental and oceanic islands. *Biodiv. Lett.*

Mindell, D. P. and Thacker, C. E. 1996. Rates of molecular evolution—phylogenetic issues and applications. *Annu. Rev. Ecol. Syst.* **27**, 279–303

Mitra, S., Landel, H. and Pruett-Jones, S. 1996. Species richness covaries with mating system in birds. *Auk* **113**, 544–551.

Mitter, C., Farrell, B. and Wiegmann, B. 1988. The phylogenetic study of adaptive zones: has phytophagy promoted insect diversification? *Am. Nat.* **132**, 107–128.

Mooers, A.Ø. and Heard, S. B. 1997. Evolutionary processes from tree shape. *Q. Rev. Biol.* **72**, 31–54.

Mooers A.Ø. and Møller, A. P. 1996. Colonial breeding and speciation in birds. *Evol. Ecol.* **10**, 375–385

Møller, A. P. and Birkhead, T. R. 1992. A pairwise comparative method as illustrated by copulation frequency in birds. *Am. Nat.* **139**, 644–656.

Morse, D. R., Lawton, J. H., Dodson, M. and Williamson, M. H. 1985. Fractal dimension of vegetation and the distribution of arthropod body lengths. *Nature* **314**, 731–733.

Nee, S. and Harvey, P. H. 1994. Getting to the roots of flowering plant diversity. *Science* **264**, 1549–1550.

Nee, S., Holmes, E. C., May, R. M. and Harvey, P. H. 1994. Extinction rates can be estimated from molecular phylogenies. *Phil. Trans. R. Soc. London Ser. B.* **349**, 25–31.

Nee, S., Barraclough, T. G. and Harvey, P. H. 1996a. Temporal changes in biodiversity: detecting patterns and identifying causes. In *Biodiversity: a biology of numbers and difference.* (Ed. K. J. Gaston). pp. 230–252. Blackwell Science, Oxford.

Nee, S., Read, A. F. and Harvey, P. H. 1996a. Why phylogenies are necessary for comparative analysis. In *Phylogenies and the comparative method in animal behaviour* (Ed. E. P. Martins). pp. 399–411. Oxford University Press, Oxford.

Orr, H. A. 1995. The population genetics of speciation: the evolution of hybrid incompatabilities. *Genetics* **139**, 1805–1813.

Orr, H. A. and Orr, L. H. 1996. Waiting for speciation—the effect of population subdivision on the time to speciation. *Evolution.* **50**, 1742–1749.

Panchen, A. L. 1992. *Classification, evolution and the nature of biology.* Cambrige University Press, Cambridge.

Purvis, A. 1995. A composite estimate of primate phylogeny. *Phil. Trans. R. Soc. London Ser. B.* **348**, 405–421.

Purvis, A. 1996. Using interspecies phylogenies to test macroevolutionary hypotheses. In *New uses for new phylogenies.* (Ed. P. H. Harvey, A. J. Leigh Brown, J. Maynard Smith and S. Nee). pp. 153–168. Oxford University Press, Oxford.

Raup, D. M. 1985. Mathematical models of cladogenesis. *Paleobiology.* **11**, 42–52.

Raup, D. M., Gould, S. J., Schopf, T. J. M. and Simberloff, D. S. 1973. Stochastic models of phylogeny and the evolution of diversity. *J. Geol.* **81**,525–42.

Read, A. F., and Nee, S. 1995. Inferences from binary comparative data. *J. Theor. Biol.* **173**, 99–108.

Ricklefs, R. E. and Schluter, D. 1993. *Species diversity in ecological communities: historical and geographical perspectives.* University of Chicago Press, Chicago.

Rosenzweig, M. L. 1995. *Species diversity in space and time.* Oxford University Press, Oxford.

Saetre, G. P., Moum, T., Bures, S., Kral, M., Adamjan, M. and Moreno, J. 1997. A sexually selected character displacement in flycatchers reinforces premating isolation. *Nature* **387**, 589–592.

Sanderson, M. J. 1990. Estimating rates of speciation and evolution: a bias due to homoplasy. *Cladistics* **6**, 387–391.

Sanderson, M. J. and Donoghue, M. J. 1994. Shifts in diversification rate with the origin of angiosperms. *Science* **264**, 1590–1593.

Sanderson, M. J. and Donoghue, M. J. 1996. Reconstructing shifts in diversification rates on phylogenetic trees. *Trends Ecol. Evol.* **11**, 15–20.

Sanderson. M. J. and Wojciechowski, M. F. 1996. Diversification rates in a temperate legume clade—are there so many species of *Astragalus* (Fabaceae). *Am. J. Bot.* **83**, 1488–1502

Sibley, C. G. and Ahlquist, J. E. 1990. *Phylogeny and classification of Birds: a study in molecular evolution.* Yale University Press, New Haven, CT.

Slatkin, M. 1996. In defense of founder-flush theories of speciation. *Am. Nat.* **147**, 493–505.

Smith, J. L. and Bush, G. L. 1997. Phylogeny of the genus *Rhagoletis* (Diptera: Tephritidae) inferred from DNA sequences of mitochondrial cytochrome oxidase II. *Mol. Phylog. Evol.* **7**, 33–43.

Southwood, T. R. E. 1973. The insect/plant relationship—an evolutionary perspective. In *Insect-plant relationships.* (Ed. H. F. van Emden). pp. 3–30. Blackwell Scientific Publications, London.

Strong, D. R., Simberloff, D. S., Abele, L. G. and Thistle, A. B. 1984. *Ecological communities: conceptual issues and the evidence.* Princeton University Press, Princeton, NJ.

Taper, M. L. and Case, T. J. 1992. Coevolution among competitors. *Oxf. Surv. Evol. Biol.* **8**, 63–109.

Templeton, A. R. 1996. Experimental evidence for the genetic transilience model of speciation. *Evolution* **50**, 909–915.

Vogler, A. P. and Barraclough, T. G. 1998. Reconstructing shifts in diversitication rate during the radiation of tiger beetles (Cicindelidae). Bulletino del Museo Turino, in press.

Vogler, A. P., Welsh, A. and Barraclough, T. G. 1998. Molecular phylogeny of the *Cicindela maritima* (Coleoptera: Cicindelidae) group indicates fast radiation in Western North America. *Ann. Entomol. Soc. Am.* **91**, 185–194.

West-Eberhard, M. J. 1983. Sexual selection, social competition and speciation. *Q. Rev. Biol.* **58**, 155–183.

Wiegmann, B. M., Mitter, C. and Farrell, B. D. 1993. Diversification of carnivorous parasitic insects: Extraordinary radiation, or specialized dead end? *Am. Nat.* **142**, 737–754

Zeh, D. W., Zeh, J. A. and Smith, R. L. 1989. Ovipositors, amnions and eggshell architecture in the diversification of terrestrial arthropods. *Q. Rev. Biol.* **64**, 147–168.

Gulliver's further travels: the necessity and difficulty of a hierarchical theory of selection

Stephen Jay Gould

11.1 Introduction

For principled and substantially philosophical reasons, based largely on his reform of natural history by inverting the Paleyan notion of overarching and purposeful beneficence in the construction of organisms, Darwin built his theory of selection at the single causal level of individual bodies engaged in unconscious (and metaphorical) struggle for their own reproductive success. But the central logic of the theory allows selection to work effectively on entities at several levels of a genealogical hierarchy, provided that they embody a set of requisite features for defining evolutionary individuality. Genes, cell lineages, demes, species, and clades—as well as Darwin's favored organisms—embody these requisite features in enough cases to form important levels of selection in the history of life.

R. A. Fisher explicitly recognized the unassailable logic of species selection, but denied that this real process could be important in evolution because, compared with the production of new organisms within a species, the origin of new species is so rare, and the number of species within most clades so low. I review this and other classical arguments against higher-level selection, and conclude (in the first part of this chapter) that they are invalid in practice for interdemic selection, and false in principle for species selection. Punctuated equilibrium defines the individuality of species and refutes Fisher's classical argument based on cycle time.

In the second part of the chapter, I argue that we have failed to appreciate the range and power of selection at levels above and below the organismic because we falsely extrapolate the defining properties of organisms to these other levels (which are characterized by quite different distinctive features), and then regard the other levels as impotent because their effective individuals differ so much from organisms. We would better appreciate the power and generality of hierarchical models of selection if we grasped two key principles: first, that levels can interact in all modes (positively, negatively, and orthogonally), and not only in the negative style (with a higher level suppressing an opposing force of selection from the lower level) that, for heuristic and operational reasons, had received almost exclusive attention in the existing literature; and second, that each hierarchical level differs from all others in substantial and interesting ways, both in the style and frequency of patterns in change and causal modes.

11.2 The validity and necessity of selection at supraorganismic levels, with emphasis on species selection

R. A. Fisher and the compelling logic of species selection

R. A. Fisher added a short section entitles 'the benefit of species' to the second edition (1958) of his founding document for the Modern Synthesis: *The Genetical Theory of Natural Selection* (first published in 1930). I do not know why he did so, but I (as a proponent of species selection) could not be more pleased by the content—for Fisher, in these few additional paragraphs, supplies the two features most needed by any healthy and controversial theory. Fisher proclaims the logic of species selection unassailable, and then denies that this genuine phenomenon could have any substantial importance in the empirical record of evolution on our planet. No promoter of species selection could possibly ask for more from a great thinker like Fisher than validation in logic and a healthy dispute about actual evidence!

Fisher begins this interpolated passage by stating that Natural Selection (his upper case letters), in the conventional organismic mode, cannot explicitly build any features for 'the benefit of the species' (though organismic adaptation may engender such an effect as a side consequence). Speaking of instinctual behaviors, Fisher writes (1958, p. 50): 'Natural Selection can only explain these instincts in so far as they are individually beneficial, and leaves entirely open the question as to whether in the aggregate they are a benefit or an injury to the species.' But Fisher then recognizes that, in principle, selection among species could occur, and could lead to higher-level adaptations directly beneficial to species. However, lest this logical imperative derail his strict Darwinian commitments to the primacy of organismic selection, Fisher then adds that species selection—though clearly valid in logic and therefore subject to realization in nature—must be far too weak (relative to organismic selection) to have any practical effect upon evolution. I regard the following lines (Fisher, 1958, p. 50) as one of the 'great quotations' in the history of evolutionary thought:

> There would, however, be some warrant on historical grounds for saying that the term Natural Selection should include not only the selective survival of individuals of the same species, but of mutually competing species of the same genus or family. The relative unimportance of this as an evolutionary factor would seem to follow decisively from the small *number* of closely related species which in fact do come into competition, as compared to the number of individuals in the same species; and from the vastly greater *duration* of the species compared to the individual.

Fisher's theoretical validation of the logic behind species selection has never been challenged. Even the most ardent gene selectionists have granted this point, and have dismissed species selection from extensive consideration only for its presumed weakness relative to their favored genic level, and not because they doubt the validity, or even the reality, of selection at this higher level. Dawkins (1982, pp. 106–107) has sharpened Fisher's point by noting that, at most, species selection might accentuate some relatively 'uninteresting' linear trends (like size increase among species in a lineage), but could not possibly 'put together complex [organismal] adaptations such as eyes and brains.' Dawkins continues:

> When we consider the species as a replicator . . . the replacement cycle time is the interval from speciation event to speciation event, and may be measured in thousands of years, tens of thousands, hundreds of thousands. In any given period of geological time, the number of selective species extinctions that can have taken place is many orders of magnitude less than the number of selective allele replacements that can have taken place . . . We shall have to make a quantitative judgment taking into account the vastly greater cycle time between replicator deaths in the species selection case than in the gene selection case.

I strongly support Dawkins's last statement, and will argue that, when we factor punctuated equilibria into the equation, species selection emerges as a powerful force in macroevolution (though not as an architect of complex organismic adaptations).

Williams has also supported Fisher's argument about the logic of higher level selection—even in his gene selectionist manifesto of 1966, where he defends the possibility, but then denies the actuality: 'If a group of adequately stable populations is available, group selection can theoretically produce biotic adaptations, for the same reason that genic selection can produce organic adaptations'. (1966, p. 110). In his later book, however, Williams becomes much more positive about the importance and reality of selection at several hierarchical levels: 'To Darwin and most of his immediate and later followers, the physical entities of interest for the theory of natural selection were discrete individual organisms. This restricted range of attention has never been logically defensible' (1992, p. 38).

The developing literature has added three 'classical' arguments against higher-level selection to Fisher's primary point that cycle times are incomparably slow relative to the lives of organisms. All these arguments share the favorable property of accepting a common logic but challenging the empirical importance of legitimate phenomena—a far happier state for productive science than the confusion about concepts and definitions that so often reigns. In the rest of this section, I shall summarize the four classical arguments; note that they can all be effectively challenged at the level—'group,' or interdemic, selection—for which they were devised; and then demonstrate that none have any strong force, in principle, against the empirical importance of the still higher level of species selection.

The classical arguments against efficacy of higher-level selection

The usual arguments against higher-level selection, all quite potent, admit that such phenomena must be possible in principle, but cannot play any meaningful role in evolution on grounds of rarity and weakness relative to ordinary natural selection upon organisms.

Weakness (based on cycle time)

R. A. Fisher's classical argument: How could species selection have any substantial effect upon evolution? Rate and effect depend upon numbers and timings of births and deaths—to provide a sufficient population of items for differential sorting. But species persist for thousands or millions of years, and clades count their 'populations' of component species in tens, or at most hundreds, and not as the millions or billions of organisms in many populations. How could species selection have any measurable effect at all (relative to ordinary organismic selection) when, on average, billions of organismic births and deaths occur for each species origin or extinction, and when populations of organisms contain orders of magnitude more members than populations of related species in a clade?

Weakness (based on variability)

Hamilton (1971), in devising arguments against interdemic selection, pointed out that variation among mean group values for genetically relevant and selected aspects of organismic phenotypes will generally be lower than variation among organisms within a population for the same features. Group selection cannot be strong if the mean phenotypes of demic individuals express such limited variation to serve as the raw material of change.

Instability, as in Dawkins's (1982) metaphors of duststorms in the desert and clouds in the sky

This argument has also been urged most strongly against interdemic selection. Demes, by definition, have porous borders because organisms in the same species can interbreed, and members of one deme can therefore, in principle, invade and join another in a reproductive role. If such invasions are frequent and numerous, the deme ceases to be a discrete entity, and cannot be called an evolutionary individual (thereby losing any status as a potential unit of selection). Moreover, many demes lack cohesion on their own account, and not only by susceptibility to incursion. Demes may be entirely temporary and adventitious aggregates of organisms, devoid of any inherent mechanism for cohesion, and defined only by the transient and clumpy nature of appropriate habits that may not even persist for a requisite generation—the deme of all mice in a haystack, or all cockroaches in a dirty urban kitchen, for example.

Invasibility from other more potent levels, usually from below

This standard argument, related to Fisher's first point about cycle time, and classically used to question the potential evolution of altruism by interdemic selection, asks how higher-level selection could possibly be effective when more powerful, lower-level invaders can cancel any result by working in the opposite direction. More particularly, suppose that interdemic selection is cranking along at its characteristic pace, increasing the overall frequency of altruistic alleles in the entire species because demes with altruists enjoy differential success in competition against demes without altruists. Fine—but as soon as a selfish mutant arises in any deme with altruists, its advantage in organismic selection against the altruistic allele should be so great that the frequency of altruistic genes must plummet within the deme, even while the deme profits in group selection from the presence of altruistic organisms. By Fisher's argument of cycle time, organismic selection of the self-serving should trump interdemic selection for altruism.

Overcoming these classical arguments, in practice for interdemic selection, but in principle for species selection

Since most modern debate about higher-level selection has addressed interdemic (or so-called 'group') selection, the classical arguments have been framed mainly at the level just above our conventional focus upon organisms (though I predict that emphasis will shift to higher levels, particularly to species selection, as macroevolutionary theory develops). All four arguments have force, and do spell impotency (or even incoherence and nonexistence) for interdemic selection in many circumstances. But, as generalities, these arguments have failed either to disprove interdemic selection as a meaningful force worthy of consideration at all, or to deny the efficacy of interdemic selection in certain important circumstances.

I shall not review this enormous literature here (as my primary concern rests at still higher levels of selection), but I wish to note that two classes of argument grant interdemic selection sufficient strength and presence to count as a potentially major force in evolution. First, much mathematical modelling (and some experimental work) has adequately shown that, under reasonable conditions of potentially frequent occurrence in nature, group selection can assert its sway against the legitimate force of the four classical objections. In the cardinal example, under several plausible models, the frequency of altruistic alleles can increase within a species, so long as the rate of differential survival and propagation of demes with altruistic members (by group selection) overcomes the admitted decline in frequency of altruists within successful demes by organismic selection. The overall frequency may rise even while the frequency within each surviving deme declines (Wilson, 1983).

Second, some well documented patterns in nature seem hard to explain without a strong component of interdemic selection. Female-biased sex ratios, as discussed by Wilson and Sober (1994, pp. 640–641), provide the classic example because two adjacent levels make opposite and easily tested predictions: conventional organismic selection should favour a 1:1 ratio by Fisher's famous argument (1930); while interdemic selection should promote strongly female-biased ratios to enhance the productivity of groups. Williams (1966) accepted this framework, which he proposed as a kind of acid test for the existence of group selection. He allowed that female-biased ratios would point to group selection, but denied that any, in fact, existed, thus validating empirically the theoretical arguments he had developed for the impotence of group selection. Williams concluded (1966, p. 151): 'Close conformity with the theory is certainly the rule, and there is no convincing evidence that sex ratios ever behave as a biotic adaptation.' But many empirical examples of female-biased ratios were soon discovered (see Colwell, 1981; also numerous references in Wilson and Sober, 1994, p. 592). Some authors (e.g. Maynard Smith, 1987) tried to interpret this evidence without invoking group selection, but I think that all major participants in the discussion now admit a strong component of interdemic selection in such results—and reported cases now number in the hundreds, so we are not talking about odd anomalies in tiny corners of nature. Williams now accepts this interpretation (1992, p. 49), writing 'that selection in female-biased Mendelian populations favors males, and that it is only the selection among such groups that can favor the female bias.'

The primary appeal of this admirably documented example lies in the discovery that female biases are usually only moderate—more than organismic selection could allow (obviously, as any bias at all would establish the point), but less than models of purely interdemic selection predict. Thus, the empirical evidence suggests a balance between adjacent and opposing levels of selection—with alleles for female-biased sex ratios reduced in frequency by organismic selection within demes, but boosted above the Fisherian balance (across species as a whole) because they increase the productivity of demes containing them at high frequency.

When we move to the level of species selection, the most important for macroevolutionary theory, we encounter an even more favorable situation. For interdemic selection, the classical contrary arguments had legitimate force, but could be overcome under conditions broad enough to grant the phenomenon considerable importance. For species selection, on the other hand, most of the classical arguments don't even apply in principle—while the one that does (weakness due to cycle time) becomes irrelevant if punctuated equilibrium prevails at a dominant relative frequency.

Proceeding through the classical objections in reverse order, the fourth argument about invasibility from below has strength only in particular con-texts—when, in principle, a favored direction of higher-level selection will usually be opposed by stronger selection at the level immediately below. (In the classic case, selfish organismal 'cheaters' derail group selection for altruism. Nonetheless, while the argument of invasibility may hold for this particular case—and while, for contingent reasons in the history of science, this example became the paradigm for interdemic selection—I see no reason in principle for thinking that organismal selection must always, or even usually, oppose interdemic selection. The two levels may operate simultaneously and in the same direction, or orthogonally.)

In any case, I cannot devise any rationale for supposing that organismic or interdemic selection should characteristically oppose species selection—and the argument of invasibility therefore collapses. Of course, organismic selection *may* operate contrary to the direction of species selection—and must frequently do so, particularly in the phenomenon that older textbooks called 'overspecialization,' or the development of narrowly focused and complex adaptations (the peacock's tail as a classic example) that enhance the reproductive success of individual organisms, but virtually guarantee a decreased geological life span for the species. But other equally common modes of organismic selection either tend to increase geological longevity (e.g. improvements in general biomechanical design) or to operate orthogonally, and therefore 'beneath the notice' of species selection. As our best examples of species selection work through differential rate of speciation rather than varying propensities for extinction, and as most organismal adaptations probably don't strongly influence a population's rate of speciation (or at least don't manifest any bias for decreasing the rate), essential orthogonality of the two levels must often prevail in evolution.

The third argument of instability, while potent for demes, clearly does not apply to species, which are as well bounded as organisms. Just as genes and cell lineages generally do not wander from organism to organism (whereas organisms often move readily from deme to deme), neither can organisms or demes wander from species to species. The reasons for such tightness of bounding differ between organism and species, but these two evolutionary individuals probably exceed all others in strength of this key criterion. Species maintain and 'police' their borders just as well as organisms do.

The tight bounding of an organism arises from functional integration among constituent parts, including an impermeable outer covering in most cases, and often an internal immune system to keep out invaders. The tight bounding of a species arises from reproductive interaction among parts (organisms), with firm exclusion of parts from any other species. Moreover, this exclusion is actively maintained, not merely passively propagated, by properties that became a favorite subject of study among founders of the Modern Synthesis, especially Dobzhansky (1951) and Mayr (1963)—the so-called 'isolating mechanisms.' Species may lack a literal skin, but they are just as well-bounded as organisms in the sense required by the theory of natural selection.

This discussion highlights one of my few unhappinesses with Wilson and Sober's (1994) excellent discussion and defense of hierarchical selection. They insist upon functional integration as the main criterion for identifying units of selection (vehicles in their terminology, interactors or evolutionary individuals for others). They insist that the following question 'is and always was at the heart of the group selection controversy—can groups be like individuals in the harmony and coordination of their parts' (1994, p. 591).

I do not object to the invocation of functionality itself, but rather to their narrow definition, too parochially based upon the kind of functionality that organisms display. The cohesion (or 'functionality') of species does not lie in the style of interaction and homeostasis that unites organisms by the integration of their tissues and organs. Rather, the cohesion of species rests upon their active maintenance of distinctive properties, achieved by interdigitating their parts (organisms) through sexual reproduction, while excluding the parts of other species by evolution of isolating mechanism.

I much prefer and support Wilson and Sober's more general definition (1994, p. 599): 'Groups are real to the extent that they become functionally organized by natural selection at the group level.' Species meet his criterion by evolving species-level properties that maintain their cohesion as evolutionary individuals. The key to a broader concept of 'functionality' (that is, ability to operate discretely as a unit of selection) lies in the evolution of active devices for cohesion, not in any particular style of accomplishment—either the reproductive barriers that maintain species, or the homeostatic mechanisms that maintain organisms.

The second argument of weakness based on lack of sufficient variability among group mean values also does not apply to species. Demes of mice from separated but adjacent haystacks may differ so little in group properties that the survival of only one deme, with replenishment of all haystacks by migrants from this successful group, might scarcely alter either allelic frequencies across the entire species, or even average differences among demes. But new species must differ, by definition, from all others—at least to an extent that prevents reproductive merging of members. Thus, the differential success of some species in a clade must alter—usually substantially—the average properties of the clade (whereas, one level down, the differential success of some demes need not change the average properties of the species very much, if at all).

We are thus left with the first argument about weakness due to long cycle time and small populations as the only classical objection with potential force against species selection. And, at first glance, Fisher's argument would seem both potent and decisive. The basic observation is surely true: billions of organism births usually occur for each species birth; and populations of organisms within a species are almost always vastly larger than populations of species in a clade. How then could species selection, despite its impeccable logic, have any measurable import-ance when conventional organismal selection must be so incomparably stronger?

The logic of Fisher's argument is undeniable, but we must also consult the empirical world. Organismic selection must overwhelm species selection when both processes operate steadily and unidirectionally—for if both levels work in the same direction, then species selection can only add a small increment to the vastly greater power of organismic selection; whereas, if the two levels work in opposite directions, organismic selection must overwhelm and cancel the effect of species selection.

But the empirical record of the great majority of well documented fossil species affirms stasis throughout the geological range (Eldredge and Gould, 1972; Stanley, 1979; Gould and Eldredge, 1993 and references therein). The causes for observed nondirectionality within species are controversial, and the phenomenon is by no means incompatible with continuous operation of strong organismic selection—for two common explanations of stasis as a central component of punctuated equilibrium include general prevalence of stabilizing selection, and fluctuating directional selection with no overall linear component due to effectively random changes of relevant environments through time. But the observation of general stasis seem well established at high relative frequency.

In this factual circumstance, since change does not generally accumulate through time within a species, organismic selection in the conventional grad-ualistic and anagenetic mode cannot contribute much to the direction of a trend within a clade. Change must therefore be concentrated in events of branching speciation, and trends must arise by the differential sorting of species with favored attributes. If new species generally arise in geological moments, as the theory of punctuated equilibrium holds (Eldredge and Gould, 1972; Gould and Eldredge, 1977, 1993), then trends owe their explanation even more clearly to higher-level sorting among species-individuals acting as firm entities with momentary births and stable durations.

Organismic selection may trump species selection in principle, but if change at speciation is 'the only game in town,' then a 'weak force' prevails while a potentially stronger force lies dormant. Nuclear bombs certainly make conventional bullets look risible as instruments of war, but if we choose not to employ the nukes, then bullets can be devastatingly effective. The empirical pattern of punctuated equilibrium therefore becomes the factual 'weapon' that overcomes Fisher's strong theoretical objection to the efficacy of species selection.

11.3 A literary statement about the two major properties of hierarchies

Our vernacular language recognizes a triad of terms for the structural description of any phenomenon that we wish to designate as a unitary item or thing. The thing itself becomes a focus, and we call it an object, an entity, an individual, an organism, or any one of a hundred similar terms, depending on the substance and circumstance. The subunits that make up the individual are then called 'parts' (or units, or organs, etc., depending upon the nature of the focal item); while any recognized grouping of similar individuals becomes a 'collectivity' (or aggregation, society, organization, etc.). In other words, and in epitome, individuals are made of parts and aggregate into collectivities.

The hierarchical theory of selection recognizes many kinds of evolutionary individuals, banded together in a rising series of increasingly greater inclusion, one within the next—genes in cells, cells in organisms, organisms in demes, demes in species, species in clades. The focal unit of each level is an *individual*, and we may choose to direct our evolutionary attention to any of the levels. Once we designate any focal level as primary in a particular study, then the unit of that level—the gene, or the organism, or the species etc.—becomes our relevant or focal individual, and its continuent units become parts, while the next higher unit becomes its collectivity. Thus, if I place my focus at the conventional organismic level, genes and cells are parts, while demes and species are collectivities. But if I need to focus on species as individuals, then organisms become parts, and clades are collectivities. In order words, the triad of part—individual—collectivity will shift bodily up and down the hierarchy, depending upon the chosen subjects and objects of any particular study.

This dry linguistic point becomes important for a fundamental reason of psychological habit. Humans are hidebound creatures of convention, particularly tied to the spatial and temporal scales most palpably familiar in our personal lives. Among nature's vastly different realms of time, from the femtoseconds of some atomic phenomena to the aeons of stellar and geological time, we really grasp, in a visceral sense, only a small span from the seconds of our incidents to the few decades of our lives. We can formulate other scales in mathematical terms; we can document their existence and the processes that unfold in their domains. But we experience enormous difficulty in trying to bring these alien scales into the guts of our understanding—largely for the parochial reason of personal inexperience.

We make frequent and legendary errors because we read the styles and modes of our own scale into the different realms of the incomprehensibly fleeting or vast. Geologists, for example, well appreciate the enormous difficulties that most people encounter in trying to visualize or understand the meaning of any ordinary statement in 'earth time'—that a landscape took millions of years to develop, or that a lineage exhibits a trend to increasing size throughout the Cretaceous period. We continue to make the damnedest mistakes, professional and laypeople alike. I have, for example struggled for more than 20 years against the conventional misreading of punctuated equilibrium as a saltational theory in the generational terms usually applied to such a concept in evolutionary studies. The theory's punctuations are saltational on geological scales—in the sense that most species arise during an unmeasurable geological moment (meaning, in operational terms, that all the evidence appears on a single bedding plane). But geological moments encompass thousands of human years—more than enough time for a continuous process that we would regard as glacially slow by the measure of our lives (e.g. see Goodfriend and Gould, 1996). Thus, punctuated equilibrium is the proper geological scaling of speciation in a few thousand years, not a slavish promotion of the ordinary human concept of instantaneity to the origins of species.

As we misunderstand the scales of time, we fail just as badly with differing realms of size. Our bodies lie in the middle of a range from the angstroms of atoms to the light years of galaxies. Individuality exists in all these domains, but when we try to understand the phenomenon at any distant scale, we fall under the thrall of the greatest of all parochialisms. We know one kind of individual so intimately and with such familiarity—our own bodies—that we impose the characteristic properties of this level upon the very different styles of 'thingness' featured at other scales of size. This inevitable bias provokes considerable trouble, for organic bodies are very peculiar kinds of individuals—and very poor models for the phenomenon of individuality at most other scales.

The 'feel' of individuality at other scales becomes so elusive that most of the best exploration has been accomplished by writers of fiction, not by scientists. The tradition goes back at least to Lemuel Gulliver, who didn't range very widely from our kind of body and our norm of size, and has been best promoted, in our generation, within the genre of science fiction. Film has also become a powerful medium of exploration, perhaps best expressed in two 'cult' films, *Fantastic Voyage* and *Inner Space*, both about humans shrunk to cellular size and injected into the body of another unaltered conspecific. This ordinary body becomes the environment of the shrunken protagonists, a 'collectivity' rather than a discrete entity—while the 'parts' of this body become individuals to the shrunken guests. When Raquel Welch fights an antibody to the death in *Fantastic Voyage*, we understand how location along the triadic continuum of part—individual—

collectivity depends upon circumstances and concern. A tiny, if crucial, part of my body at my true size becomes an entire and ultimately dangerous individual to Ms. Welch at a fraction of a millimeter.

The parochiality of time has served us badly enough, but this parochiality of bodily size has, for two reasons, placed even more imposing barriers in our path to an improved and generalized evolutionary theory—a formulation well within our grasp if we can learn how to expand the Darwinian perspective to all levels of nature's hierarchy. First, we understand (often viscerally) what our bodies do best as Darwinian agents—and we then grant universal importance to these properties, both by denying interest to the different key features of individuals at other levels, and by assuming that our 'best' properties must, by extension, power Darwinian systems wherever they work. Our bodies are best at developing adaptations in the complex and coordinated form that we call 'organic.' Many evolutionists therefore argue, in the most constraining parochialism of all, that only adaptations matter as an explanatory goal of Darwinism, and that such adaptations must therefore drive evolution at all levels. I don't even think that such a perspective works well for organisms—surely the locus of most promising application (Gould and Lewontin, 1979)—but this attitude will surely stymie any understanding of individuality at other levels, where complex adaptations do not figure so prominently. How will Dawkins ever appreciate the different individuality of species, where exaptive effects hold at least equal sway with adaptations, if he continues to view sequelae and side consequences as 'the boring by-product theory' (1982, p. 215).

Second, we don't comprehend the scale-bound realities of other size domains, or we err by imposing our own perceptions when we try to think about the world of a gene, or of a species. In one of the most famous statements of twentieth century biology, D'Arcy Thompson (1942, p. 77) ended his chapter 'On Magnitude' (in his classic work, *On Growth and Form*) by noting how badly we misread the world of similar organisms because we are big and therefore live in gravity's domain (a result of falling surface/volume ratios as creatures become large, but not a significant feature in other realms of size). If we have so much trouble for extremes within our own level of organismic individuality, how will we grasp the even more distant worlds of other kinds of evolutionary individuals? D'Arcy Thompson wrote:

> Life has a range of magnitude narrow indeed compared to that with which physical science deals; but it is wide enough to include three such discrepant conditions as those in which a man, an insect, and a bacillus have their being and play their several roles. Man is ruled by gravitation, and rests on mother earth. A water-beetle finds the surface of a pool a matter of life and death, a perilous entanglement or an indispensable support. In a third world, where the bacillus lives, gravitation is forgotten, and the viscosity of liquid, the resistance defined by Stokes's law, the molecular shocks of the Brownian movement, doubtless also the electric charges of the ionized medium, make up the physical environment and have their potent and immediate influence on the organism. The predominant factors are no longer those of our scale; we have come to the edge of a world of which we have no experience, and where all our preconceptions must be recast.

Once we become mentally prepared to seek and appreciate (and not to ignore or devalue) the structural and causal differences among nature's richly various scales, we can formulate more fruitfully the two cardinal properties of hierarchies that make the theory of hierarchical selection both so interesting and so different from the conventional single-level Darwinism of organismal selection. The key to both properties lies in 'interdependence with difference,' for the hierarchical levels of causality are bonded in interaction, but also (for some attributes) fairly independent in modality. Moreover, the levels all operate differently, one from the other, despite unifying principles, like selection, applicable to all.

Selection at one level may enhance, counteract, or just be orthogonal to, selection at any adjacent level. All modes of interaction prevail among levels and have strong impact upon nature

I emphasize this crucial point because many students of the subject have focused almost exclusively on negative interaction between levels, albeit for a sensible and eminently practical reason. In so doing, they verge on the serious error of equating an operational advantage with a theoretical restriction, and almost seem to deny the other modes of positive (synergistic) and orthogonal (independent) interaction. Negative interaction wins primary heuristic attention because this mode provides our most cogent evidence, not merely for simultaneous action of two levels, but even for the very existence of a controversial or unsuspected level. If two levels work in synergism, then we easily miss the one we do not expect to see, and attribute the full effect to a greater strength than expected for the level we know. But if the controversial level yields an unexpected effect contrary to the known direction of selection at a familiar level, then we may specify and measure the disputed phenomenon.

In the example cited previously (p. 225), individual selection favors a balanced sex ratio, while interdemic selection leads to female bias in many circumstances. Our best evidence for the reality of interdemic selection arises from the discovery of such biases—not so strong as purely interdemic selection would produce (for organismic selection operates simultaneously in the other direction), but firm enough to demonstrate the existence of a controversial phenomenon. But if interdemic selection also worked towards a 1:1 ratio, we could attribute such an empirical finding exclusively to the conventional operation of organismic selection.

Negative interaction, however, does yield a distinguishing consequence to highlight this mode of interaction as especially important in the revisions to evolutionary theory that the hierarchical model will engender. In the conventional Darwinism of organismal selection alone, stabilities generally receive interpretation as adaptive peaks or optima, thus enhancing the functionalist bias inherent in the theory, and are unfortunate in my view. The only structuralist intrusion into this theme ordinarily occurs when we have been willing to allow that natural selection can't surmount a constraint—elephants too heavy to fly even if genetic variability for wings existed; insects confined to small sizes by the inherited *bauplan* of an exoskeleton that must be molted, and a respiratory system of skeletal invaginations that would need to become too extensive at the surface/volume ratio of large organisms. But constraints in these cases are passive walls, not active agents.

The hierarchical theory of selection suggests a strikingly different and dynamic reason for many of nature's stabilities: an achieved balance, at an intermediary point optimal for neither, between two levels of selection working in opposite directions. Several important phenomena may be so explained: weak female bias as the negative interaction of organismal and interdemic selection (see above); restriction of multiple copy number in 'selfish DNA' as a balance between positive selection at the gene level, suppressed by negative selection (based, perhaps, on energetic coasts of producing so many copies irrelevant to the phenotype) at the organismic level (Doolittle and Sapienza, 1980; Orgel and Crick, 1980). I also suspect that stable and distinctive features of species and clades often represent balances between positive organismic selection that would drive a feature to further elaboration, and negative species selection to limit the geological longevity of such 'overspecialized' forms. In any case, a world of conceptual difference exists between stabilities read as optima of a single process, and stabilities interpreted as compromises between active and opposed forces.

As an example of overemphasis upon negative interaction, Wilson and Sober (1994, p. 592) ask: 'Why aren't examples of within-individual [organism] selection more common?' They mention the most familiar case of meiotic drive, and then give the conventional argument for rarity of such phenomena: the integrity of complex organisms implies strong balance and homeostasis among parts; any part that begins to proliferate independently will threaten this stability, and must therefore be opposed by organismic selection, a force generally strong enough to eliminate such a threat from below (though terminal cancer may represent a pyrrhic victory for cell-lineage over organismic selection).

If selection within bodies always opposes the organismic level, then we must predict a low frequency for the phenomenon, since evolution has endowed the organismic level with a plethora of devices for resisting such dysfunctional invasion from within. While I accept this argument for a low frequency of selection *contrary* to the interests of enclosing organisms, selection within bodies may be common when we include the other modalities of *synergistic* and *orthogonal* interaction. The most interesting hypothesis for extensive selection at the gene level, the notion originally dubbed 'selfish DNA' (Orgel and Crick, 1980; Doolittle and Sapienza, 1980), explains the observed copy number of much middle-repetitive DNA as a result of orthogonal gene-level selection initially 'unnoticed' by the organism, although eventually suppressed by negative selection from above when copies reach sufficient numbers to exact an energetic drain upon construction of the phenotype. In fact, organismic complexity might never have evolved at all without extensive gene-level selection in this orthogonal (or synergistic) mode. For if we accept the common argument that freedom to evolve new phenotypic complexity requires genetic duplication to 'liberate' copies for modification in novel directions, then how could such redundancy ever arise if organismic selection worked with such watchdog efficiency that even a single 'extra' copy, initially unneeded by the organismic phenotype, induced strong negative selection from above, and got immediately flushed out of a population by a kamikaze-like organism?

Leo Buss (1987), in a fascinating book on the role of hierarchical selection in the phylogenetic history of development, makes a compelling case for the vital importance of *both* synergistic and negative selection between levels in the history of life, which he views largely as a tale of sequential addition in hierarchical levels—so that nature's current hierarchy becomes a problem for historical explanation, and not an inherent system fully present throughout time. Buss argues that synergism must fuel the first steps in adding a new level atop a preexisting hierarchy (for initial negativity with the previous highest level would preclude the origin of a new level). But once the new level achieves a tentative foothold, it stabilizes best by imposing negative selection against differential proliferation of individuals at the level just below—for these individuals have now become part of the new level's integrity, and selection at the new level will tend to check any dysfunctional imbalance caused by differential proliferation below.

Each hierarchical level differs from all others in substantial and interesting ways, both in the style and frequency of patterns in change and causal modes

Nature's hierarchy, for all the commonality of unifying principles (selection, for example, acting at each level), does not display fractal structure with self-similarity across level.

As the theory of hierarchical selection develops, I predict that no subject within its aegis will prove more fascinating than the varying strengths and modalities among levels. Just as the study of allometry has discovered characteristic and predictable scale-dependent differences in the structure and function of organisms at strongly contrasting sizes—a prominent subject in biology ever since Galileo formulated the principle of surfaces and volumes in the late 1630s, and beautifully codified in D'Arcy Thompson's masterpiece of prose and concept, *On Growth and Form*—so too does individuality as a tiny gene imply substantially different properties for a unit of selection than 'personhood' as a large species or an even larger clade. Allometric effects across hierarchical levels should greatly exceed the familiar (and extensive) differences between tiny and gigantic organisms for two reasons. First, the size ranges among levels are far greater still. Second, organisms share many common properties just by occupying a common level of evolutionary individuality despite an immense range of size; but the levels themselves differ strongly in basic modes of individuality, and therefore develop far greater disparity.

This problem of allometric effects strongly impacts our struggle to formulate a hierarchical theory of selection. Human beings are both evolutionary individuals and organisms—yet the equally well defined evolutionary individuals of other hierarchical levels are not organisms. Unfortunately, organisms represent a special and rather 'funny' kind of evolutionary individual, imbued with distinctive properties absent from, or much weaker in, other individuals (at other levels) that are equally potent as evolutionary agents. But if we mistakenly view our own unique properties as indispensable traits for *any* kind of evolutionary individual— the classic error of parochialism—then we will devalue, or even fail to identify, other individuals defined by different properties, and resident at other levels.

Consider, for example, the difficulty we experience—despite our preferences for reductionism in science—in trying to visualize the world of genes, where nucleotides function as active and substitutable evolutionary parts, and where chromosomes build a first encasement, followed by nuclei and cells, with our body becoming a collectivity, whose death will also destroy any gene still resident within. Think of the initial resistance that most of us felt for Kimura's neutralist theory—largely because we falsely 'downloaded' our adaptationist views about organisms into this different domain, where high frequencies of neutral substitution become so reasonable once we grasp the disparate nature of life at such smallness. And if we fare so badly in grasping this world of the tiny and immediate, supposedly so valued by our reductionist preferences, how can we comprehend an opposite extension into the longer life, the larger size, and the markedly different character of species-individuals—a world that we have usually viewed exclusively as a collectivity, an aggregation of our bodies, and not as a different kind of individual in any sense at all?

I like to play a game of 'science fiction' by imagining myself as an individual of another scale (not just as a human being reduced or enlarged for a visit to such a *terra incognita*). But I do not know how far I can succeed. We are organisms and tend to see the world of selection and adaptation as expressed in the good design of wings, legs, and brains. But randomness may predominate in the world of genes—and we might interpret the universe very differently if our primary vantage point resided at this lower level. We might then see a world of largely independent items, drifting in and out by the luck of the draw—but with little islands dotted about here and there, where selection reins in tempo and embryology ties things together. What, then, is the still different order of a world much larger and longer than ourselves? If we missed the strange domain of genic neutrality because we are too big, then what are we not seeing because we are too small? Our mortal bodies are like genes in some larger world of change among species in the vastness of geological time. What are we missing in trying to read this world by the inappropriate scale of our small bodies and minuscule lifetimes?

References

Buss, L. W. 1987. *The Evolution of Individuality*. Princeton University Press.
Colwell, R. K. 1981. Group selection is implicated in the evolution of female-biased sex ratios. *Nature* **290**:401–404.
Dawkins, R. 1982. *The Extended Phenotype*. Oxford University Press.

Dobzhansky, Th. 1951. *Genetics and the Origin of Species* (Third Edition). New York: Columbia University Press.

Doolittle, W. F. and Sapienza, C. 1980. Selfish genes, the phenotype paradigm and genome evolution. *Nature* **284**: 601–603.

Eldredge, N. and Gould, S. J. 1972. Punctuated equilibria: An alternative to phyletic gradualism. In: T. J. M. Schopf (ed.), *Models in Paleobiology*, pp. 82–115. Freeman, Cooper & Co., San Francisco CA.

Fisher, R. A. 1930. *The Genetical Theory of Natural Selection.* Oxford University Press.

Fisher, R. A. 1958. *The Genetical Theory of Natural Selection (Second Edition).* New York: Dover Publications.

Goodfriend, G. A. and Gould, S. J. 1996. Paleontology and chronology of two evolutionary transitions by hybridization in the Bahamian land snail *Cerion*. *Science* **274**: 1894–1897.

Gould, S. J. and Eldredge, N. 1977. Punctuated equilibria: The tempo and mode of evolution reconsidered. *Paleobiology* **3**: 115–151.

Gould, S. J. and Eldredge, N. 1993. Punctuated equilibrium comes of age. *Nature* **366**: 223–227.

Gould, S. J. and Lewontin, R. C. 1979. The spandrels of San Marco and the Panglossian paradigm: A critique of the adaptationist programme. *Proceedings of the Royal Society of London Series B* **205**: 581–598.

Hamilton, W. D. 1971. Selection of selfish and altruistic behavior in some extreme models. In: J. S. Eisenberg and W. S. Dillon (eds), *Man and Beast: Comparative Social Behavior.* Washington, DC: Smithsonian Institution Press.

Maynard Smith, J. 1987. How to model evolution. In: J. Dupré (ed.), *The Latest on the Best: Essays on Evolution and Optimality.* Cambridge, MA: MIT Press.

Mayr, E. 1963. *Animal Species and Evolution.* Cambridge, MA: Harvard University Press.

Orgel, L. E. and Crick, F. 1980. Selfish DNA: the ultimate parasite. *Nature* **284**: 604–607.

Stanley, S. M. 1979. *Macroevolution: Pattern and Process.* San Francisco: W. H. Freeman.

Thompson, D. W. 1942. *On Growth and Form.* London: MacMillan.

Williams, G. C. 1966. *Adaptation and Natural Selection.* Princeton University Press.

Williams, G. C. 1992. *Natural Selection: Domains, Levels, and Challenges.* Oxford University Press.

Wilson, D. S. 1983. The group selection controversy: history and current status. *Annual Review of Ecology and Systematics* **14**: 159–187.

Wilson, D. S. and Sober, E. 1994. Reintroducing group selection to the human behavioral sciences. *Behavioral and Brain Sciences* **17**: 585–654.

12

Geographic range size and speciation

Kevin J. Gaston and Steven L. Chown

12.1 Introduction

No species is distributed ubiquitously across the Earth. Indeed, the sizes of global geographic ranges vary by at least 12 orders of magnitude (Brown *et al.* 1996). At one extreme lie those species constrained, by ecology or by history, to occupy small, isolated islands of habitat or very scarce sets of environmental conditions. At the other extreme lie those species which are distributed across multiple biogeographic regions. This interspecific variance in geographic range sizes has stimulated a host of investigations of the constraints on the occurrences of more narrowly distributed species, and of the mechanisms which enable the more widely distributed to become so (e.g. for references and reviews see Woodward 1987; Hengeveld 1990). On the contrary, little interest has been directed toward broad interspecific patterns in the determinants and consequences of this variation (for reviews see Kunin and Gaston 1993, 1997; Gaston 1994a; Brown 1995).

Perhaps the most basic summary of, and pattern in, the variation of geographic range sizes exhibited by a taxonomic assemblage is the species-range size distribution and its associated statistics. The frequency distribution of the geographic range sizes of species in a taxonomic assemblage tends to be unimodal with a strong right-skew. That is, most species have relatively small range sizes while a few have relatively large ones (Fig. 12.1), an observation that has been made both for paleontological (e.g. Jablonski 1986b, 1987; Jablonski and Valentine 1990; Roy 1994) and extant assemblages (e.g. Willis 1922; Freitag 1969; Anderson 1977, 1984a,b, 1985; Rapoport 1982; McAllister *et al.* 1986; Schoener 1987; Russell and Lindberg 1988a; Pomeroy and Ssekabiira 1990; Pagel *et al.* 1991; Gaston 1994a, 1996a; Brown 1995; Roy *et al.* 1995; Blackburn and Gaston 1996; Brown *et al.* 1996; Hughes *et al.* 1996). Indeed, in virtually all published species-range size distributions in which range size is untransformed, the most left-hand range size class is also the modal class. These distributions tend toward an approximately normal distribution when geographic range sizes are subject to a logarithmic transformation (e.g. Anderson 1984a,b; McAllister *et al.* 1986; Pagel *et al.* 1991; Gaston 1994a, 1996a; Blackburn and Gaston 1996; Brown *et al.* 1996; Hughes *et al.* 1996). However, they also appear consistently to acquire a mild to moderate left skew under such a transformation (Gaston 1998).

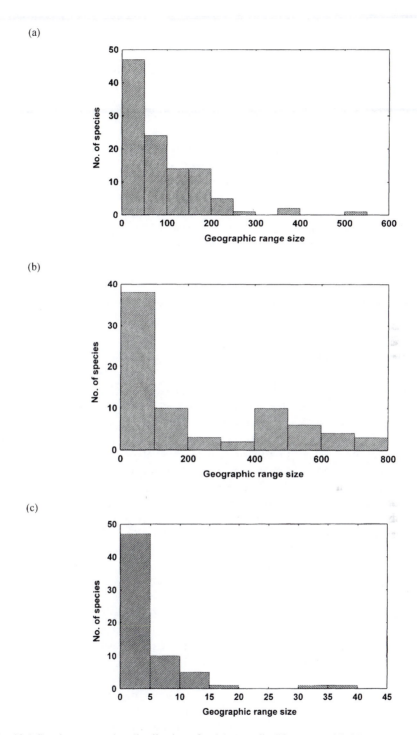

Fig. 12.1 Species–range size distributions for (a) procellariiforms worldwide (K. J. G. and S. L. C., unpubl. analyses), (b) cetaceans worldwide (from estimated distributions in Watson 1981) and (c) trogid beetles in Africa (S. L. C. and C. H. Scholtz, unpubl. data). In all cases ranges sizes are crudely measured as the number of 611 000 km² cells occupied.

Whatever its shape, the species-range size distribution must ultimately be a product of three processes: speciation, extinction, and the temporal dynamics of the range sizes of species between speciation and extinction (hereafter termed 'range transformation'). This begs a number of important, and in the main remarkably little explored, questions. These include the form of the relationship between likelihood of speciation and geographic range size, the likelihood of different patterns of subdivision of range sizes at speciation, whether the lifetime dynamics of the range sizes of species conform to one or more simple models, and the inter-relations of the three processes (Gaston and Blackburn 1997a,b; Gaston 1998). In this chapter we explore the first, and surprisingly tortuous, of these issues. That is, how the probability of speciation might change with differences in the geographic range sizes of species. Because range transformations and the implications of range size for species persistence influence the likelihood of speciation, we also touch on these issues, but only insofar as they interact with the latter process.

12.2 Speciation: modes and models

Speciation generates new species and hence additional geographic ranges, adding to a species-range size distribution. Its influence on such a distribution is, however, more complex, and depends fundamentally on the geographic mode of speciation (allopatric, sympatric, etc., see Brooks and McLennan 1991), the form of the speciation event (or speciation model—anagenetic, cladogenetic, etc., see Wagner and Erwin 1995) and to some extent, the interaction of the two (e.g. Archibald 1993; Wagner and Erwin 1995). To render the issue more tractable, we assume that speciation is predominantly allopatric, while acknowledging that other forms are probably more frequent or important than has often been claimed in the past (Tauber and Tauber 1989; Ripley and Beehler 1990; Schliewen *et al.* 1994; Lazarus *et al.* 1995; Rosenzweig 1995). We also ignore the multitude of models that result from various combinations of mode, form and speciation rate (gradual vs. punctuated, see Vrba 1980; Erwin and Anstey 1995); a total of eight in Archibald (1993). Thus we distinguish between three major speciation models, of which two may be modified by the mode of speciation (allopatric mode I or vicariant speciation, and allopatric mode II or peripheral isolate speciation) (Fig. 12.2). These models are: (i) anagenesis—an entire ancestral species (i.e. all populations or one surviving population) evolves into a descendant species; (ii) cladogenesis without ancestral persistence—an ancestral species gives rise to two daughter species by vicariance but does not itself subsequently persist (bifurcation of Wagner and Erwin 1995); and (iii) cladogenesis with ancestral persistence—an ancestral species gives rise to one or more daughter species by vicariance or peripheral isolation and does persist beyond these events (cladogenesis of Wagner and Erwin 1995). Again, we ignore mixed models, while accepting that in principle these are possible.

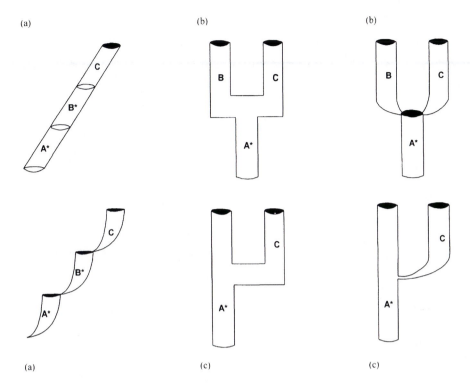

Fig. 12.2 The three major speciation models recognised in this treatment (after Archibald 1993). (a) anagenesis, (b) cladogenesis without ancestral persistence, and (c) cladogenesis with ancestral persistence. Note that in the context of geographic range sizes, the two cladogenetic models are profoundly influenced by the mode of speciation (allopatric mode I or vicariant speciation, and allopatric mode II or peripheral isolate speciation). This is illustrated by the width of the branches. Likewise, anagenesis may involve all populations or a single surviving one (see text for further details).

12.3 Taking a time slice: instantaneous assessments

Anagenesis

Consideration of the relationship between the instantaneous rate of speciation and geographic range size under an anagenetic model begs a number of contentious issues, relevant both to cladogenetic and anagenetic models, but which are more familiar in the context of the latter. Essentially these are concerned with the relationships between selection, evolution (in the strict sense of Endler 1986), and speciation, and the influence that population size and structure has on these processes. These topics are the subject of an impressive literature (for modern treatments see Stanley 1979; Vrba 1985; Endler 1986; Levinton 1988; Eldredge 1989; King 1993; Holt 1997), and certainly cannot be treated here to any large extent. None the less it is worth noting that the way in which speciation actually takes place and the way we view the process (and the factors affecting it), are likely to vary, respectively, with the real relationships between these processes, and our theoretical perceptions of them.

In addition, these considerations point to a recurrent theme throughout this chapter, which is the extent to which theory and empirical findings for populations of different sizes and for geographic ranges of different sizes are interchangeable. Much of the literature discussing the interaction between range size and speciation, and related issues, implicitly assumes that larger populations can be equated with larger ranges and smaller populations with smaller ranges. At a global scale this is probably not unreasonable, insofar as there is a broad positive correlation between the total population sizes and the total geographic range sizes of species in a taxonomic assemblage (Mace 1994; Gaston and Blackburn 1996). Indeed, this reflects a more general observation that, almost universally, taxonomic assemblages exhibit positive interspecific relationships between mean local densities (at occupied sites) and regional occurrences or geographic range sizes (Hanski 1982; Brown 1984; Gaston and Lawton 1990; Hanski *et al.* 1993; Gaston 1994a, 1996b; Gotelli and Graves 1996; Gaston *et al.* 1997). That is, not only do population sizes increase with the extent of the distribution of a species, but abundances increase at a faster rate than would be predicted on the basis of increasing distribution alone. This makes it seem highly likely that local populations of more widespread species also tend to be larger than those of more restricted species. However, these statements must be tempered by recognition that the variance about all of these relationships is typically very large. For example, those between local density and regional occurrence or geographic range size tend to have explained variances in the order of about 20–30% (Gaston 1996b).

Largely ignoring all of these issues, it at first seems probable that under an anagenetic model the instantaneous likelihood of speciation will decline with range size because small range sizes (and populations) are thought to be more vulnerable to the spread of marked evolutionary changes (Levinton 1988; King 1993; but see Otto and Whitlock 1997 for a more complex model incorporating effects of population size changes). However, recent work suggesting that very small populations may be unlikely to undergo change (e.g. Rice and Hostert 1993; Moya *et al.* 1995; Walsh 1995; Holt 1997; see also Vrba 1987), means that speciation may well be a peaked, rather than a declining, function of geographic range size under an anagenetic model.

Cladogenesis with and without ancestral persistence

Although the two cladogenetic models are likely to differ markedly with regard to the asymmetry of range division (see below), they should behave similarly with regard to the relationship between instantaneous rate of speciation and geographic range size, and will be considered together.

The idea that species with larger geographic range sizes have a greater probability of speciation dates at least to Darwin (1859) and continues to attract support (e.g. Terborgh 1973; Marzluff and Dial 1991; Budd and Coates 1992; Roy

1994; Wagner and Erwin 1995; Tokeshi 1996; Maurer and Nott in press), apparently on the premise that larger geographic ranges present more opportunities for subdivision (Endler 1977, p. 175 makes a similar argument for parapatric speciation). Rosenzweig (1975, 1978, 1995) argues that on a purely probabilistic basis species with larger geographic range sizes are more likely to undergo speciation, because the likelihood of their ranges being bisected by a barrier is greater than for a small range size. Differentiating, as Rosenzweig does, between two kinds of barriers, 'knives' (which have beginnings and ends) and 'moats' (which surround their isolates), then strictly this assertion is only true of moats. Very large geographic ranges will tend to engulf knives, such that they do not engender speciation, and the probability of division will have a peak at intermediate range sizes. Rosenzweig (1995) argues that this is unlikely to occur because there are no, or virtually no, species with geographic ranges so large that reducing them would make them an easier target for barriers. However, this view depends critically on the frequency distribution of barrier sizes. If most barriers are small to intermediate in size, relative to the range sizes of widespread species, then intermediate-sized ranges may indeed have a higher probability of speciation. Such an effect would be enhanced because barriers seem far more likely to take the form of knives than of moats.

Data with which to ascertain the frequency distribution of the sizes of barriers are difficult to assemble. None the less, the distances between spatial features of landscapes at a variety of scales seem commonly to be distributed according to a right-skewed function. Thus the full peaked function for the relationship between likelihood of speciation and geographic range size may actually be the more appropriate.

In support of the view that species with small to intermediate range sizes are more likely to speciate than are those with larger, it has been argued that widespread and abundant taxa may often possess well-developed dispersal abilities (perhaps associated with them becoming widespread) and should as a consequence have a strong proclivity to maintain gene flow among populations, which will tend to inhibit speciation (e.g. Mayr 1963, 1988; Stanley 1979; Flessa and Thomas 1985; Lieberman et al. 1993; García-Ramos and Kirkpatrick 1997). Narrowly distributed and locally rare taxa with poor dispersal abilities (and patchy populations which may tend to form isolates), will tend to have higher speciation rates. Although the extent to which widely distributed species do indeed tend to have greater dispersal abilities remains debatable (e.g. Levinton 1988; Palumbi 1994; Gaston and Kunin 1997), recent models have demonstrated that both enhanced gene flow and weak environmental gradients are liable to promote genetic cohesion (Rice and Hostert 1993; Kirkpatrick and Barton 1997).

In a similar vein, Chown (1997) proposes that rare, but not the rarest, species have the highest probability of speciation. Based on Stanley's (1986) 'fission effect' model, Chown envisages that speciation is a peaked function of geographic range size which rises rapidly at small range sizes and then progressively subsides towards large range sizes. This is a substantial modification of the original fission effect model, which was intended to capture how the relative rates of speciation and extinction varied with the *mean* population size across all the species in an assemblage. However, equating geographic range size with population size, the underlying relationship between speciation rate and range size is not dissimilar to the full relationship modelled by Rosenzweig (1978, 1995) when the truncation effects he postulates (i.e. the absence of ranges so large that they are a reduced target for barriers) are ignored.

In sum, it seems reasonable to argue, on present evidence, that for both cladogenetic models the instantaneous rate of speciation is a peaked function of geographic range size, or at least that the greatest likelihood of speciation is probably associated with small to intermediate range sizes rather than large ones.

12.4 Dynamic assessments: variable persistence, range transformations and speciation

A thought-provoking complication to assertions that geographic ranges of small to intermediate sizes are more likely to undergo speciation arises from the relationship between geographic range size and persistence. Although most are relatively short-lived, there is substantial variance in the duration of evolutionary or geological time for which species persist (Raup 1994). Moreover, it is generally accepted that species with larger range sizes are more resistant to global extinction, and thus persist for longer, than do those with larger range sizes (an observation which dates at least from Lamarck; see McKinney 1997). If species with larger range sizes are likely to persist for longer, this may enhance their probability of speciation, although there have been suggestions that the likelihood of speciation is low when taxa are very old (Stanley 1979; Eldredge 1989). Thus, even if species with smaller range sizes have a greater likelihood of speciation per unit time, species with larger ranges could potentially still be as, or more, likely to leave descendants. In this context, the form of the relationship between range size and persistence is crucial, as is the likelihood and form of the transformation of the geographic range sizes of species through time.

There appears to be abundant empirical evidence for a positive relationship between time to extinction and range size for various paleontological species assemblages (Jackson 1974; Hansen 1978, 1980; Stanley 1979; Koch 1980; Flessa and Thomas 1985; Jablonski 1986a,b; Buzas and Culver 1991; Jablonski and Raup 1995; Erwin 1996; Flessa and Jablonski 1996; such relationships have also been argued to underpin the observed greater persistence of ecological generalists compared with specialists—e.g. Kammer *et al.* 1997). Furthermore, the correlation appears very general, although it is not always especially strong and may break down during periods of mass extinction (Jablonski 1986b; Norris 1991). A component of this pattern is doubtless an artefact resulting from the fact that more widely distributed species tend also to be locally more abundant (the pattern has been documented for paleontological assemblages as well as the contemporary ones for which references were provided earlier; Buzas *et al.* 1982, McKinney 1997; but see Koch 1980) and thus are more likely to be recorded in the fossil record (Russell and Lindberg 1988a,b); persistence will, of course, always tend to be underestimated to some extent, regardless of abundance. However, it is generally held that this effect is not sufficient to explain the observed pattern in its entirety (Jablonski 1988; Smith 1994), although it remains a moot point how much of the variation in observed relationships between persistence and range size can be explained as a product of sampling (a number of techniques have been developed to improve the reliability of estimated stratigraphic durations; e.g. Benton 1994; Marshall 1991, 1994).

Potentially of far more significance for this purported relationship between range size and persistence is the taxonomic level at which investigations thereof are undertaken, and the way in which geographic range sizes are assessed. Even if a positive relationship is demonstrated between the range size of a higher taxon and its likelihood of persistence (e.g. Westrop 1991; Jablonski and Raup 1995; Flessa and Jablonski 1996), the relationship may well be of limited relevance at the species level. This could be a consequence of different extinction patterns for species and higher taxa during extinction events of different magnitudes (Jablonski 1986b; Westrop 1989), or because range size variance is partitioned to a large extent at the species level (Gaston 1998). In the latter case, summing species ranges to provide those for generic (or other higher) taxa may provide insufficient insight into relationships at the species level (but see counter-arguments in Flessa and Thomas 1985). Essentially, these complications mean that with regard to the relationship between geographic range size and the likelihood of speciation, most confidence should be placed in those investigations undertaken at the species level.

Such species-level investigations as have been performed have, to a large extent, provided support for a relationship between geographic range size and persistence (Jackson 1974; Hansen 1978, 1980; Koch 1980; Jablonski et al. 1985; Jablonski 1986a,b, 1987; but see also Stanley 1986; Stanley et al. 1988). However, where studies are explicit about the way range size is calculated, it appears that in some cases this is done by summing the spatial extent of localities (often number of provinces) over the total geological duration of the species, or geological period under investigation (e.g. Westrop and Lundvigsen 1987; Stanley et al. 1988; see also Koch 1987; Koch and Morgan 1988; Budd and Coates 1992; Roy 1994). Such an 'allochronic extent of occurrence' estimate (see Gaston 1991, 1994b for discussion of 'extent of occurrence') would, of course, provide a reliable indication of the geographic range size of a species if it could be demonstrated that subsequent to speciation, species rapidly attain the range size that will characterize them for the bulk of their geological duration (ranges decline only as species go extinct), or that species range sizes remains static throughout a species duration. The latter model is clearly unrealistic (Gaston 1998), but based on Late Cretaceous molluscs, Jablonski (1987) has suggested that species show range size dynamics conforming to the former one. However, in many instances this may not be the case. The range transformation scenario outlined by Jablonski (1987) is only one of at least seven such scenarios (Fig. 12.3).

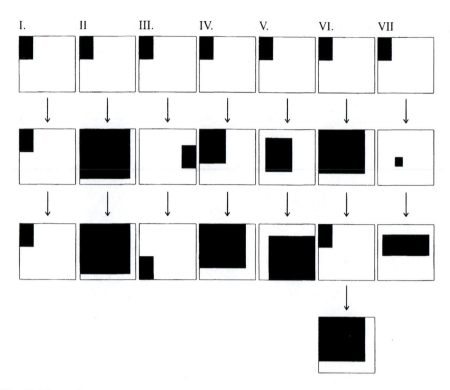

Fig. 12.3 Potential ways in which geographic range size and position can vary through time. I, stasis; II, stasis post-radiation; III, stasis with positional shift; IV, age and area; V, age and area with positional shift; VI, cyclic; VII, idiosyncratic. In each case the shaded box represents the geographic range and the arrows the transitions between different time periods (see text for details).

(I) Stasis—an unrealistic model.

(II) Stasis post-radiation—the Jablonski model discussed above (see also Rosen 1988). This model requires that species spread very rapidly post-speciation. An explicit test of this model is difficult, but the rates of spread of invasive species suggests that dispersal rates can be very high (e.g. Wing 1943; Hengeveld 1989; Lynch 1989; Grosholz 1996; Veit and Lewis 1996; Williamson 1996).

(III) Stasis with positional shift—this appears to be characteristic of at least some taxa during the Quaternary (Woods and Davis 1989; Dennis 1993; Burton 1995; Coope 1995). It may be far from typical of all species, and it is frequently not clear to what extent such dynamism is reflected in changes in range size rather than simply range shifts (Hengeveld 1990).

(IV) Age and area—Willis (1922) argued that the geographic ranges of species followed a trajectory of steadily increasing size the longer they persisted, presumably culminating with a rapid decline to extinction or disappearance through cladogenesis. This hypothesis underwent vigorous debate, and was widely rejected (e.g. Gleason 1924; Stebbins and Major 1965; Stebbins 1978). In its simplest form it cannot be correct, because there are ample examples of young species which are widely distributed and old species which are very restricted in their distribution. None the less, it continues to be maintained in some circles that there is a broad positive interspecific correlation between age and area (although such a pattern need not follow from an intraspecific age and area relationship), albeit not necessarily of strong predictive value (McLaughlin 1992), and some recent analyses have reported such a pattern, at least among small numbers of closely related species (Taylor and Gotelli 1994) and at the generic level (Miller 1997).

(V) Age and area with positional shift—this is essentially a combination of models (III) and (IV), and in many ways conforms to the taxon pulse concept (Erwin 1985; Liebherr and Hajek 1990), which was derived from a cyclic model of range size transformation.

(VI) Cyclic—this model is very similar to the previous one, with the exception that in most instances the range size of the ancestral species eventually fragments as the species divides up into subspecies and eventually species. Various authors have regarded the range size dynamics of species as essentially cyclic (e.g. Dillon 1966), and this is most explicitly expressed in the concept of the taxon cycle (Wilson 1961; Ricklefs and Cox 1972, 1978). Evidence has been produced both in support of (Ricklefs and Cox 1972; Glazier 1980; Rummel and Roughgarden 1985; Roughgarden and Pacala 1989) and against (Pregill and Olson 1981; Liebherr and Hajek 1990; Losos 1992) the existence of taxon cycles for particular assemblages; the observed pattern is likely to be determined by the geographic pattern of isolation. Exceptions to a taxon cycle concept of range contraction and expansion may include range size changes associated with contraction of species ranges to refugia and their subsequent expansion (see e.g. Kavanaugh 1979a; Lynch 1988), and the processes that have given rise to bipolar distributions (e.g. Crowson 1981; Crame 1993). However, the nature of refugia and their geographic location, and the processes leading to bipolar distributions, remain polemical (Lynch 1988; Crame 1993; Bush 1994).

(VII) Idiosyncratic—there need, of course, be no general pattern of change in the geographic range sizes of species between speciation and extinction. Different species may exhibit entirely idiosyncratic trajectories. This would seem in keeping with the continual adaptation and change that may result from responding to the demands of the Red Queen (Van Valen 1973; Ricklefs and Latham 1992; Vrba 1993), and with the changes in the distributions of some species over the past few decades (e.g. Frey 1992; Burton 1995; Parmesan 1996). If such a pattern prevails, then one might expect to see little similarity in the geographic range sizes of closely related species, unless the traits which influence range size are strongly phylogenetically conserved in which case in climatically and ecologically similar regions the distributions of close relatives might be expected to fluctuate in parallel (Ricklefs and Latham 1992). There are two lines of evidence regarding the similarity of geographic range sizes of closely related species. First, a significant positive correlation has been documented between the geographic range sizes of closely related species of Late Cretaceous molluscs, from which Jablonski (1987) inferred that range size is heritable at the species level (see also Ricklefs 1989; Ricklefs and Latham 1992). However, examination of a few contemporary data sets based on sister-species does not uphold this as a generality (Gaston 1998). These results could potentially be reconciled if the geographic range sizes of species are highly labile over their lifetimes such that at any one time closely related species do not have very similar range sizes, but that when this variation is effectively summed over periods of evolutionary or geological time (as inevitably occurs in the fossil record) then strong phylogenetic patterns of interspecific variation in range sizes become apparent. Equally, the results for Late Cretaceous molluscs simply may not generalise. Second, it also appears that most of the variation in the geographic range sizes of species is explained at low taxonomic levels (Arita 1993; Brown 1995; Gaston and Blackburn 1997b; Gaston 1998; for similar results for range sizes at meso-scales see Hodgson 1993; Peat and Fitter 1994; Kelly and Woodward 1996).

Under models I and II, summing geographic range sizes through time is unlikely to make a significant difference to the relationship that is finally demonstrated between range size and likelihood of persistence. However, under all of the other models, the change in the position of the range of a species and/or the change of its size through time may lead to overestimates of the range size of a species in the geological record, *if* the range sizes of species are summed over time for all (or an ecologically significant portion of) the records of the particular species. Given that the range sizes of species can change over timescales that are poorly resolved within the fossil record (see above and also Erwin and Anstey 1995; Johnson *et al.* 1995; Jablonski and Sepkoski 1996; Roy *et al.* 1996; Jablonski 1997), all but models I and II cast some doubt on the relationships between range size and persistence so far described for fossil taxa (see also McKinney and Allmon 1995).

Paleontologists have sought to overcome this problem in three ways. First, determination of geographic range sizes has been restricted to a particular geological period, and persistence calculated by looking at the occurrence of species before and after this time (e.g. Koch 1980). Second, determination of geographic ranges has been restricted to that geological period in which any given species had its maximum geographic spread, and persistence calculated by looking at the occurrence of species before and after this time (e.g. Hansen 1980). Third, the geographic range sizes of Recent taxa have been determined (based on museum collections or published occurrence data) and the persistence of these taxa back through the fossil record (e.g. Jackson 1974; Buzas and Culver 1989; Flessa and Jablonski 1996). Although these methods largely resolve the dilemma of calculation of range sizes, range transformations throughout the duration of the species concerned are still likely to bedevil the conclusions of such studies. In fact, studies based on the latter method may at best do little more than confirm the weak, interspecific age and area hypothesis. Clearly, considerably greater attention needs to be paid to the range transformations of species. Not only do they have implications for the relationship between range size and speciation, but they may confound any cladistic and/or phylogenetic biogeographic reconstructions that are based on the modern distributions of species (e.g. see Cracraft 1982, 1986; Cracraft and Prum 1986; Myers and Giller 1987; Brooks and McLennan 1991), as well as scenarios developed for species level evolution (discussed by Sites and Moritz 1987; King 1993); some recent cladistic techniques aim to address problems associated with dispersal (Alroy 1995; Lieberman and Eldredge 1996; Ronquist 1997).

If the problems raised here are ignored, attempts are made to correct for them (e.g. Koch and Morgan 1988) or they are treated as unimportant, and it is assumed that species with the largest range sizes persist for longest (see Jablonski and Valentine 1990 for some evidence supporting the assumption), such a pattern may exist for three possible reasons, although often only the first is explicitly stated: (i) species with larger geographic range sizes may be less likely to walk randomly to extinction, and thus they persist for longer (see Rosenzweig 1995; Gaston and Blackburn 1996; Mace and Kershaw 1997); (ii) species with traits which make them less prone to local extinction, and hence able to persist for longer, may also be enabled to maintain larger range sizes because of this extinction resistance; and (iii) species with larger range sizes may have them because they have persisted for longer (there is an age and area relationship; see above). Of course, these are not mutually exclusive. A large range size may intrinsically buffer a species against extinction, but this range size may be associated with traits which enable the species also to attain a high local abundance (because there is a positive interspecific abundance-range size relationship; Gaston et al. 1997) which will buffer the species from local extinction, and the species may progressively expand its range size through time.

Such considerations highlight another issue, namely the relative contributions of differences in geographic range size and population size to extinction risk. While the bulk of paleontological studies have emphasised the importance of range size to persistence, perhaps because this is more readily measured than population size, Stanley (1986; see also Stanley et al. 1988) argues that '. . . population size has been of primary importance in determining the vulnerability of species to extinction, whereas geographic range has been of second-order importance'. This assertion is founded upon analyses of Lyellian percentages (i.e. the proportion of species in a fossil fauna which survive to the present) for Neogene marine bivalve molluscs. In virtually all cases the tests are indirect, based on relationships between measures of survivorship and variables which are claimed, though not demonstrated, to be correlated with abundance. Indeed, others have failed to find relationships between abundance and persistence (e.g. Koch 1980; but see Johnson et al. 1995). An explicit analysis of the relative contributions of population size and range size to the risks thought to be faced by extant species has shown that population size is more important, but this difference may not have been independent of the basis on which risks were themselves evaluated (Gaston and Blackburn 1996).

Irrespective of the reasons for the, albeit poorly tested, positive relationship between geographic range size and persistence, its effect on the relationship between range size and speciation is liable to be quite straightforward. In the case of the anagenetic model in which the instantaneous likelihood of speciation declines with range size, it will tend to reduce the differences in lifetime likelihood of speciation between species with different sized ranges, possibly to the extent that species of all range sizes are equally likely to leave a descendant. In the case of both cladogenetic models, and the second anagenetic one, in which the relationship between the instantaneous probability of speciation and range size is a peaked function, the greater persistence of species with larger range sizes will tend to reduce or obviate the decline in speciation rate from intermediate to large range sizes. This effect is expected to be stronger in the case of cladogenesis with ancestor persistence, because there is more variance in the number of ancestors that it is possible to leave (without ancestral persistence there are only two daughters). These effects might be expected to be further accentuated because small ancestral range sizes are more likely to give rise to potential descendant species with very small range sizes, which may rapidly become extinct and not attain full species status (or if they do attain such status they may persist for only a short period and go unrecorded in the fossil record).

We are aware of only one explicit attempt empirically to test the relationship between geographic range size and likelihood of speciation. Wagner and Erwin (1995) find that in analyses of two Neogene clades of Foraminifera (the Globigerinidae and Globorotaliidae) and an Ordovician family of gastropods (the Lophospiridae), species with larger geographic ranges are likely to leave more descendants in two cases but not in the third, and in all three cases species which have persisted for longer are likely to leave more descendants (geographic range sizes appear to be calculated as allochronic extents of occurrence). Importantly, the patterns are not consistent between cladogenetic and anagenetic modes of speciation. Thus, for the globigerinids, although there was an overall relationship between leaving descendants and possessing wider geographic ranges, this was not significant for anagenetic speciation events alone. In the gastropod case (for which data are provided) the partial correlation between number of descendants (dependent variable) and range size, controlling for differences in longevity can be calculated. It is not statistically significant (Gaston 1998). More such studies are required.

12.5 Asymmetry of range division

Thus far we have ignored the possible consequences of one speciation event on the subsequent likelihood of speciation by the ancestral species. Plainly this differs for the anagenetic and cladogenetic models. Although range division as a consequence of speciation is obviously not of relevance in an anagenetic model, the form of the speciation event itself is. As Archibald (1993) illustrated, anagenesis could be the consequence of an entire species (all populations) changing through time (anagenesis model I), or as the consequence of the extinction of most populations and the gradual change of the remaining few (anagenesis model II) (Fig. 12.2). If the instantaneous rate of speciation is either a peaked or decreasing function of range size (see above), then as the total population size (or number of linked metapopulations) declines, as a consequence of an extinction event, speciation probability will increase; the idea is elaborated by Vrba (1993) as the turnover pulse hypothesis, and also discussed by others (e.g. Kauffman 1993; Chown 1997). Hence, anagenesis model II will lead to a new species with a small range size, which, in turn has a high probability of speciation and extinction. If anagenesis follows model I, then range size, and hence the instantaneous like-lihood of speciation, are liable to remain unchanged. In our view, the scenario described for anagenesis model II seems most likely because if anagenesis is given any support at all, it is this model that is preferred (e.g. Allmon 1994).

In the cladogenetic models, if a speciation event has a marked effect on the range size of an ancestral species, and the instantaneous likelihood of speciation is a function of range size, then that event will change this likelihood. If it is ancestral species with relatively small geographic range sizes which are most likely to speciate, then the importance of the asymmetry of range division for the probability of subsequent speciation is markedly lessened. The products of any range division can only be two small ranges (be they two daughters, or a daughter and its ancestor if the latter persists). If species with large geographic range sizes are more likely to speciate, depending on the asymmetry of division, the outcome may span one large and one small range size through to two ranges each half the size of that of the ancestor at speciation. These have very different consequences for the likelihood of subsequent speciation. If speciation patterns are thought of in terms of simple random events then the most likely immediate products of speciation are two species of which one is more widely distributed than is the other. A perfect 50–50 split is highly improbable for this reason, and because subsequent genotypic change (speciation) may be rather slow in two large, widespread populations (García-Ramos and Kirkpatrick 1997).

The question of the degree of asymmetry in the range sizes of sister-groups, immediately post-speciation, is closely related to the issue of whether allopatric speciation is best typified by a peripheral isolation model or by a vicariance model (see Brooks and McLennan 1991; King 1993). In the former, peripheral isolates form by waif dispersal (establishment of a new population through long-distance movement across a barrier), microvicariance (physical division of a previously continuous distribution) or range retraction (causing peripheral populations to become isolated; Frey 1993). Here, the relatively widespread ancestral species is likely to change little while the peripheral isolate diverges (Glazier 1987). In the vicariance model, a subdivision of the range of an ancestral species occurs, such that there is cessation of contact between the two subpopulations, giving rise to new species. Here, both daughters of the ancestral range are likely to diverge, and the ancestral species will cease to exist. The two models are plainly very closely related, and may in some sense be seen as constituting points on a continuum of speciation processes. However, vicariant speciation may potentially result in any degree of asymmetry in the initial range sizes of daughter species, and is often portrayed as generating very similar sized ranges. In contrast, peripheral isolation results, immediately post-speciation, in a highly asymmetrical split.

Both peripheral isolation (e.g. Kavanaugh 1979b; Ripley and Beehler 1990; Levin 1993; Chesser and Zink 1994) and vicariance models (e.g. Cracraft 1982, 1986; Cracraft and Prum 1988; Lynch 1989) have significant support. The relative frequency of the two modes of speciation remains a point of some contention (e.g. see Bush 1975; Barton and Charlesworth 1984; Mayr 1988; Lynch 1989; Ripley and Beehler 1990; Brooks and McLennan 1993; Frey 1993; King 1993; Chesser and Zink 1994; Taylor and Gotelli 1994; Wagner and Erwin 1995). Resolution of the issue rests, however, in major part on the extent of post-speciational change in geographic range sizes (see above) and hence the extent to which the present-day distributions of species can be used to reconstruct past patterns of speciation (Sites and Moritz 1987; Lynch 1989; Brooks and McLennan 1991; King 1993).

The range division symmetry issues explored above raise a familiar, but in this context, rather important matter. Modern cladists, and many other biologists, give short shrift to anagenesis and cladogenesis with ancestral persistence as models of speciation, preferring cladogenesis without ancestral persistence (for discussion see above and Eldredge and Cracraft 1980; Nelson and Platnick 1981; Nelson and Rosen, 1981; Wiley 1981; Harvey and Pagel 1991; Cracraft 1992; Smith 1994; Mooers and Heard 1997). On the contrary, many paleontologists are comfortable with models of speciation that include either cladogenesis with ancestral persistence (Gould and Eldredge 1977; Vrba 1980, 1985; Jablonski and Lutz 1983), or anagenesis (Archibald 1993; Allmon 1994; Wagner and Erwin 1995), and their evidence certainly seems to provide some support for this (e.g. Geary 1990, 1995; Lazarus *et al.* 1995; Wagner and Erwin 1995). Based on our analyses we are of the opinion that the anagenesis identified by paleobiol-ogists may simply be cladogenesis without ancestral persistence, but where one of the nascent sister species has declined to extinction immediately, the likely fate of most small populations (Glazier 1987; Gorodkov 1992). Such 'aborted species' have been discussed in a different, but related context, by Mayr (1963), Stanley (1979), Vrba (1985, 1993) and King (1993). None the less, judging by the fossil evidence, it does appear that at least in some taxa, ancestral persistence is real. Hence it should be treated as such, rather than as a cladistic conundrum.

12.6 Conclusions

Charles Darwin (1856–1858) described the study of geographic ranges as 'a grand game of chess with the world for a board'. The game is not a simple one. The pieces, the species, occupy different numbers of squares on the board at different times, they appear and then disappear, and many pieces may occupy the same square at the same time. Moreover, we can only gain glimpses of the past moves.

None the less, this is also no idle game. The pattern of moves has resulted in the patterns of biodiversity which we observe today.

In the context of the relationship between geographic range size and the likelihood of, and asymmetry of range division during, speciation it is clear that the rules of the game have scarcely been comprehended; much of the evidence required for choosing among the many possible moves is lacking. None the less we have shown that while instantaneous speciation rates are likely to be a peaked function of geographic range size under virtually all allopatric speciation models, the overall function may be an increasing one because of a positive relationship between range size and persistence. Furthermore, we have also provided a strong argument in favour of asymmetrical range size divisions. Thus speciation appears to be rather biased with regard to range size. This may at least partly account for the left-skew to species-range size distributions observed under logarithmic transformation, and is a potentially important dynamic in understanding why most species are narrowly distributed.

Acknowledgements

K. J. G. is a Royal Society University Research Fellow; S. L. C. is funded by the South African Department of Environmental Affairs and Tourism and the Foundation for Research Development. We are very grateful to Tim Blackburn, Bob May, Mike McKinney, Terry Robinson, John Spicer, Phil Warren, Paul Williams and Mark Williamson for helpful discussion and comments, to Bob May for his encouragement to pursue the ideas outlined herein, and to Anne Magurran and Bob May for the invitation to contribute to this volume.

References

Allmon, W. D. 1994 Taxic evolutionary paleoecology and the ecological context of macroevolutionary change. *Evol. Ecol.* **8**, 95–112.

Alroy, J. 1995 Continuous track analysis: a new phylogenetic and biogeographic method. *Syst. Biol.* **44**, 152–178.

Anderson, S. 1977 Geographic ranges of North American terrestrial mammals. *Am. Mus. Novitates* **2629**, 1–15.

Anderson, S. 1984a Geographic ranges of North American terrestrial birds. *Am. Mus. Novitates* **2785**, 1–17.

Anderson, S. 1984b Areography of North American fishes, amphibians and reptiles. *Am. Mus. Novitates* **2802**, 1–16.

Anderson, S. 1985 The theory of range-size (RS) distributions. *Am. Mus. Novitates* **2833**, 1–20.

Archibald, J. D. 1993 The importance of phylogenetic analysis for the assessment of species turnover: a case history of Paleocene mammals in North America. *Paleobiology* **19**, 1–27.

Arita, H. T. 1993 Rarity in Neotropical bats: correlations with phylogeny, diet, and body mass. *Ecol. Appl.* **3**, 506–517.

Barton, N. H. and Charlesworth, B. 1984 Genetic revolutions, founder effects, and speciation. *Annu. Rev. Ecol. Syst.* **15**, 133–164.

Benton, M. J. 1994 Palaeontological data and identifying mass extinctions. *Trends Ecol. Evol.* **9**, 181–185.

Blackburn, T. M. and Gaston, K. J. 1996 Spatial patterns in the geographic range sizes of bird species in the New World. *Phil. Trans. R. Soc. London Ser. B* **351**, 897–912.

Brooks, D. R. and McLennan, D. A. 1991 Phylogeny, ecology, and behavior: a research program in comparative biology. Chicago: University of Chicago Press.

Brooks, D. R. and McLennan, D. A. 1993 Comparative study of adaptive radiations with an example using parasitic flatworms (Platyhelminthes: Cercomeria). *Am. Nat.* **142**, 755–778.

Brown, J. H. 1984 On the relationship between abundance and distribution of species. *Am. Nat.* **124**, 255–279.

Brown, J. H. 1995 *Macroecology*. Chicago: University of Chicago Press.

Brown, J. H., Stevens, G. C. and Kaufman, D. M. 1996 The geographic range: size, shape, boundaries, and internal structure. *Annu. Rev. Ecol. Syst.* **27**, 597–623.

Budd, A. F. and Coates, A. G. 1992 Nonprogressive evolution in a clade of Cretaceous *Montastraea*-like corals. *Paleobiology* **18**, 425–446.

Burton, J. F. 1995 *Birds and climate change*. London: Christopher Helm.

Bush, G. L. 1975 Modes of animal speciation. *Annu. Rev. Ecol. Syst.* **6**, 334–364.

Bush, M. B. 1994 Amazonian speciation: a necessarily complex model. *J. Biogeog.* **21**, 5–17.

Buzas, M. A. and Culver, S. J. 1989 Biogeographic and evolutionary patterns of continental margin benthic foraminifera. *Paleobiology* **15**, 11–19.

Buzas, M. A. and Culver, S. J. 1991 Species diversity and dispersal of benthic foraminifera. *BioScience* **41**, 483–489.

Buzas, M. A., Koch, C. F., Culver, S. J. and Sohl, N. F. 1982 On the distribution of species occurrence. *Paleobiology* **8**, 143–150.

Chesser, R. T. and Zink, R. M. 1994 Modes of speciation in birds: a test of Lynch's method. *Evolution* **48**, 490–497.

Chown, S. L. 1997 Speciation and rarity: separating cause from consequence. In *The biology of rarity: causes and consequences of rare-common differences* (ed. W. E. Kunin and K. J. Gaston), pp. 91–109. London: Chapman and Hall.

Coope, G. R. 1995 Insect faunas in ice age environments: why so little extinction? In *Extinction rates* (ed. J. H. Lawton and R. M. May), pp. 55–74. Oxford: Oxford University Press.

Cracraft, J. 1982 Geographic differentiation, cladistics, and vicariance biogeography: reconstructing the tempo and mode of evolution. *Am. Zool.* **22**, 411–424.

Cracraft, J. 1986 Origin and evolution of continental biotas: speciation and historical congruence within the Australian avifauna. *Evolution* **40**, 977–996.

Cracraft, J. 1992 Explaining patterns of biological diversity: integrating causation at different spatial and temporal scales. In *Systematics, ecology, and the biodiversity crisis* (ed. N. Eldredge), pp. 59–76. New York: Columbia University Press.

Cracraft, J. and Prum, R. O. 1988 Patterns and processes of diversification: speciation and historical congruence in some neotropical birds. *Evolution* **42**, 603–620.

Crame, J. A. 1993 Bipolar molluscs and their evolutionary implications. *J. Biogeog.* **20**, 145–161.

Crowson, R. A. 1981 *The biology of the Coleoptera*. London: Academic Press.

Darwin, C. 1856–1858 Charles Darwin's natural selection: being the second part of his big species book written from 1856–1858 (ed. R. C. Stauffer). publ. 1975. Cambridge: Cambridge University Press.

Darwin, C. 1859 *On the origin of species by means of natural selection, or the preservation of favoured races in the struggle for life.* London: John Murray.

Dennis, R. L. H. 1993 *Butterflies and climate change.* Manchester: Manchester University Press.

Dillon, L. S. 1966 The life cycle of the species: an extension of current concepts. *Syst. Zool.* **15**, 112–126.

Eldredge, N. 1989 *Macroevolutionary dynamics. Species, niches and adaptive peaks.* New York, McGraw-Hill.

Eldredge, N. and Cracraft, J. 1980 *Phylogenetic patterns and the evolutionary process. Method and theory in comparative biology.* New York: Columbia University Press.

Endler, J. A. 1977 *Geographic variation, speciation, and clines.* Princeton: Princeton University Press.

Endler, J. A. 1986 *Natural selection in the wild.* Princeton: Princeton University Press.

Erwin, D. H. 1996 Understanding biotic recoveries: extinction, survival and preservation during the End-Permian mass extinction. In *Evolutionary paleobiology* (ed. D. Jablonski, D. H. Erwin and J. H. Lipps), pp. 398–418. Chicago: University of Chicago Press.

Erwin, D. H. and Anstey, R. L. 1995 Speciation in the fossil record. In *New approaches to speciation in the fossil record* (ed. D. H. Erwin and R. L. Anstey), pp. 11–38. New York: Columbia University Press.

Erwin, T. L. 1985 The taxon pulse: a general pattern of lineage radiation and extinction among carabid beetles. In *Taxonomy, phylogeny and zoogeography of beetles and ants* (ed. G. E. Ball), pp. 3437–3472. Dordrecht: Junk.

Flessa, K. W. and Jablonski, D. 1986 The geography of evolutionary turnover: a global analysis of extant bivalves. In *Evolutionary paleobiology* (ed. D. Jablonski, D. H. Erwin and J. H. Lipps), pp. 376–397. Chicago: University of Chicago Press.

Flessa, K. W. and Thomas, R. H. 1985 Modelling the biogeographic regulation of evolutionary rates. In *Phanerozoic diversity patterns. Profiles in macroevolution* (ed. J. W. Valentine), pp. 355–376. Princeton: Princeton University Press.

Freitag, R. 1969 A revision of the species of the genus *Evarthrus* LeConte (Coleoptera: Carabidae). *Quaest. Entomol.* **5**, 89–212.

Frey, J. K. 1992 Response of a mammalian faunal element to climatic changes. *J. Mamm.* **73**, 43–50.

Frey, J. K. 1993 Modes of peripheral isolate formation and speciation. *Syst. Biol.* **42**, 373–381.

García-Ramos, G. and Kirkpatrick, M. 1997 Genetic models of adaptation and gene flow in peripheral populations. *Evolution* **51**, 21–28.

Gaston, K. J. 1991 How large is a species' geographic range? *Oikos* **61**, 434–438.

Gaston, K. J. 1994a *Rarity.* London: Chapman and Hall.

Gaston, K. J. 1994b Measuring geographic range sizes. *Ecography* **17**, 198–205.

Gaston, K. J. 1996a Species-range size distributions: patterns, mechanisms and implications. *Trends Ecol. Evol.* **11**, 197–201.

Gaston, K. J. 1996b The multiple forms of the interspecific abundance-distribution relationship. *Oikos* **75**, 211–220.

Gaston, K. J. 1998. Species-range size distributions: products of speciation, extinction and transformation. *Phil. Trans. R. Soc. London Ser. B* **353**, 219–230.

Gaston, K. J. and Blackburn, T. M. 1996 Global scale macroecology: interactions between population size, geographic range size and body size in the Anseriformes. *J. Anim. Ecol.* **65**, 701–714.

Gaston, K. J. and Blackburn, T. M. 1997a Evolutionary age and risk of extinction: the global avifauna. *Evol. Ecol.* **11**, 557–565.

Gaston, K. J. and Blackburn, T. M. 1997b Age, area and avian diversification. *Biol. J. Linn. Soc.* **62**, 239–253.

Gaston, K. J. and Kunin, W. E. 1997 Rare-common differences: an overview. In *The biology of rarity: causes and consequences of rare-common differences* (ed. W. E. Kunin and K. J. Gaston), pp. 12–29. London: Chapman and Hall.

Gaston, K. J. and Lawton, J. H. 1990 Effects of scale and habitat on the relationship between species local abundance and large scale distribution. *Oikos* **58**, 329–335.

Gaston, K. J., Blackburn, T. M. and Lawton, J. H. 1997 Interspecific abundance-range size relationships: an appraisal of mechanisms. *J. Anim. Ecol* **66**, 579–601.

Geary, D. H. 1990 Patterns of evolutionary tempo and mode in the radiation of *Melanopsis* (Gastropoda; Melanopsidae). *Paleobiology* **16**, 492–511.

Geary, D. H. 1995 The importance of gradual change in species-level transitions. In *New approaches to speciation in the fossil record* (ed. D. H. Erwin and R. L. Anstey), pp. 67–86. New York: Columbia University Press.

Glazier, D. S. 1980 Ecological shifts and the evolution of geographically restricted species of North American *Peromyscus* (mice). *J. Biogeog.* **7**, 63–83.

Glazier, D. S. 1987 Toward a predictive theory of speciation—the ecology of isolate selection. *J. Theor. Biol.* **126**, 323–333.

Gleason, H. A. 1924 Age and area from the viewpoint of phytogeography. *Am. J. Bot.* **11**, 541–546.

Gorodkov, K. B. 1992 Dynamics of range: General approach. II Dynamics of range and evolution of taxa (qualitative or phyletic changes of range). *Entomol. Rev.* **70**, 81–99.

Gotelli, N. J. and Graves, G. R. 1996 *Null models in ecology*. Washington: Smithsonian Institution.

Gould, S. J. and Eldredge, N. 1977 Punctuated equilibria: The tempo and mode of evolution reconsidered. *Paleobiology* **3**, 115–151.

Grosholz, E. D. 1996 Contrasting rates of spread for introduced species in terrestrial and marine systems. *Ecology* **77**, 1680–1686.

Hansen, T. A. 1978 Larval dispersal and species longevity in Lower Tertiary gastropods. *Science* **199**, 886–887.

Hansen, T. A. 1980 Influence of larval dispersal and geographic distribution on species longevities in neogastropods. *Paleobiology* **6**, 193–207.

Hanski, I. 1982 Dynamics of regional distribution: the core and satellite species hypothesis. *Oikos* **38**, 210–221.

Hanski, I., Kouki, J. and Halkka, A. 1993 Three explanations of the positive relationship between distribution and abundance of species. In *Historical and geographical determinants of community diversity* (ed. R. Ricklefs and D. Schluter), pp. 108–116. Chicago: University of Chicago Press.

Harvey, P. H. and Pagel, M. D. 1991 *The comparative method in evolutionary biology*. Oxford University Press.

Hengeveld, R. 1989 *Dynamics of biological invasions*. London: Chapman and Hall.

Hengeveld, R. 1990 *Dynamic biogeography*. Cambridge: Cambridge University Press.

Hodgson, J. G. 1993 Commonness and rarity in British butterflies. *J. Appl. Ecol.* **30**, 407–427.

Holt, R. D. 1997 Rarity and evolution: some theoretical considerations. In *The biology of rarity: causes and consequences of rare-common differences* (ed. W. E. Kunin and K. J. Gaston), pp. 209–234. London: Chapman and Hall.

Hughes, L., Cawsey, E. M. and Westoby, M. 1996 Geographic and climatic range sizes of Australian eucalypts and a test of Rapoport's rule. *Global Ecol. Biogeog. Lett.* **5**, 128–142.

Jablonski, D. 1986a Causes and consequences of mass extinctions: a comparative approach. In *Dynamics of extinction* (ed. D. K. Elliot), pp. 183–229. New York: Wiley.

Jablonski, D. 1986b Background and mass extinctions: the alternation of macroevolutionary regimes. *Science* **231**, 129–133.

Jablonski, D. 1987 Heritability at the species level: analysis of geographic ranges of Cretaceous mollusks. *Science* **238**, 360–363.

Jablonski, D. 1988 Response [to Russell and Lindberg]. *Science* **240**, 969.

Jablonski, D. 1997 Body-size evolution in Cretaceous molluscs and the status of Cope's rule. *Nature* **385**, 250–252.

Jablonski, D. and Lutz, R. A. 1983 Larval ecology of marine benthic invertebrates: paleobiological implications. *Biol. Rev.* **58**, 21–89.

Jablonski, D. and Raup, D. M. 1995 Selectivity of end-Cretaceous bivalve extinctions. *Science* **268**, 389–391.

Jablonski, D. and Sepkoski, J. J. Jr. 1996 Paleobiology, community ecology, and scales of ecological pattern. *Ecology* **77**, 1367–1378.

Jablonski, D. and Valentine, J. W. 1990 From regional to total geographic ranges: testing the relationship in Recent bivalves. *Paleobiology* **16**, 126–142.

Jablonski, D., Flessa, K. W. and Valentine, J. W. 1985. Biogeography and paleobiology. *Paleobiology* **11**, 75–90.

Jackson, J. B. C. 1974 Biogeographic consequences of eurytopy and stenotopy among marine bivalves and their evolutionary consequences. *Am. Nat.* **108**, 541–560.

Johnson, K. G., Budd, A. F. and Stemann, T. A. 1995 Extinction selectivity and ecology of Neogene Caribbean reef corals. *Paleobiology* **21**, 52–73.

Kammer, T. W., Baumiller, T. K. and Ausich, W. I. 1997 Species longevity as a function of niche breadth: evidence from fossil crinoids. *Geology* **25**, 219–222.

Kauffman, S. A. 1993 *The origins of order. Self-organization and selection in evolution.* Oxford: Oxford University Press.

Kavanaugh, D. H. 1979a Rates of taxonomically significant differentiation in relation to geographical isolation and habitat: examples from a study of the Nearctic *Nebria* fauna. In *Carabid beetles: their evolution, natural history and classification* (ed. T. L. Erwin, G. E. Ball and A. L. Halpern), pp. 35–57. The Hague: Junk.

Kavanaugh, D. H. 1979b Investigations on present climatic refugia in North America through studies on the distributions of carabid beetles: concepts, methodology and prospectus. In *Carabid beetles: their evolution, natural history and classification* (ed. T. L. Erwin, G. E. Ball and A. L. Halpern), pp. 371–381. The Hague: Junk.

Kelly, C. K. and Woodward, F. I. 1996 Ecological correlates of plant range size: taxonomies and phylogenies in the study of plant commonness and rarity in Great Britain. *Phil. Trans. R. Soc. London Ser. B* **351**, 1261–1269.

King, M. 1993. *Species evolution. The role of chromosome change.* Cambridge: Cambridge University Press.

Kirkpatrick, M. and Barton, N. H. 1997 Evolution of a species' range. *Am. Nat.* **150**, 1–23.

Koch, C. F. 1980 Bivalve species duration, areal extent and population size in a Cretaceous sea. *Paleobiology* **6**, 184–192.

Koch, C. F. 1987 Prediction of sample size effects on measured temporal and geographic distribution patterns of species. *Paleobiology* **13**, 100–107.

Koch, C. F. and Morgan, J. P. 1988 On the expected distribution of species' ranges. *Paleobiology* **14**, 126–138.

Kunin, W. E. and Gaston, K. J. 1993 The biology of rarity: patterns, causes, and consequences. *Trends Ecol. Evol.* **8**, 298–301.

Kunin, W. E. and Gaston, K. J. (ed.) 1997 *The biology of rarity: causes and consequences of rare-common differences.* London: Chapman and Hall.

Lazarus, D., Hilbrecht, H., Spencer-Cervato, C. and Thierstein, H. 1995 Sympatric speciation and phyletic change in *Globorotalia truncatulinoides*. *Paleobiology* **21**, 28–51.

Levin, D. A. 1993 Local speciation in plants: the rule not the exception. *Syst. Bot.* **18**, 197–208.

Levinton, J. 1988 *Genetics, paleontology, and macroevolution*. Cambridge: Cambridge University Press.

Lieberman, B. S. and Eldredge, N. 1996 Trilobite biogeography in the Middle Devonian: geological processes and analytical methods. *Paleobiology* **22**, 66–79.

Lieberman, B. S., Allmon, W. D. and Eldredge, N. 1993 Levels of selection and macroevolutionary patterns in turritellid gastropods. *Paleobiology* **19**, 205–215.

Liebherr, J. K. and Hajek, A. E. 1990 A cladistic test of the taxon cycle and taxon pulse hypothesis. *Cladistics* **6**, 39–59.

Losos, J. B. 1992 A critical comparison of the taxon-cycle and character-displacement models for size evolution of *Anolis* lizards in the Lesser Antilles. *Copeia* **1992**, 279–288.

Lynch, J. D. 1988 Refugia. In *Analytical biogeography. An integrated approach to the study of animal and plant distributions* (ed. A. A. Myers and P. S. Giller), pp. 311–342. London: Chapman and Hall.

Lynch, J. D. 1989 The gauge of speciation: on the frequencies of modes of speciation. In *Speciation and its consequences* (ed. D. Otte and J. A. Endler), pp. 527–553. Sunderland, Massachusetts: Sinauer.

Mace, G. M. 1994 An investigation into methods for categorizing the conservation status of species. In *Large-scale ecology and conservation biology* (ed. P. J. Edwards, R. M. May and N. R. Webb), pp. 293–312. Oxford: Blackwell Science.

Mace, G. M. and Kershaw, M. 1997 Extinction risk and rarity in an ecological timescale. In *The biology of rarity: causes and consequences of rare-common differences* (ed. W. E. Kunin and K. J. Gaston), pp. 130–149. London: Chapman and Hall.

Marshall, C. R. 1991 Estimation of taxonomic ranges from the fossil record. In *Analytical paleobiology* (ed. N. Gilinsky and P. Signor), pp. 19–38. Knoxville, Tennessee: Paleontological Society.

Marshall, C. R. 1994 Confidence intervals on stratigraphic ranges: partial relaxation of the assumption of randomly distributed fossil horizons. *Paleobiology* **20**, 459–469.

Marzluff, J. M. and Dial, K. P. 1991 Life history correlates of taxonomic diversity. *Ecology* **72**, 428–439.

Maurer, B. A. and Nott, M. P. Geographic range fragmentation and the evolution of biological diversity. In *Biodiversity dynamics: turnover of populations, taxa, and communities* (ed. M. L. McKinney). New York: Columbia University Press. In press.

Mayr, E. 1963 *Animal species and evolution*. Cambridge, Massachusetts: Harvard University Press.

Mayr, E. 1988 *Toward a new philosophy of biology: observations of an evolutionist*. Cambridge, Massachusetts: Harvard University Press.

McAllister, D. E., Platania, S. P., Schueler, F. W., Baldwin, M. E. and Lee, D. S. 1986 Ichthyofaunal patterns on a geographical grid. In *Zoogeography of freshwater fishes of North America* (ed. C. H. Hocutt and E. D. Wiley), pp. 17–51. New York: Wiley.

McKinney, M. L. 1997 How do rare species avoid extinction? A paleontological view. In *The biology of rarity: causes and consequences of rare-common differences* (ed. W. E. Kunin and K. J. Gaston), pp. 110–129. London: Chapman and Hall.

McKinney, M. L. and Allmon W. D. 1995 Metapopulations and disturbance: From patch dynamics to biodiversity dynamics. In *New approaches to speciation in the fossil record* (ed. D. H. Erwin and R. L. Anstey), pp. 123–183. New York: Columbia University Press.

McLaughlin, S. P. 1992 Are floristic areas hierarchically arranged? *J. Biogeog.* **19**, 21–32.

Miller, A. I. 1997 A new look at age and area: the geographic and environmental expansion of genera during the Ordovician Radiation. *Paleobiology* **23**, 410–419.

Mooers, A. Ø. and Heard, S. B. 1997 Inferring evolutionary processes from phylogenetic tree shape. *Q. Rev. Biol.* **72**, 31–54.

Moya, A., Galiana, A. and Ayala, F. J. 1995. Founder-effect speciation theory: Failure of experimental corroboration. *Proc. Natl Acad. Sci. USA* **92**, 3983–3986.

Myers, A. A. and Giller, P. S. 1988 Process, pattern and scale in biogeography. In *Analytical biogeography: an integrated approach to the study of animal and plant distributions* (ed. A. A. Myers and P. S. Giller), pp. 3–21. London: Chapman and Hall.

Nelson, G. and Platnick, N. 1981 *Systematics and biogeography. Cladistics and vicariance.* New York: Columbia University Press.

Nelson, G. and Rosen, D. E. 1981 *Vicariance biogeography. A critique.* New York: Columbia University Press.

Norris, R. D. 1991 Biased extinction and evolutionary trends. *Paleobiology* **17**, 388–399.

Otto, S. P. and Whitlock, M. C. 1997 The probability of fixation in populations of changing size. *Genetics* **146**, 723–733.

Pagel, M. P., May, R. M. and Collie, A. R. 1991 Ecological aspects of the geographic distribution and diversity of mammalian species. *Am. Nat.* **137**, 791–815.

Palumbi, S. R. 1994 Genetic-divergence, reproductive isolation, and marine speciation. *Annu. Rev. Ecol. Syst.* **25**, 547–572.

Parmesan, C. 1996 Climate and species' range. *Nature* **382**, 765–766.

Peat, H. J. and Fitter, A. H. 1994 Comparative analyses of ecological characteristics of British angiosperms. *Biol. Rev.* **69**, 95–115.

Pomeroy, D. and Ssekabiira, D. 1990 An analysis of the distributions of terrestrial birds in Africa. *Afr. J. Ecol.* **28**, 1–13.

Pregill, G. K. and Olson, S. L. 1981 Zoogeography of West Indian vertebrates in relation to Pleistocene climatic cycles. *Annu. Rev. Ecol. Syst.* **12**, 75–98.

Rapoport, E. H. 1982 *Areography: geographical strategies of species.* Oxford: Pergamon.

Raup, D. M. 1994. The role of extinction in evolution. *Proc. Natl Acad. Sci. USA* **91**, 6758–6763.

Rice, W. R. and Hostert, E. E. 1993 Laboratory experiments on speciation: what have we learned in 40 years? *Evolution* **47**, 1637–1653.

Ricklefs, R. E. 1989 Speciation and diversity: integration of local and regional processes. In *Speciation and its consequences* (ed. D. Otte and J. Endler), pp. 599–622. Sunderland, Massachusetts: Sinauer.

Ricklefs, R. E. and Cox, G. W. 1972 Taxon cycles in the West Indies avifauna. *Am. Nat.* **106**, 195–219.

Ricklefs, R. E. and Cox, G. W. 1978 Stage of taxon cycle, habitat distribution and population density in the avifauna of the West Indies. *Am. Nat.* **122**, 875–895.

Ricklefs, R. E. and Latham, R. E. 1992 Intercontinental correlation of geographical ranges suggests stasis in ecological traits of relict genera of temperate perennial herbs. *Am. Nat.* **139**, 1305–1321.

Ripley, S. D. and Beehler, B. M. 1990 Patterns of speciation in Indian birds. *J. Biogeog.* **17**, 639–648.

Ronquist, F. 1997 Dispersal-vicariance analysis: A new approach to the quantification of historical biogeography. *Syst. Biol.* **46**, 195–203.

Rosen, B. R. 1988 Biogeographic patterns: a perceptual overview. In *Analytical biogeography. An integrated approach to the study of animal and plant distributions* (ed. A. A. Myers and P. S. Giller), pp. 23–55. London: Chapman and Hall.

Rosenzweig, M. L. 1975 On continental steady states of species diversity. In *Ecology and evolution of communities* (ed. M. L. Cody and J. M. Diamond), pp. 124–140. Cambridge, Massachusetts: Harvard University Press.

Rosenzweig, M. L. 1978 Geographical speciation: on range size and the probability of isolate formation. In *Proceedings of the Washington State University Conference on Biomathematics and Biostatistics* (ed. D. Wollkind), pp. 172–194. Washington: Washington State University.

Rosenzweig, M. L. 1995 *Species diversity in space and time.* Cambridge: Cambridge University Press.

Roughgarden, J. and Pacala, S. 1989 Taxon cycle among *Anolis* lizard populations: review of evidence. In *Speciation and its consequences* (ed. D. Otte and J. A. Endler), pp. 403–432. Sunderland, Massachusetts: Sinauer.

Roy, K. 1994 Effects of the Mesozoic Marine Revolution on the taxonomic, morphologic, and biogeographic evolution of a group: aporrhaid gastropods during the Mesozoic. *Paleobiology* **20**, 274–296.

Roy, K., Jablonski, D. and Valentine, J. W. 1995 Thermally anomalous assemblages revisited: patterns in the extraprovincial latitudinal range shifts of Pleistocene marine mollusks. *Geology* **23**, 1071–1074.

Roy, K., Valentine, J. W., Jablonski, D. and Kidwell, S. M. 1996 Scales of climatic variability and time averaging in Pleistocene biotas: implications for ecology and evolution. *Trends Ecol. Evol.* **11**, 458–463.

Rummel, J. D. and Roughgarden, J. 1985 A theory of faunal buildup for competition communities. *Evolution* **39**, 1009–1033.

Russell, M. P. and Lindberg, D. R. 1988a Real and random patterns associated with molluscan spatial and temporal distributions. *Paleobiology* **14**, 322–330.

Russell, M. P. and Lindberg, D. R. 1988b Estimates of species duration. *Science* **240**, 969.

Schliewen, U. K., Tautz, D. and Pääbo, S. 1994 Sympatric speciation suggested by monophyly of crater lake cichlids. *Nature* **368**, 629–632.

Schoener, T. W. 1987 The geographical distribution of rarity. *Oecologia* **74**, 161–173.

Sites, J. W. and Moritz, C. 1987 Chromosomal evolution and speciation revisited. *Syst. Zool.* **36**, 153–174.

Smith, A. B. 1994 *Systematics and the fossil record: documenting evolutionary patterns.* Oxford: Blackwell Science.

Stanley, S. M. 1979 *Macroevolution: patterns and process.* San Francisco: W. H. Freeman.

Stanley, S. M. 1986 Population size, extinction, and speciation: the fission effect in Neogene Bivalvia. *Paleobiology* **12**, 89–110.

Stanley, S. M., Wetmore, K. L. and Kennett., J. P. 1988 Macroevolutionary differences between the two major clades of Neogene planktonic foraminifera. *Paleobiology* **14**, 235–249.

Stebbins, G. L. 1978 Why are there so many rare plants in California? II. Youth and age of species. *Fremontia* **6**, 17–20.

Stebbins, G. L. and Major, J. 1965 Endemism and speciation in the California flora. *Ecol. Monogr.* **35**, 1–35.

Tauber, C. and Tauber, M. J. 1989 Sympatric speciation in insects: perception and perspective. In *Speciation and its consequences* (ed. D. Otte and J. A. Endler), pp. 307–344. Sunderland, Massachusetts: Sinauer.

Taylor, C. M. and Gotelli, N. J. 1994 The macroecology of *Cyprinella*: correlates of phylogeny, body size, and geographical range. *Am. Nat.* **144**, 549–569.

Terborgh, J. 1973 On the notion of favorableness in plant ecology. *Am. Nat.* **107**, 481–501.

Tokeshi, M. 1996 Power fraction: a new explanation of relative abundance patterns in species-rich assemblages. *Oikos* **75**, 543–550.

Van Valen, L. 1973 A new evolutionary law. *Evol. Theory* **1**, 1–30.

Veit, R. R. and Lewis, M. A. 1996 Dispersal, population growth, and the Allee effect: dynamics of the house finch invasion of eastern North America. *Am. Nat.* **148**, 255–274.

Vrba, E. 1980 Evolution, species and fossils: How does life evolve? *S. Afr. J. Sci.* **76**, 61–84.

Vrba, E. 1985 Environment and evolution: alternative causes of the temporal distribution of evolutionary events. *S. Afr. J. Sci.* **81**, 228–236.

Vrba, E. 1987 Ecology in relation to speciation rates: some case histories of Miocene-Recent mammal clades. *Evol. Ecol.* **1**, 283–300.

Vrba, E. 1993 Turnover-pulses, the red queen, and related topics. *Am. J. Sci.* **293A**, 418–452.

Wagner, P. J. and Erwin, D. H. 1995 Phylogenetic patterns as tests of speciation models. In *New approaches to speciation in the fossil record* (ed. D. H. Erwin and R. L. Anstey), pp. 87–122. New York: Columbia University Press.

Walsh, J. B. 1995 How often do duplicated genes evolve new functions? *Genetics* **139**, 421–428.

Watson, L. 1981 *Whales of the World.* London: Hutchinson.

Westrop, S. R. 1989 Macroevolutionary implications of mass extinction—evidence from an Upper Cambrian stage boundary. *Paleobiology* **15**, 46–52.

Westrop, S. R. 1991 Intercontinental variation in mass extinction patterns: influence of biogeographic province. *Paleobiology* **17**, 363–368.

Westrop, S. R. and Lundvigsen, R. 1987 Biogeographic control of trilobite mass extinction at an Upper Cambrian 'biomere' boundary. *Paleobiology* **13**, 84–99.

Wiley, E. O. 1981 *Phylogenetics. The theory and practice of phylogenetic systematics.* New York: John Wiley.

Williamson, M. 1996 *Biological invasions.* London: Chapman and Hall.

Willis, J. C. 1922 *Age and area: a study in geographical distribution and origin of species.* Cambridge: Cambridge University Press

Wilson, E. O. 1961 The nature of the taxon cycle in the Melanesian ant fauna. *Am. Nat.* **95**, 169–193.

Wing, L. 1943 Spread of the starling and English sparrow. *The Auk* **60**, 74–87.

Woods, K. D. and Davis, M. B. 1989 Paleoecology of range limits: beech in the upper peninsula of Michigan. *Ecology* **70**, 681–696.

Woodward, F. I. 1987 *Climate and plant distribution.* Cambridge: Cambridge University Press.

13

Rates of speciation in the fossil record

J. John Sepkoski, Jr.

13.1 Introduction

A rate of speciation is a measurement of how many new species appear in an interval of time within a given taxon and a given habitat, region, or ecosystem. A rate of speciation can also be a comparative measure, less dependent on time, relating to how one taxon seems to diversify relative to another.

The principal processes of speciation remain inadequately understood, despite a huge literature of field, laboratory, and theoretical studies. Still, this does not preclude empirical studies of the magnitude of speciation and the variation of rates observed in historical records. In this contribution, I briefly examine empirical rates of speciation, and of origination of taxa above the species level, from a palaeontological perspective. I first consider the average global rate of speciation and then discuss several sources of variation in rate at scales ranging from the whole of the Phanerozoic down to individual taxa and climatic fluctuations.

13.2 Is there a canonical speciation rate?

Is there an average speciation rate for the biota as a whole? If there is, this might provide a benchmark rate when considering specific taxa or evolutionary situations. A canonical rate might also provide insight into processes of diversification through study of deviations from this average and the forcing factors involved.

A mean rate of global speciation can be calculated on the back of the envelope by considering the opposite process—extinction. Raup (1991*a*) noted that more than 99% of species that have ever lived are now extinct; May *et al.* (1995) calculated a slightly smaller percentage but using an older estimate of the average duration of a species. In either case, the vast diversity of the modern biota must reflect only a slight excess of speciations over extinctions. Therefore, if we can estimate a historical extinction rate, we know the approximate speciation rate.

Raup (1991*b*) calculated that the average species of marine animals has a duration of about four million years. This translates into approximately 25% of species becoming extinct per million years. If we assume that the biota has a diversity on order of 10^7 species (May 1988; Hammond 1995), then approximately 2.5 species should become extinct each year. Assuming a slight excess of speciation, and rounding, this means that the canonical rate of speciation is *three species per year*.

A rate of three new species per year suggests one reason why processes of speciation are so difficult to study: speciation is a rare event. Of the estimated 10^7 living species, fewer than 20% are known to science and, of the approximately 1.5×10^6 described species, a trivial number is monitored annually. In fact, the discovery of *any* newly formed species might suggest that the calculated canonical rate is too low. Raup (1991*b*) calculated the average duration of a fossil species using an assumption of exponential decay of species with increasing longevity (see also Raup 1985), akin to Van Valen's (1973) law of extinction. The observation of numerous endemic species in ephemeral environments, such as lakes, might suggest there have been excess numbers of very short-lived species (for examples, see Foster *et al.* 1998; for fossil examples, see Geary 1990; McCune 1996). But even if the true average duration of species is halved to two million years, the canonical rate grows only to about six new species per year, or roughly one daughter for all species now known. (It is doubtful that the average species longevity could fall much lower than two million years; species of insects seem to have long geologic durations; see Labandeira and Sepkoski 1993; Coope 1995.)

Now, there are any number of reasons to believe calculation of a canonical speciation rate is absurd. Extinction is a discrete event, defined by the death of the last individual of a species. Speciation is more nebulous, having to do with relative reproductive isolation and degrees of genetic, morphological, ecological, and behavioral differentiation, as reviewed in this volume. Mayr's (1963) classic *Animal Species and Evolution* summarizes numerous examples of hybrid zones, ring species, introgressions in disturbed habitats, etc., which reflect differentiation short of speciation. Other workers have expanded on these observations, and students of plant species have added numerous examples of ambiguous speciation in the modern world (for example, Grant 1981; Niklas 1997). So, perhaps, the time scale for a canonical speciation rate should not be a year but more like a century or even a millennium. This merely involves multiplying the calculation by the appropriate order of magnitude.

From a palaeontological perspective, what is interesting in not the mean speciation rate but the variance about it. I suspect that if we had good historical data, we would find that most years, or even decades, witnessed no new species (absent of human disturbance) and that a few witnessed bursts of new species (see, for example, Bennett 1997). In other words, even a canonical speciation rate calculated over an appropriate time scale may have limited meaning. Through the rest of this chapter, I will consider some of the components of variance in speciation rates seen in the fossil record.

13.3 Methodological note

Much of the numerical information gleaned from knowledge of the fossil record on rates of 'speciation' and extinction is not for species but for taxa higher in the Linnean hierarchy, usually genera and families. Species are not used for several reasons.

1. Palaeontological species are morphospecies recognized on limited aspects of anatomy, specifically hard parts. The relationship between fossil morphospecies and extant biospecies is not always certain (although see Jackson and Cheetham 1990; Budd *et al.* 1994). In older parts of the Phanerozoic record, criteria used to define fossil species have varied among generations, cultures, and individual researchers (perhaps not unheard of in neontology).
2. Sampling of fossil species is very uneven in time and space. Many rock formations are disappointingly unfossiliferous whereas a few are spectacularly rich in fossils. This includes not only such renowned examples as the Burgess Shale and Solnhofen Plattenkalke but also units rich in 'normal' fossils, such as the Triassic St. Cassian beds of the Alps and the Cretaceous Coon Creek deposits of the southern United States. These fossil faunas have received appropriate but still special monographic attention at the expense of less yielding rock formations.
3. Species of particular interest for economic utility and stratigraphic correlation have received magnitudes more attention than fossils of lesser utility (see Pearson 1998). Thus, species in groups such as planktonic foraminifera, ammonoids, graptolites, and conodonts have an integrity that is not comparable with other taxa such as benthic foraminifera or sponges.

Palaeontologic genera and families, as a whole, have greater taxonomic stability and, because they represent collections of fossil morphospecies, are less subject to the vicissitudes of preservation, sampling, and study (Raup 1979*a*). It is true that most genera and families erected in the palaeontologic literature have not been scrutinized by modern phylogenetic analysis. Nonetheless, when large samples are used, this does not seem to be a major impediment.

1. Most studies of diversity patterns have found that genera and families correlate reasonably well with underlying species patterns, both in the fossil record (for example, Sepkoski *et al.* 1981) and in the Recent (for example, Gaston and Williams 1993; Flessa and Jablonski 1995; Roy *et al.* 1996*a*). Differences do exist but are often predictable from the behavior of collections versus individuals (for example, dampening of the magnitudes of mass extinctions).
2. Modeling studies of the efficacy of paraphyletic versus monophyletic taxa (Sepkoski and Kendrick 1993) suggest that traditionally defined paraphyletic taxa can track species patterns in the fossil record. In fact, if palaeontologic sampling of species is poor, traditional higher taxa can reflect patterns of

diversity and underlying rates of lineage speciation and extinction better than the poorly sampled species themselves.

The last point is important: use of higher taxa in palaeontologic studies of origination and extinction is not about higher taxa as ontologically real biological entities. Rather, these palaeontological taxa are proxies, standing in where fossil species fail.

13.4 Variation in origination rates over Phanerozoic time

The idea that there is a canonical speciation rate is compromised by the vast variation in rate observed over geologic time. Figure 13.1 illustrates a time series of magnitude of marine originations over the Phanerozoic, measured as the percent of new genera relative to standing diversity per interval of time. Intervals of geologic time have been variously amalgamated and subdivided to provide as equal intervals as possible, averaging about 5.5 million years duration, given present understanding of the chronometric time scale.

Figure 13.1 shows that there are two intervals of high rates of origination in the Phanerozoic oceans. The first includes the initial radiation of animals during the Cambrian Period (Figure 13.2) which exceeds most subsequent rates by at least a factor of two. This probably reflects rapid speciation as animals exploited the ecological roles available to large motile organisms capable of capturing water currents laden with food, processing algae and microbial mats, burrowing through sediment to ingest bacteria and other food items, and preying upon animals successful in the other three endeavors.

Subsequent to the 'Cambrian explosion', origination rates decline in a general manner. This is exhibited in Figure 13.1 by the exponential fit to the data. There is no reason to believe that origination rates should decay exponentially; the fitted curve is presented merely to draw one's eye to the general decline.

The end-Permian mass extinction (Figure 13.2) disrupted the secular decline of the Palaeozoic Era at the genus level (more so than at the family level; see Sepkoski 1993). In the wake of the end-Permian event, the Triassic witnessed levels of origination comparable to the Ordovician, a few tens of million years after the Cambrian explosion. The Triassic burst of genus origination reflects the phenomenon of rebound–rapid diversification in the aftermath of mass extinctions, as will be discussed below. The exceptional magnitude of the Triassic burst of origination reflects in large part the extreme magnitude of the end-Permian mass extinction: approximately half of marine families, 80% of genera and perhaps 96% of species were extinguished (Raup 1979b; Sepkoski 1989), leaving vacant ecospace available for survivors to radiate into during the subsequent Triassic (Erwin *et al.* 1987)

The fitted curve over the Mesozoic and Cenozoic in Figure 13.1 simply draws one's eye to the continuing decline in origination rates, which will be discussed below. Note, however, that magnitudes of origination in most recent geologic

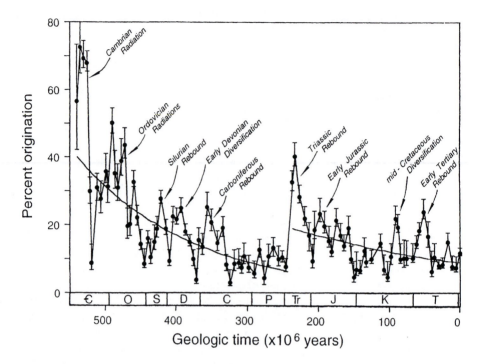

Fig. 13.1 Percent origination (100 × number of originations/standing diversity) for marine animal genera in 100 time intervals (stratigraphic stages and substages). 19 291 fossil genera spanning two or more of these intervals were used (for discussion of the data base and rationale for excluding genera known only from a single interval, see Sepkoski 1996). Error bars were calculated by the method of Sepkoski and Koch (1996). The two curves are least-squares exponentials fitted to emphasize long-term secular declines in magnitudes of origination. Major peaks above the curves are labeled. The geologic time scale is indicated along the abscissa (Є, Cambrian; O, Ordovician; S, Silurian; D, Devonian, C, Carboniferous; P, Permian; Tr, Triassic; J, Jurassic; K, Cretaceous; T, Tertiary).

times are comparable with rates seen during the comparable icehouse intervals of the late Palaeozoic.

13.5 Variation in speciation and origination among major taxa

The two phases of secular decline in average rates of genus origination are not unlike what Van Valen (1974) found for rates of extinction among families: Per-family extinction rates appear to decline through the Palaeozoic, accelerate in the wake of the end-Permian mass extinction, and then decline again through the Mesozoic and Cenozoic Eras. Raup and Sepkoski (1982) had previously analysed *total* rates of extinction and argued that there was a steady deceleration of extinction that encompassed the whole of the Phanerozoic, except when punctuated by events of mass extinction.

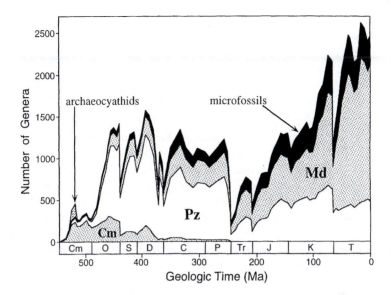

Fig. 13.2 Standing diversity of marine animal genera through the Phanerozoic, using the same data on genera as in Figure 1. The three great evolutionary faunas (Sepkoski 1984) are indicated by the fields (Cm, Cambrian Fauna; Pz, Palaeozoic fauna; Md, Modern fauna). 'microfossils' indicates the diversity of animal-like protists (Radiolaria and Foraminifera). 'archaeocyathids' indicates the enigmatic diversity of the Archaeocyatha, which are in large part considered sponges. From Sepkoski (1997).

Both sets of authors explained secular declines in terms of continuous processes, either taxa slowly evolving resistance to extinction-causing perturbations (Raup and Sepkoski 1982) or evolution and reorganization of community-level interactions (Van Valen 1974) (see review in Gilinsky 1994).

An alternative hypothesis is that secular changes in Phanerozoic origination and extinction rates reflect sorting of higher taxa. If higher taxa with rapid rates of origination and extinction (that is, are volatile in the sense of Raup and Boyajian 1988) dominate early ecosystems and are replaced by clades with relatively low rates, then secular declines in both origination and extinction becomes, in effect, epiphenomena. Testing this hypothesis with rates of extinction is actually easier than with rates of origination for several reasons.

1. Rates of extinction within major taxa through the Phanerozoic appear much more constant than rates of origination. Origination rates tend to decline somewhat over the history of major taxa, as will be discussed below; in contrast, rates of extinction tend to be more stable (Van Valen 1985a). Thus, a characteristic rate of extinction has more meaning than an average rate of origination.

2. Analytical models for calculating taxic rates of evolution are more fully developed for extinction than for origination. These models vary from the

simple models of Stanley (1979) to the branching models of Raup (1985), Gilinsky and Good (1991), and Foote and Raup (1996).

3. Extinction in the fossil record has received much more empirical study since the K/T hypothesis of Alvarez *et al.* (1980). Thus, data on times of extinction in any synoptic database have probably received more scrutiny from the community than have times of origination.

Despite these problems with calculating characteristic rates of origination, I have rashly computed rates for genera in 16 large taxonomic classes of marine animals. The rates were calculated as per-taxon rates of origination (that is originations per genus per million years) for geologic stages variously subdivided and amalgamated to provide as even as possible a 5.5 million-year sampling interval (based on a modified version of the time scale in Harland *et al.* 1990).

Figure 13.3 illustrates the average rates of genus origination with the taxa divided into the three great evolutionary faunas of the Phanerozoic marine fossil record (Figure 13.2). The rates of origination appear to vary substantially, as found at the family level for orders and classes by Holman (1989). The Cambrian evolutionary fauna encompasses the major taxa that radiated rapidly during the Cambrian explosion and dominated diversity throughout that period. Their average rates of origination tend to be higher than taxonomic groups subsequently important (although not without overlap). Thus, the huge peak in origination rates over the Cambrian in Figure 13.1 is a function of the dominance of groups in this fauna.

The taxonomic groups of the Palaeozoic evolutionary fauna illustrated in Figure 13.2 diversified during the Ordovician radiation and dominated the world ocean through the remainder of the Palaeozoic Era. Their average rates of origination tend to be lower than taxa of the Cambrian fauna, which is reflected in the smaller magnitude of percent origination over the Ordovician Period in Figure 13.1, despite the fact that three times more diversity was emplaced than during the Cambrian explosion. The waning origination rates through the post-Ordovician reflect two factors.

1. Continuing decline of members of the Cambrian evolutionary fauna (Figure 13.2), with their very high rates of origination.
2. Growing importance of the slowly-originating Modern evolutionary fauna, especially during the late Palaeozoic Era (that is, Carboniferous and Permian).

The major groups of the Modern evolutionary fauna tend to have the lowest rates of origination in Figure 13.3 and certainly have the lowest among-group variance. These are the groups that came to dominate global diversity in the wake of the end-Permian extinction event. The high rates of origination over the Triassic are partially a function of their accelerated diversification during the rebound from the Permian mass extinction. The high rates are also partially a function of a huge radiation of ammonoid cephalopods, a group more characteristic of the Palaeozoic evolutionary fauna which exhibits average origination rates

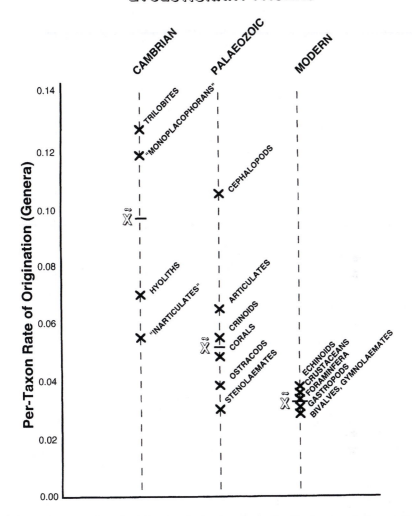

Fig. 13.3 Average rates of origination for major classes in the three evolutionary faunas. Rates were calculated as the per-taxon rate of origination (number of originations/standing diversity interval duration per million years, not 5.5 million-year sampling intervals). Averages are arithmetic means of rates weighted by standing diversity in each interval such that intervals with low diversity of a class get low weight. Median values ($\tilde{x}$) of the rates within each fauna are illustrated, showing the general decline in magnitude of origination rates from the Cambrian to the Modern fauna.

rivaling members of the Cambrian fauna (Figure 13.3). Ammonoids were a major component of marine diversity in the Triassic fossil record but later became proportionally less important as members characteristic of the Modern fauna diversified.

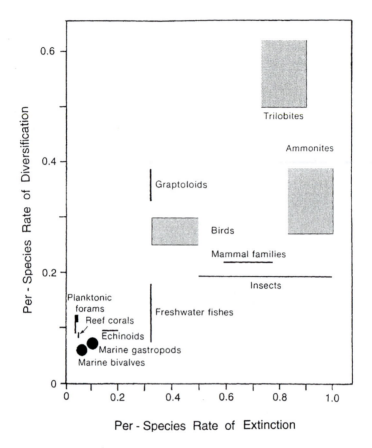

Fig. 13.4 The average rate of diversification plotted against rate of extinction for several major groups of taxa, after Stanley (1979, p. 231). Because diversification rate equals speciation rate minus extinction rate, the positive trend of the points indicates magnitudes of speciation and extinction must be correlated. (If not, the graph would show a *negative* trend.) Lengths of lines and areas of boxes indicate uncertainties in the rates. The per-species rate of extinction was calculated as the reciprocal of mean species durations (in millions of years). See Stanley (1979) for details of the calculation of mean duration and diversification rate.

This explanation for secular decline in rates of genus origination is actually quite similar to mathematical treatments of decline in the extinction rates of families presented by Sepkoski (1984, 1991c) and Gilinsky (1994). But if origination rates tend to be variable over time, why should there be a correspondence? It is a palaeontologic observation that rates of origination and extinction tend to be correlated, over nearly an order of magnitude of differences in rates. Stanley (1979, p. 258) formalized this observation as his first rule of macroevolution: the 'rate of speciation in adaptive radiation and rate of extinction are strongly correlated in the animal world' (see also Van Valen 1985b; Stanley 1990; Gilinsky

1994). Figure 13.4 illustrates Stanley's (1979) data, which are measured for *species*, using indirect calculations of extinction and diversification rates.

Stanley (1979, 1990) presented several explanations for the correlation between speciation and extinction rates. The first is almost trivial, but still important: a taxon with high extinction but low speciation will not survive long enough to be considered important. More substantively, Stanley argued that species' characteristics that might promote speciation, such as numerous small, geographically isolated populations, might also leave species prone to extinction (see also Stanley 1986a). For example, species with low dispersal ability should tend to be endemic, with comparatively small population sizes more subject to extinction than widespread species, but low-dispersing species should also have more subdivided populations prone to speciation. Jablonski (1986, 1995) expanded upon this with statistical studies of North American molluscs of the Late Cretaceous (see also Hansen 1978, 1982 for early Tertiary molluscs). Jablonski demonstrated that molluscs with planktotrophic larvae–larvae that spend weeks to months feeding in marine surface waters–have wider geographic distributions, longer geological durations, and lower speciation rates than nonplanktotrophic species. Molluscs that reproduced by either direct development or non-feeding, short-duration planktonic larvae had opposite characteristics: narrower geographic distributions and higher speciation rates (although see also Lieberman *et al.* 1993).

What is perhaps most significant in Jablonski's work is that these different rates of speciation and longevity are heritable at the level of clades; that is, molluscan species with planktotrophic larvae tended to give rise to daughter species with similarly low rates of speciation whereas molluscan species with nonplanktotrophic larvae gave rise to species with similarly restricted geographic ranges and high rates of speciation (Jablonski 1986). In Jablonski's analyses, the underlying cause of characteristic speciation rates is evident from modes of reproduction. In other cases, characteristic speciation rates may relate to ecologic factors, such as physiology and function, that affect population structure and density and that are shared by common descent within higher taxa (for example, Martin 1992; Lawton *et al.* 1994). But why characteristic rates should remain heritable over 10s to 100s of million years within taxa at the level of orders and classes as in Figure 13.3 remains unclear.

This problem of why evolutionary rates are conserved within disparate taxonomic groups is one of the great unsolved problems of macroevolution. Yet it is incredibly important since much of the faunal turnover we observe at higher taxonomic levels in the marine fossil record can be attributed to these characteristic and evidently heritable rates of speciation and extinction.

13.6 Rebounds of origination rates after mass extinctions

The decay curves plotted in Figure 13.1 fit the data in only the roughest of ways. There are many peaks and troughs in the magnitude of origination that represent much more than random variation about best-fit curves. Some of the peaks reflect acceleration of origination in the aftermath of mass extinctions, as labeled in

Figure 13.1. Rebound, discussed with respect to the end-Permian mass extinction, is a general feature of recoveries from all of the larger extinction events. What is surprising, though, is that acceleration of origination is not immediate after mass extinctions (Hallam 1991). Measured on time scales of geologic stages (averaging about five million years duration), there is a lag of several stages between the initial rise of diversity after the mass extinctions and the maximum rate of origination. The end-Ordovician mass extinction and subsequent Silurian rebound provide a typical example, illustrated in Figure 13.5. The mass extinction involves a sharp peak in extinction rate (although distributed over two geologically short intervals of the latest Ordovician; see Brenchley 1990; Brenchley *et al.* 1995). With the start of the Silurian Period, extinction rates drop to levels characteristic of the background rates of earlier times. However, measured magnitudes of origination do not jump immediately; rather, there is a protracted increase over 20 million years, with maximum rates reached only at the end of the Wenlockian Series. (Note that magnitudes in Figure 13.5 are measured as percentages; the delayed peak of origination is not simply a function of increasing diversity.)

There are various explanations for this recurring pattern.

1. The protracted rise in origination rates after mass extinctions is an artefact of imperfect knowledge of the fossil record. Incomplete preservation of fossil taxa and incomplete sampling of their preserved temporal ranges means that documented first and last occurrences are only minimum estimates of true times of origination and extinction. This problem has received most attention with respect to the often apparent gradual declines of species toward mass extinctions (Signor and Lipps 1982) and is often referred to as the 'Signor–Lipps' effect'. The same attenuation of documented times of origination should occur during rebounds from mass extinctions, and this has been referred to as the 'Lipps–Signor' or even 'Sppil–Rongis' effect. Erwin (1996) discusses evidence for this phenomenon in data on the recovery from the end-Permian mass extinction.

2. The protracted rise in origination is an artefact of mixing data for taxa with very different characteristic rates of origination. Because rates of speciation and extinction are correlated, rapidly speciating taxa should suffer more at mass extinctions than slowly speciating groups (see, for example, Sepkoski 1984). Thus, the aftermath of mass extinction initially will see disproportionate numbers of slowly-speciating taxa, with rapidly-speciating groups only later recovering and becoming important components of the biota. The data for marine animal genera, however, display no such pattern: when divided into evolutionary faunas, the more slowly speciating Modern fauna exhibits the same prolonged rise in origination rates as the more rapidly speciating Palaeozoic and, especially, Cambrian faunas.

3. The protracted increase in origination could have a biological basis, involving the slow re-establishment of stable ecological communities (see also Schindler 1990; Hallam 1991; Harries 1993; Erwin 1996). The immediate aftermaths of mass extinctions are characterized by outbreaks of what have been termed 'disaster species' (Kauffman and Erwin 1995; Hart 1996; Walliser 1996). These

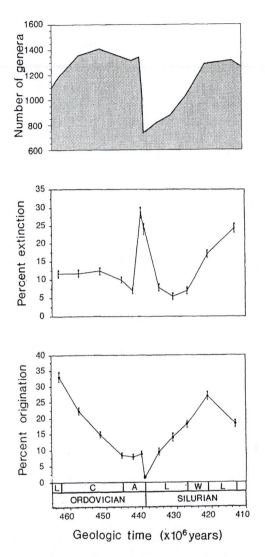

Fig. 13.5 Diversity, extinction, and origination around the end-Ordovician mass extinction and Silurian rebound. The upper panel illustrates the sharp drop in standing diversity of marine animal genera at the end of the Ordovician Period and then the long recovery of diversity to pre-extinction levels through the first half of the Silurian Period. The middle panel shows that the peak of extinction at the end of the Ordovician was sharp. The lower panel shows, in contrast, that originations did not increase abruptly after the pulse of extinction, even though diversity began to recover; rather, origination rates increased to the middle Silurian. Data, calculations, and error bars as in Figures 13.1 and 13.2. Letters on the abscissa indicate stratigraphic series (from left to right: L, Llandeilo; C, Caradoc; A, Ashgill; L, Llandovery; W, Wenlock; L, Ludlow).

are marine species that have nearly cosmopolitan distributions, are hugely abundant in local fossil deposits, and have remarkably short durations, often lasting on order of 10^5 years. A recurring pattern is that 'disaster species' blossom immediately after a mass extinction and are quickly replaced (on geological time scales) by subsequent disaster species. Thus, it appears as if the world ecosystem undergoes a turbulence in the wake of mass extinctions before the biological system settles down and rediversification begins in earnest. Increased speciation and rediversification apparently ensues on both geographic scales, with the re-establishment of biogeographic provinciality, and on environmental scales, with the re-establishment of community differentiation (i.e. beta diversity). The latter has best been documented for the Silurian rebound after the end-Ordovician mass extinction (Sheehan 1975), although much more work needs to be done with respect to that rebound and others.

13.7 Variation of speciation and origination rates within taxa

Most of the considerations above were directed toward large-scale secular variations in rates of origination. Much of this seems to involve either sorting among higher taxa over long intervals of time or response to historical events such as mass extinctions. However, there is variation within major taxonomic groups that can be related to time, geography, climate, and life history strategies. The last has already been mentioned with respect to studies by Hansen (1982), Jablonski (1986) and others. Below, I discuss several other sources of large-scale variation in speciation rates within major taxa.

Time

Few major taxa exhibit constant rates of origination across long intervals of geologic time. Gilinsky and Bambach (1987) found that a great majority of higher taxa of marine animals exhibit regular decline in magnitudes of origination through their histories. They studied 99 major taxa, both extant and extinct, and used Linnean families as proxies for species-level originations. They subdivided Phanerozoic time into coarse intervals of roughly 20 million years' duration (in order to avoid small-number variation and to smooth effects of short-termed, unique events), and they calculated several metrics of origination and extinction for each interval.

Figure 13.6 illustrates a typical example from Gilinsky and Bambach's (1987) study. It illustrates percent origination and extinction for the Rugosa, an extinct clade of Palaeozoic corals. Gilinsky and Bambach found that among a majority of the 99 taxa analysed, magnitudes of origination declined significantly when regressed on time but extinction tended to be stationary or, in a minority of taxa, increase only slightly through history.

Gilinsky and Bambach's (1987) results are probably not an artefact of taxonomy. Although studies at lower levels in the Linnean hierarchy are sparse,

RUGOSA (Rugose Corals):

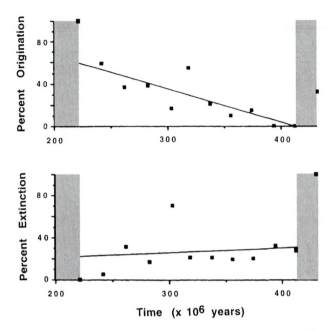

Fig. 13.6 Proportion of originations and extinctions relative to standing diversity through time for families of rugose corals, after Gilinsky and Bambach (1987). The fitted lines show that rate of origination declined rather regularly over time whereas average extinction increased very mildly. Points represent data for stratigraphic series of the Palaeozoic Era. The lines are linear regressions on time (measured from the approximate start of the Vendian Period). First and last points in the stippled fields were excluded from the fits in order to avoid the necessary 100% origination in the first time interval and 100% extinction in the last.

a few anecdotal examples exhibit similar patterns. For example, Foote (1988) used cohort analysis (Raup 1978, 1985) to estimate species-level rates of origination and extinction from genera of early Palaeozoic trilobites. He found a substantial decline in calculated rates of speciation from the Cambrian to the Ordovician Period. Maurer (1989) examined actual measured rates of speciation for North American horses and found marked decline with respect to diversity; since diversity of horses expanded through much of the Tertiary, speciation within the group must have declined through time.

Gilinsky and Bambach (1987) offered several hypotheses for secular declines in origination rates within major taxa. One was in the flavor of Van Valen's (1973) Red Queen Hypothesis: the basic adaptations of a successful group are established early in its history, as seen, a posteriori, in its diversification; however, when a group's biotic context changes as a result of continuing evolutionary turnover among surrounding taxa, the initial adaptations become increasingly less beneficial and overall speciation rate declines. A second scenario presented was that 'genetic and developmental systems may become resistant to change as they age'

(Gilinsky and Bambach 1987, p. 444). This speculation, bolstered by some reference to genetic studies, was also suggested to reduce possibilities of speciation over the history of a major taxon.

A simpler explanation for secular decline in origination rates is offered in Maurer's (1989) work: as a group radiates, it exerts a crowding effect (or, alternatively, it fills available niches), slowly decreasing opportunities for further speciation (see also Rosenzweig 1975; Sepkoski 1978, 1991a). This effect in marine higher taxa may be further enhanced by the general succession of evolutionary faunas, during which older faunas are replaced in sequence by faunas with new ecological organizations. This replacement could lead to origination rates actually declining below extinction rates within taxa of the older evolutionary fauna (Sepkoski 1991b).

Geography

Speciation rates vary in space as well as time. The role of geography in speciation is central to many hypotheses of population differentiation, from the shifting-balance model of Wright (1931, 1982), peripatric model of Mayr (1963), vicariance model of Croizat et al. (1974), to the clinal model of Endler (1977), among others. However, measurements of variations in speciation rates at grand scales remain limited.

High diversity in equatorial oceans and land masses suggests that rates of speciation may be higher in tropical than in temperate, boreal, or polar regions. However, high diversity could also result from low rates of extinction in the tropics (Jablonski 1993; Flessa and Jablonski 1996). Stehli et al. (1972) calculated mean geological ages of extant species of planktonic foraminifera along a transect from the equatorial Pacific to the Arctic Ocean and found little difference. Since mean age of extant species relates to mean duration (Stanley 1979; Pease 1987) and extinction rate is the reciprocal of mean duration (Raup 1985), this would suggest near constant extinction rate across latitude for this group. Thus, the high diversity of foraminifers near the equator (but not on; see Rutherford and D'Hondt 1996) would suggest higher rates of speciation, at least at some point in time, at lower latitudes.

At taxonomic levels of genera and families, mean geological ages tend to decrease toward the tropics. This has been shown for marine corals (Stehli and Wells 1971; Veron 1995), bivalves (Stehli et al. 1969; Hecht and Agan 1972; Flessa and Jablonski 1996), and barnacles (Newman 1986). Flessa and Jablonski (1996) present a detailed statistical analysis for genera of extant bivalves and argue compellingly that smaller mean ages of low-latitude (within 30° N–S) marine bivalves relative higher latitude bivalves reflects higher speciation and not reduced extinction.

At taxonomic levels even above the family, there is evidence that the tropics experience enhanced origination, or at least more frequent appearance, of groups with novel morphologies. Jablonski (1993) analysed the palaeolatitudes of first documented fossil occurrences of 42 taxonomic orders that originated in the oceans during the Mesozoic and Cenozoic Eras. He divided these orders into two sets: 26 that he assessed to have good fossil records (and thus less likely to reflect merely the vagaries of preservation and palaeontologic sampling) and 18 that were lightly sclerotized or otherwise lacked potential to be widely sampled. Jablonski found that first appearances of the former set were significantly more numerous in tropical latitudes than expected at random whereas the latter set conformed to expectations of palaeontologic sampling. Again, the evidence suggests that tropics are cauldrons of evolution, with high rates of evolutionary activity.

Climate

One could hypothesize that latitudinal variation in rates of origination is simply a correlate of climatic variation and the forced responses of effected populations. But even in the tropics, climate has changed considerably over time scales of 10^5 years (see for example, Colinvaux 1989, 1996; Guilderson et al. 1994). Variation in the Earth's orbit about the Sun affects solar insulation with characteristic periodicities of approximately 2×10^5, 4×10^5, and 10×10^5 years—the Milankovitch cycles. Bennett (1990) was one of the first workers to recognize the effects that climatic oscillations on these time scales might have for speciation rates (although see also Cronin 1987). Bennett (1990, 1997) argued that disruption and reassembly of terrestrial and marine communities (Coope 1987, 1995; Overpeck et al. 1992; Valentine and Jablonski 1993; Elias 1994; FAUNMAP Working Group 1996) on Milankovitch time scales might cancel microevolutionary processes affecting phyletic evolution, leading to the appearance of widespread evolutionary stasis. On the other hand, disruption of communities during climatic oscillations might also isolate small populations for time periods of 10^5 years, long enough for some surviving isolates to diverge from their ancestral species.

Vrba (1992) expanded upon these ideas in her 'habitat theory' of evolutionary change (see also Vrba 1985, 1987). Figure 13.7 is her illustration of how hypothetical populations might respond to Milankovitch oscillations in climate. Each cell in the figure is meant to represent a map view of a species' geographic range, and differences among columns illustrate different scenarios for response of geographic range, population structure, and genetic constitution as climate oscillates. Except for a, in which there is no change, species are shown to respond by range fragmentation and (usually) re-expansion with each oscillation. Figure 13.7b shows what may be the most common case: fragmentation of range into a number of isolated populations but without genetic differentiation so that the species later re-expands unchanged. The alternative is illustrated in Figure 13.7d, in which the isolates are so small and limited that the species becomes extinct.

Figure 13.7c, e, f and g are variants in which populations isolated by climatic change develop sufficient differentiation to be considered distinct species. In Figure 13.7c, the differentiated population expands parallel to the merged, undifferentiated isolates of the ancestral species to establish a parapatric pattern; f is similar but the species remain allopatric. Figure 13.7e illustrates a situation in which the differentiated population survives the climatic oscillation, but all undifferentiated isolates of the ancestral species do not; the new species expands over the ancestor's original range. Finally, Figure 13.7g illustrates a scenario in

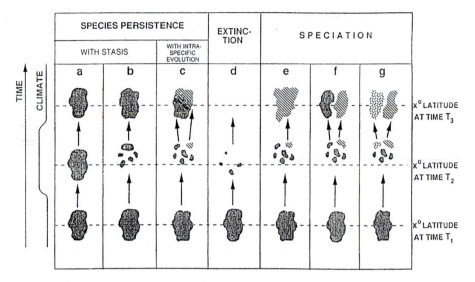

Fig. 13.7 Hypothetical map views of a species range at three times through a climatic oscillation (left-hand side), from Vrba (1992). Different patterns of shading at times T_2 and T_3 indicate evolution of new species from isolates of the parental species. See text for further explanation.

which more than one isolated population differentiates and survives as remaining populations of the ancestor disappear.

These are hypothetical situations currently with very limited documentation. Vrba (1985) has presented some evidence for clustered speciations and extinctions ('turnover pulses') among diverse groups of Neogene mammals during rapid climatic shifts over Africa. Similar pulses of rapid faunal extinction and speciation have been recorded among marine animals at the onset of major glaciation in high northern latitudes, about 2.5 million years ago (Stanley 1986b; Allmon et al. 1993; Jackson et al. 1996). Repeated episodes of speciation during Milankovitch fluctuations have yet to be demonstrated, and, in fact, there is some preliminary evidence that rates of speciation may have been slow during the glacial oscillations of the Pleistocene (Jackson 1995; Zink and Slowinski 1995; Roy et al. 1996b; Sheldon 1996; Bennett 1997; although see also Chiba 1998).

Even if pulses of speciation are not obvious at time scales of 10^4 to 10^5 years, there is suggestion in the fossil record that pulses such as seen in the Pliocene may have been recurrent, especially in the marine realm. Brett and Baird (1995) documented high levels of persistence of benthic species over intervals on order of 5×10^6 years in mid-Palaeozoic rocks in the northern Appalachian Basin of the United States. These intervals of stability were interrupted by short periods of extinction and appearance of many newly evolved species and immigrant species. This pattern of pulses of speciation followed by prolonged stability of faunas has been termed 'coordinated stasis' (see Brett et al. 1996). Miller (1997) presents a very useful review of the phenomenon and concludes that, like the habitat theory, coordinated stasis needs far more empirical scrutiny (see also Patzkowsky and Holland 1997).

13.8 Prospectus

This review has examined a few aspects of large-scale patterns of speciation and taxon origination from the perspective of the fossil record. Although it is possible to use palaeontologic data to calculate a canonical rate of speciation–three species per year–it is clear that this has limited meaning. There is large variation about this average rate on time scales of 10^4–10^8 years. A variety of explanations have been offered for this variation, but many uncertainties and unsolved problems persist. Below, I re-emphasize three of these.

1. Why do major taxa vary over approximately an order of magnitude in characteristic rates of speciation (and rates of extinction)? And why are these characteristic rates maintained to some extent through the histories of the taxa? Although some insight has been provided by consideration of life-history strategies (for example, Jablonski 1986), differences among taxonomic classes and orders and evident heritability of rates have never been adequately explained. Yet, if difference and heritability are taken as empirical fact, a seemingly disparate variety of palaeontologic observations can be modeled, including the succession of evolutionary faunas (Sepkoski 1984), the secular decline of origination and extinction rates (Sepkoski 1984; Gilinsky 1994), and even the onshore–offshore expansion of diversity seen in so many marine taxa (Sepkoski 1991c).

2. Is there a consistent structure to patterns of speciation during rebounds from mass extinctions? The rapid turnover of species in low-diversity communities prevalent after mass extinctions and the delayed onset of rapid rates of speciation also do not have adequate understanding. Here, the solution to the problem may begin with much better palaeontologic documentation of pattern, investigated with far greater temporal resolution, taxonomic discrimination, and geographic detail than represented in Figure 13.5. Better biological models of speciation in extensive, disturbed habitats may also contribute. Such study may have relevance reaching far beyond life's past history and enhance understanding of what may happen if present-day biodiversity continues to decline and if humankind can ever stop its devastation.

3. Is most speciation clustered during times of rapid change in climate? The hypotheses of Bennett (1990, 1997), Vrba (1992), and Brett and Baird (1995) remain largely untested. This is in part because it is very difficult to resolve spatial patterns, such as conjectured in Figure 13.6, on the necessary time scales of 10^4–10^5 years in the fossil record. Yet, such data are again important for questions about modern biodiversity: if projections of global warming are true, we will need comparative data to predict consequences for the biota, how

ranges might become fragmented, and how rates of speciation, as well as extinction, might change.

Acknowledgements

I thank C. M. Janis and R. Lupia for assistance with illustrations. Collection and analysis of data received support from the National Aeronautics and Space Administration (USA) under grant NAGW-1693.

References

Allmon, W. D., Rosenberg, G., Portell, R. W. and Schindler, K. S. 1993. Diversity of Atlantic Coastal Plain mollusks since the Pliocene. *Science (Washington)* **260**, 1626–1629.

Alvarez, L. W., Alvarez, W., Asaro, F. and Michel, H. V. 1980. Extraterrestrial cause for the Cretaceous-Tertiary extinction. *Science (Washington)* **208**, 1095–1108.

Bennett, K. D. 1990. Milankovitch cycles and their effects on species in ecological and evolutionary time. *Paleobiology* **16**, 11–21.

Bennett, K. D. 1997. *Evolution and ecology: the pace of life.* Cambridge: Cambridge University Press.

Brenchley, P. J. 1990. End-Ordovician. In *Palaeobiology: a synthesis* (ed. D. E. G. Briggs and P. R. Crowther), pp. 181–184. Oxford: Blackwell.

Brenchley, P. J., Carden, G. A. F. and Marshall, D. 1995. Environmental changes associated with the 'first strike' of the Late Ordovician mass extinction. *Modern Geology* **20**, 69–82.

Brett, C. E. and Baird, G. C. 1995. Coordinated stasis and evolutionary ecology of Silurian to Middle Devonian faunas in the Appalachian Basin. In *New approaches to speciation in the fossil record* (ed. D. H. Erwin and R. L. Anstey), pp. 285–315. New York: Columbia University Press.

Brett, C. E., Ivany, L. C. and Schopf, K. M. 1996. Coordinated stasis: an overview. *Palaeogeography, Palaeoclimatology, Palaeoecology* **23**, 1–20.

Budd, A. F., Johnson, K. G. and Potts, D. C. 1994. Recognizing morphospecies in colonial corals: I. Landmark-based methods. *Paleobiology* **20**, 484–505.

Chiba, S. 1998. Synchronized evolution in lineages of land snails in oceanic islands. *Paleobiology* **24**, 99–108.

Colinvaux, P. A. 1989. Ice-age Amazon revisited. *Nature* **340**, 188–189.

Colinvaux, P. A. 1996. Quaternary environmental history and forest diversity in the neotropics. In *Evolution and environment in tropical America* (ed. J. B. C. Jackson, A. G. Coates and A. F. Budd), pp. 359–405. Chicago: University of Chicago Press.

Coope, G. R. 1987. The response of late Quaternary insect communities to sudden climatic changes In *Organization of communities: past and present* (ed. J. H. R. Gee and P. S. Giller), pp. 421–438. Oxford: Blackwell.

Coope, G. R. 1995. Insect faunas in ice age environments: why so little extinction? In *Extinction rates* (ed. J. H. Lawton and R. M. May), pp. 55–74. Oxford: Oxford University Press.

Croizat, L., Nelson, G. and Rosen, D. E. 1974. Centers of origin and related concepts. *Systematic Zoology* **23**, 265–287.

Cronin, T. M. 1987. Speciation and cyclic climate change. In *Climate: history, periodicity, and predictability* (ed. M. R. Rampino, J. E. Sanders, W. S. Newman and C. K. Königsson), pp. 333–342. New York: Van Nostrand.

Elias, S. A. 1994. *Quaternary insects and their environments.* Washington, DC: Smithsonian Institution Press.

Endler, J. A. 1977. *Geographic variation, speciation, and clines.* Princeton, New Jersey: Princeton University Press.

Erwin, D. H. 1996. Understanding biotic recoveries: extinction, survival, and preservation during the end-Permian mass extinction. In *Evolutionary paleobiology* (ed. D. Jablonski, D. H. Erwin and J. H. Lipps), pp. 398–418. Chicago: University of Chicago Press.

Erwin, D. H., Valentine, J. W. and Sepkoski, J. J., Jr. 1987. A comparative study of diversification events: the early Paleozoic versus the Mesozoic. *Evolution* **41**, 1177–1186.

FAUNMAP Working Group. 1996. Spatial response of mammals to Late Quaternary environmental fluctuations. *Science (Washington)* **272**, 1601–1606.

Flessa, K. W. and Jablonski, D. 1995. Biogeography of Recent marine bivalve molluscs and its implications for paleobiogeography and the geography of extinction: a progress report. *Historical Biology* **10**, 25–48.

Flessa, K. W. and Jablonski, D. 1996. The geography of evolutionary turnover: a global analysis of extant bivalves. In *Evolutionary paleobiology* (ed. D. Jablonski, D. H. Erwin and J. H. Lipps), pp. 376–397. Chicago: University of Chicago Press.

Foote, M. 1988. Survivorship analysis of Cambrian and Ordovician trilobites. *Paleobiology* **14**, 258–3271.

Foote, M. and Raup, D. M. 1996. Fossil preservation and the stratigraphic ranges of taxa. *Paleobiology* **22**, 121–140.

Foster, S. A., Scott, R. J. and Cresko, W. A. 1998. Nested biological variation and speciation. *Philosophical Transactions of the Royal Society, London B*, **353**, 207–218.

Gaston, K. J. and Williams, P. H. 1993. Mapping the world's species. The higher taxon approach. *Biodiversity Letters* **1**, 2–8.

Geary, D. H. 1990. Patterns of evolutionary tempo and mode in the radiation of *Melanopsis* (Gastropoda; Melanopsidae). *Paleobiology* **16**, 492–511.

Gilinsky, N. L. 1994. Volatility and the Phanerozoic decline of background extinction intensity. *Paleobiology* **20**, 445–458.

Gilinsky, N. L. and Bambach, R. K. 1987. Asymmetrical patterns of origination and extinction in higher taxa. *Paleobiology* **13**, 427–445.

Gilinsky, N. L. and Good, I. J. 1991. Probabilities of origination, persistence, and extinction of families of marine invertebrate life. *Paleobiology* **17**, 145–166.

Grant, V. 1981. *Plant speciation.* New York: Columbia University Press.

Guilderson, T. P., Fairbanks, R. G. and Rubenstone, J. L. 1994. Tropical temperature variations since 20,000 years ago: modulating interhemispheric climate change. *Science (Washington)* **263**, 663–665.

Hallam, A. 1991. Why was there a delayed radiation after the end-Palaeozoic extinctions? *Historical Biology* **5**, 257–262.

Hammond, P. M. 1995. The current magnitude of biodiversity. In *Global biodiversity assessment* (ed. V. H. Heywood), pp. 113–138. Cambridge: Cambridge University Press.

Hansen, T. A. 1978. Larval dispersal and species longevity in lower Tertiary gastropods. *Science (Washington)* **199**, 885–887.

Hansen, T. A. 1982. Modes of larval development in early Tertiary neogastropods. *Paleobiology* **8**, 367–377.

Harland, W. B., Armstrong, R., Cox, A. V., Craig, L. E., Smith, A. G. and Smith, D. G. 1990. *A geologic time scale 1989.* Cambridge: Cambridge University Press.

Harries, P. J. 1993. Dynamics of survival following the Cenomanian-Turonian (Upper Cretaceous) mass extinction event. *Cretaceous Research* **14**, 563–583.

Hart, M. B. (ed.) 1996. Biotic recovery for mass extinction events. *Geological Society of London Special Paper* **102**.

Hecht, A. D. and Agan, B. 1972. Diversity and age relationships in Recent and Miocene bivalves. *Systematic Zoology* **21**, 308–312.

Holman, E. W. 1989. Some evolutionary correlates of higher taxa. *Paleobiology* **15**, 357–363.

Jablonski, D. 1986. Larval ecology and macroevolution in marine invertebrates. *Bulletin of Marine Science* **39**, 565–587.

Jablonski, D. 1993. The tropics as a source of evolutionary novelty through geological time. *Nature* **364**, 142–144.

Jablonski, D. 1995. Extinction in the fossil record. In *Extinction rates* (ed. R. M. May and J. H. Lawton), pp. 25–44. Oxford: Oxford University Press.

Jackson, J. B. C. 1995. Constancy and change of life in the sea. In *Extinction rates* (ed. J. H. Lawton and R. M. May), pp. 45–54. Oxford: Oxford University Press.

Jackson, J. B. C. and Cheetham, A. H. 1990. Evolutionary significance of morphospecies: a test with cheilostome Bryozoa. *Science (Washington)* **248**, 579–583.

Jackson, J. B. C., Budd, A. F. and Pandolfi, J. M. 1996. The shifting balance of natural communities? In *Evolutionary paleobiology* (ed. D. Jablonski, D. H. Erwin and J. H. Lipps), pp. 89–122. Chicago: University of Chicago Press.

Kauffman, E. G. and Erwin, D. H. 1995. Surviving mass extinctions. *Geotimes* **March 1995**, 14–17.

Labandeira, C. C. and Sepkoski, J. J., Jr. 1993. Insect diversity in the fossil record. *Science (Washington)* **261**, 310–315.

Lawton, J. H., Nee, S., Letcher, A. and Harvey, P. 1994. Animal distributions: patterns and process. In *Large-scale ecology and conservation biology* (ed. P. J. Edwards, R. M. May and N. Webb), pp. 41–58. Oxford: Blackwell.

Lieberman, B. S., Allmon, W. D. and Eldredge, N. 1993. Levels of selection and macroevolutionary patterns in the turritellid gastropods. *Paleobiology* **19**, 205-215.

Martin, R. A. 1992. Generic richness and body mass in North American mammals: support for the inverse relationship of body size and speciation rate. *Historical Biology* **6**, 73–90.

Maurer, B. A. 1989. Diversity-dependent species dynamics incorporating the effects of population-level processes on species dynamics. *Paleobiology* **15**, 133–146.

May, R. M. 1988. How many species on Earth? *Science (Washington)* **241**, 1441–1449.

May, R. M., Lawton, J. H. and Stork, N. E. 1995. Assessing extinction rates. In *Extinction rates* (ed. J. H. Lawton and R. M. May), pp. 1–24. Oxford: Oxford University Press.

Mayr, E. 1963. *Animal species and evolution*. Cambridge, Massachusetts: Belknap Press.

McCune, A. R. 1996. Biogeographic and stratigraphic evidence for rapid speciation in semionotid fishes. *Paleobiology* **22**, 34–48.

Miller, A. I. 1997. Coordinated stasis or coincident relative stability? *Paleobiology* **23**, 155–164.

Newman, W. A. 1986. Origin of the Hawaiian marine fauna: dispersal and vicariance as indicated by barnacles and other organisms. In *Crustacean biogeography* (ed. R. H. Gore and K. L. Heck), pp. 21–49. Rotterdam: Balkema.

Niklas, K. J. 1997. *The evolutionary biology of plants*. Chicago: University of Chicago Press.

Overpeck, J. T., Webb, R. S. and Webb, T., III. 1992. Mapping eastern North American vegetation changes of the past 18 ka: non-analogs and the future. *Geology* **20**, 1071–1074.

Potzkowsky, M. E. and Holland, S. M. 1997. Patterns of turnover in Middle and Upper Ordovician brachiopods in the eastern United States: a test of coordinated stasis. *Paleobiology* **23**, 420–443.

Pearson, P. N. 1998. Speciation and extinction asymmetries in paleontological phylogenies: evidence for evolutionary progress? *Paleobiology* **24**, 204–335.

Pease, C. M. 1987. Lyellian curves and mean taxonomic duration. *Paleobiology* **13**, 484–487.

Raup, D. M. 1978. Cohort analysis of generic survivorship. *Paleobiology* **4**, 1–5.

Raup, D. M. 1979*a*. Biases in the fossil record of species and genera. *Bulletin of the Carnegie Museum of Natural History* **13**, 85–91.

Raup, D. M. 1979*b*. Size of the Permo-Triassic bottleneck and its evolutionary implications. *Science (Washington)* **206**, 217–218.

Raup, D. M. 1985. Mathematical models of cladogenesis. *Paleobiology* **11**, 42–52.

Raup, D. M. 1991*a*. *Extinction: bad genes or bad luck?* New York: Norton.

Raup, D. M. 1991*b*. A kill curve for Phanerozoic marine species. *Paleobiology* **17**, 37–48.

Raup, D. M. and Boyajian, G. E. 1988. Patterns of generic extinction in the fossil record. *Paleobiology* **14**, 109–125.

Raup, D. M. and Sepkoski, J. J., Jr. 1982. Mass extinctions in the marine fossil record. *Science* **215**, 1501–1503.

Rosenzweig, M. L. 1975. On continental steady states of species diversity. In *Ecology and evolution of communities* (ed. M. L. Cody and J. M. Diamond), pp. 121–140. Cambridge, Massachusetts: Belknap.

Roy, K., Jablonski, D. and Valentine, J. W. 1996*a*. Higher taxa in biodiversity studies: patterns from eastern Pacific marine molluscs. *Philosophical Transactions of the Royal Society, London Series B* **351**, 1605–1613.

Roy, K., Jablonski, D., Valentine, J. W. and Kidwell, S. K. 1996*b*. Scales of climate variability and time averaging in Pleistocene biotas: implications for ecology and evolution. *Trends in Ecology and Evolution* **11**, 458–463.

Rutherford, S. and D'Hondt, S. 1996. Re-evaluating gradients in planktic foraminiferal diversity. *Geological Society of America Abstracts with Program* **28(7), A296.**

Schindler, E. 1990. The late Frasnian (Upper Devonian) Kellwasser crisis. In *Extinction events in Earth history* (ed. E. G. Kauffman and O. H. Walliser), pp. 153–159. Berlin: Springer-Verlag.

Sepkoski, J. J., Jr. 1978. A kinetic model of Phanerozoic taxonomic diversity I. Analysis of marine orders. *Paleobiology* **4**, 223–251.

Sepkoski, J. J., Jr. 1984. A kinetic model of Phanerozoic taxonomic diversity III. Post-Paleozoic families and mass extinctions. *Paleobiology* **10**, 246–267.

Sepkoski, J. J., Jr. 1989. Periodicity in extinction and the problem of catastrophism in the history of life. *Journal of the Geological Society, London* **146**, 7–19.

Sepkoski, J. J., Jr. 1991*a*. Diversity in the Phanerozoic oceans: a partisan review. In *The unity of evolutionary biology* (ed. E. C. Dudley), pp. 210–236. Portland, Oregon: Dioscorides Press.

Sepkoski, J. J., Jr. 1991*b*. Population biology models in macroevolution. In *Analytical paleobiology* (Short Courses in Paleontology) (ed. N. L. Gilinsky and P. W. Signor), pp. 136–156. The Paleontological Society.

Sepkoski, J. J., Jr. 1991*c*. A model of onshore-offshore change in faunal diversity. *Paleobiology* **17**, 58–77.

Sepkoski, J. J., Jr. 1993. Ten years in the library: new data confirm paleontological patterns. *Paleobiology* **19**, 43–51.

Sepkoski, J. J., Jr. 1996. Patterns of Phanerozoic extinctions: a perspective from global data bases. In *Global events and event stratigraphy* (ed. O. H. Walliser), pp. 35–52. Berlin: Springer-Verlag.

Sepkoski, J. J., Jr. 1997. Biodiversity: past, present, and future. *Journal of Paleontology*, **71**, 533–539.

Sepkoski, J. J., Jr. and Kendrick, D. C. 1993. Numerical experiments with model monophyletic and paraphyletic taxa. *Paleobiology* **19**, 168–184.

Sepkoski, J. J., Jr. and Koch, C. F. 1996. Evaluating paleontologic data relating to bioevents. In *Global events and event stratigraphy* (ed. O. H. Walliser), pp. 21–34. Berlin: Springer-Verlag.

Sepkoski, J. J., Jr., Bambach, R. K., Raup, D. M. and Valentine, J. W. 1981. Phanerozoic marine diversity and the fossil record. *Nature* **293**, 435–437.

Sheehan, P. M. 1975. Brachiopod synecology in a time of crisis (Late Ordovician-Early Silurian). *Paleobiology* **1**, 205–212.

Sheldon, P. R. 1996. Plus ça change—a model for stasis and evolution in different environments. *Palaeogeography, Palaeoclimatology, Palaeoecology* **127**, 209–227.

Signor, P. W. and Lipps, J. H. 1982. Sampling bias, gradual extinction patterns, and catastrophes in the fossil record. *Geological Society of America Special Paper* **190**, 291–296.

Stanley, S. M. 1979. *Macroevolution: pattern and process.* San Francisco: W. H Freeman.

Stanley, S. M. 1986a. Population size, extinction, and speciation: the fission effect in Neogene Bivalvia. *Paleobiology* **12**, 89–110.

Stanley, S. M. 1986b. Anatomy of a regional mass extinction: Plio-Pleistocene decimation of the western Atlantic bivalve fauna. *Palaios* **1**, 17–36.

Stanley, S. M. 1990. The general correlation between rate of speciation and rate of extinction: fortuitous causal linkages. In *Causes of evolution: a paleontological perspective* (ed. R. M. Ross and W. D. Allmon), pp. 103–127. Chicago: University of Chicago Press.

Stehli, F. G. and Wells, J. W. 1971. Diversity and age patterns in hermatypic corals. *Systematic Zoology* **20**, 115–126.

Stehli, F. G., Douglas, R. G. and Newell, N. D. 1969. Generation and maintenance of gradients in taxonomic diversity. *Science (Washington)* **164**, 947–949.

Stehli, F. G., Douglas, R. G. and Kafescioglu, I. A. 1972. Models for the evolution of planktonic foraminifera. In *Models in paleobiology* (ed. T. J. M. Schopf), pp. 116-128. San Francisco: Freeman, Cooper and Co.

Valentine, J. W. and Jablonski, D. 1993. Fossil communities: compositional variation at many time scales. In *Species diversity in ecological communities: historical and geographical perspectives* (ed. R. E. Ricklefs and D. Schluter), pp. 341–348. Chicago: University of Chicago Press.

Van Valen, L. 1973. A new evolutionary law. *Evolutionary Theory* **1**, 1–30.

Van Valen, L. 1974, Two modes of evolution. *Nature* **252**, 298–300.

Van Valen, L. M. 1985a. How constant is extinction? *Evolutionary Theory* **7**, 93–106.

Van Valen, L. M. 1985b. A theory of origination and extinction. *Evolutionary Theory* **7**, 133–142.

Veron, J. E. N. 1995. *Corals in space and time: the biogeography and evolution of the Scleractinia.* Ithaca, New York: Cornell University Press.

Vrba, E. S. 1985. Environment and evolution: alternative causes of the temporal distribution of evolutionary events. *South African Journal of Science* **81**, 229–236.

Vrba, E. S. 1987. Ecology in relation to speciation rates: some case histories of Miocene-Recent mammal clades. *Evolutionary Ecology* **1**, 283–300.

Vrba, E. S. 1992. Mammals as a key to evolutionary theory. *Journal of Mammalogy* **73**, 1-28.

Walliser, O. H. 1996. Patterns and causes of global events. In *Global events and event stratigraphy* (ed. O. H. Walliser), pp. 7–19. Berlin: Springer-Verlag.

Wright, S. 1931. Evolution in Mendelian populations. *Genetics* **16**, 97–159.

Wright, S. 1982. Character change, speciation, and the higher taxa. *Evolution* **36**, 427–443.

Zink, R. M. and Slowinski, J. B. 1995. Evidence from molecular systematics for decreased avian diversification in the Pleistocene Epoch. *Proceedings of the National Academy of Sciences, USA* **92**, 5832–5835.

The evolution of diversity in ancient ecosystems: a review

S. Conway Morris

14.1 Introduction

It is John Phillips (1860) who is generally credited with first plotting a history of diversity as revealed by the fossil record (Figure 14.1, upper left). Although schematic his diagram has several features of interest. It depicts the triple peaks of Palaeozoic, Mesozoic and Cainozoic diversity, divided by the caesuras of the end-Permian and end-Cretaceous events. Interestingly, the respective dips in the highs of Palaeozoic and Mesozoic diversity were attributed to taphonomic failure, a problem that still dogs any analysis of ancient diversity. Phillips (1860) also recognized the precipitous increase in Cainozoic diversity. Thus, within a year of the first edition of Darwin's *Origin of Species* the basic information on the history of organic diversity was available, with Phillips (1860) not only drawing attention to taphonomic lacunae but also in the pages preceding his figure 4 making some preliminary attempts to quantify his analysis.

It took more than a hundred years, however, for serious interest in this history to be rekindled. First were the pioneering formulations of Newell (1967), Valentine (1969, 1973) and Tappan (1969), but the subsequent enterprise has been strongly influenced by the massive compilations, in part unpublished, by Sepkoski (1992) (Figure 14.1) and an important series of analytical papers (Sepkoski 1978, 1979, 1981, 1988, 1993; see also Sepkoski 1997). The analysis of diversity and the resurgence of interest in mass extinctions remain a principal focus for palaeobiology (e.g. Signor 1990; Benton 1990; Valentine 1990; Erwin 1996a; Foote 1996), and has reached a certain stage of maturity. In addition to analysis of marine invertebrate diversity, comparable compilations for land plants (Niklas *et al.* 1983) and non-marine tetrapods (Benton 1985) are also available. Earlier concerns that the patterns of diversity were largely artefacts, controlled by such factors as areas or volumes of fossiliferous sediment, diversity 'hot-spots' centred on monographic treatment and fossil Lagerstätten, and a bias in favour of the preservation of younger sediments (the 'Pull of the Recent') have been quelled if not dismissed (see Sepkoski *et al.* 1981). Even so the shape of the curve of

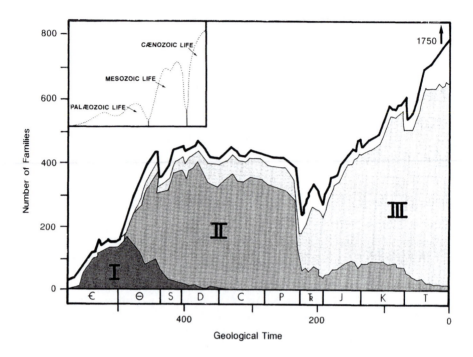

Fig. 14.1 The diversity curve of marine metazoan families, and their distinction into the three great evolutionary faunas, identified by factor analysis, and referred to as Cambrian (I), Palaeozoic (II) and Mesozoic-Cainozoic or Modern (III). The residual area above the main curves corresponds to that part of the diversity that cannot be accommodated by the three main statistical factors. The figure of 1750 (right) is the estimated number of marine families in Recent oceans, many of which have a low fossilization potential. The smaller diversity curve inserted top-left is redrawn from Phillips (1860). [Main figure is redrawn from Sepkoski (1981, fig. 5). Reprinted with permission from *Paleobiology* and the Paleontological Society.]

organic diversity remains contentious. Thus Benton (1995), employing the data from his edited compilation of *Fossil Record 2* (Benton 1993), identified in general an exponential increase in Phanerozoic diversity. This, however, was refuted by the critical reanalysis of Courtillot and Gaudemer (1996) who reaffirmed the basic pattern identified earlier by J. J. Sepkoski of diversity curves governed by logistical growth and interrupted on occasion by mass extinctions of varying severity.

There is a continuing shift in research emphasis in universities and happily to a lesser extent in natural history museums away from primary systematics and the monographic treatment of fossil groups. It is perhaps questionable, therefore, whether any updated databases produced in the next few years based on the precepts of Linnean taxonomy of generic or familial diversity will lead to a radical revision of current depictions of the histories of diversity, of which marine life remains the best known (Figure 14.1). The one exception to this trend in data accumulation might appear to be the substantial growth in the study of fossil

Lagerstätten, especially those displaying soft-part preservation. Paradoxically, in recent years, the research programme in this area has been largely on the taphonomy of Lagerstätten and the possible reasons for exceptional preservations. The documentation of primary systematics of many of these extraordinarily well-preserved biotas has lagged far behind.

The reader will realize that this review makes no pretence to completeness. So far as it has an aim it is to touch on areas that perhaps have suffered from either relative neglect or are particularly topical. To forestall justifiable criticism I should say that among the many potential topics that will not be considered here are: the problem of selectivity in extinctions (e.g. Bennett and Owens 1997), apart from some passing comments at the end of this paper any detailed discussion of the mathematical treatment of diversity (e.g. Patterson and Fowler 1996; Solé and Bascompte 1996; Solé et al. 1996; see also Kirchner and Weil 1998), the questions of stratigraphic completeness and the sampling of the fossil record (e.g. Foote and Raup 1996; Marshall 1997), variable species longevity (e.g. Levinton and Ginzburg 1984; Kammer et al. 1997), biogeographical changes engendered by continental splitting, the influence of seawater chemistry (Grotzinger 1990; Harper et al. 1997), the patterns of onshore-offshore diversity (e.g. Sepkoski 1991), the merits (or otherwise) of Linnean taxa versus cladistic methodologies (e.g. Sepkoski and Kendrick 1993; Foote 1996), the stability of taxonomic concepts (Hughes and Labandeira 1995), and the important role of refugia as 'ecological bunkers' in times of adversity and also 'evolutionary museums' of archaic diversity (see, e.g. Oji 1996). Finally, no serious attempt is made to cover the influence of climate on diversity in terms of either originations (e.g. Cronin 1985), extinctions (see Clarke 1993; Coope 1987, 1994), or faunal replacements (e.g. Janis 1989; Jackson 1994). At the risk of an eclectic journey the reader is invited to continue.

14.2 Ancient diversity: potential pitfalls

In assessing ancient diversity there appear to be two fundamental problems, which need to be reviewed before I comment, with equal brevity, on four particular topics (sections 14.3–14.6). The first problem concerns the divergence in approach between those who seek regularities and, if not laws, at least a degree of predictability in the behaviour and fate of ancient ecosystems. While the antithesis is by no means absolute, an alternative view prefers to emphasize the contingencies of history and often by implication the lack of discernible trends or patterns.

At first sight the attempt to seek regularities in the history of diversity appears to be frustrated by the obvious changes in the biosphere in the last four billion years. Most obviously this applies to the Archaean (3.8–2.5 Byr), when ecosystems may have been only occupied by prokaryotes (see Schopf 1983). The emergence of eukaryotes, which in terms of the fossil record apparently can be traced back to about 2.1 Byr (Han and Runnegar 1992; but see Doolittle et al. 1996), presumably defined the onset of more complex ecosystems but direct evidence for significant reorganization of microbial communities is at present effectively restricted to two areas. One is the documentation in the changes in the diversity of the microbially generated structures known as stromatolites, and in particular a dramatic decline in their diversity in the later Proterozoic (Walter and Heys 1985). This decline is widely attributed to the rise of grazing metazoans (Garrett 1970; Awramik 1971; Walter and Heys 1985), but abiotic explanations such as changes in the carbonate saturation of seawater seem as reasonable (Grotzinger 1990). There is a further complication in as much as the relationships between macroscopic expression of the stromatolite and the complexity or otherwise of the microbial communities remain obscure. The other area where an understanding of Precambrian diversity has advanced considerably is the depiction of protistan diversity, especially in terms of acritarchs (Knoll 1994). Even with the appearance of the first animals, whose origins may be deeper in the Proterozoic than the fossil record at present indicates (Wray *et al.* 1996; see also Conway Morris 1997), the ecology of the planet seems to have operated at an effectively microbial level for at least three billion years.

In this regard the subsequent Cambrian 'explosion' (ca. 545–520 Myr) remains one of the fulcrum points in the history of diversity with the origin and radiation of macroscopic metazoans. It is, nevertheless, a matter of contention as to the extent to which the Cambrian radiation is an 'explosion' of fossils whose newly acquired skeletal hard-parts breached a taphonomic threshold as against a genuine revolution in diversity. This potential dilemma was explored by Runnegar (1982*a*), but the recent reviews of the fossil record in this interval (e.g. Lipps and Signor 1992; Brasier 1992; Conway Morris 1998) treat the Cambrian 'explosion' as an evolutionary event of the first magnitude. That the trace fossils show a parallel increase in diversity (e.g. Jensen 1997) is a strong argument in favour of the latter view, although it would still be possible to argue that the Cambrian 'explosion' is more the result of a scaling-up in size, perhaps due to changes in the concentration of atmospheric oxygen, rather than a genuine series of innovations. The problem of cryptic diversity is returned to below (Section 14.6), but here we need only note that at least so far as the immediately preceding Ediacaran faunas are concerned evidence continues to grow for phyletic links with life in the Cambrian (e.g. Gehling 1991; Conway Morris 1993*a*; Fortey *et al.* 1996; Gehling and Rigby 1996; Waggoner 1996; Fedonkin and Waggoner 1997).

The impact of the Cambrian 'explosion' in other ways is perhaps still less fully appreciated. For example Logan *et al.* (1995) drew attention to a significant shift in the ratio of carbon isotopes ($\delta^{13}C$) in certain hydrocarbons extracted from sediments spanning the Proterozoic–Phanerozoic boundary. This they interpreted as a consequence of the evolution of planktonic grazers, possibly arthropods (see Butterfield 1994, 1997) adept at harvesting suspended organic matter which after passage through the gut was encapsulated into faecal pellets that then rapidly descended to the seafloor a kilometre or more below. Prior to the evolution of these grazers the descent of organic matter through the water column would have been markedly slower because much of it would have been as minute particles of marine 'snow'. The transition from a microbial ecosystem to one with metazoan harvesters probably had a profound effect on ocean chemistry, especially in terms of oxygen demand mediated by bacterial utilization of the slowly sinking particulate 'snow' (Logan *et al.* 1995).

The literature is replete with many other examples of subsequent changes in Phanerozoic diversity and their ecological consequences. Of these the invasions of the land (e.g. Labandeira and Beall 1990; Kenrick and Crane 1997) and air (Crepet 1979; Feduccia 1980; Labandeira and Sepkoski 1993) perhaps are the most familiar, but others e.g. increasing degrees of sediment bioturbation (e.g. Droser 1991; Droser and Bottjer 1993), taphonomic feedback of substrate types (e.g. Sprinkle and Guensburg 1995; Kidwell and Brenchley 1996), the recruitment to and the reorganization of planktonic ecosystems (e.g. Rigby 1997) are also highly significant. Ironically, those workers who emphasize the many historical contingencies in the history of diversity are often particularly vociferous in their denial of what to others appears to be strong evidence for increases in the complexity (e.g. Valentine *et al.* 1994; see also McShea 1996) and sophistication of both biotas and ecosystems. Such trends seem to deserve the name of progress (Rosensweig and McCord 1991; see also Jackson and McKinney 1990), which need not be equated with orthogenesis.

The second general problem concerns the question of scales of analysis in terms of geological time. A good example is that given by McGhee (1996) in his discussion of the late Devonian (Frasnian–Fammenian) mass extinction. The type and style of extinction differs according to whether the analysis is conducted at the level of epoch, stage, substage or subzone (Figure 14.2). Which is the 'true' pattern, or at least the one with greatest explanatory value? The crux of this problem seems to be the problem of deciding whether the pattern of biotic change viewed on short time-scales ('ecological time') may be extrapolated seamlessly to the long-term intervals of 'evolutionary time', or whether emergent properties occur that would remain invisible to observers, such as ourselves, if a fossil record happened to be unavailable. While strong claims have been made for such emergence, especially in terms of species selection (but see Budd and Johnson 1991; Lieberman 1995) and mass extinctions (but see Jackson and McKinney 1990), it is not yet clear the extent to which such principles actually govern the history of diversity.

The main purpose of this chapter is to review four specific areas to further our understanding of the changes in diversity.

14.3 The problem of rapid diversification

A survey of the marine fossil record (Sepkoski 1981, 1997) reveals three notable episodes of major diversification: the early Cambrian, the mid Ordovician and the Cainozoic (Figure 14.1). At present, the first of these, the Cambrian 'explosion' has attracted considerable attention, in part because of the realization of the significance of Burgess Shale-type faunas (Conway Morris 1998) and the interest in the advances in the molecular biology of bodyplan development and its relevance to the origin of metazoan bodyplans (Conway Morris 1994*a*;

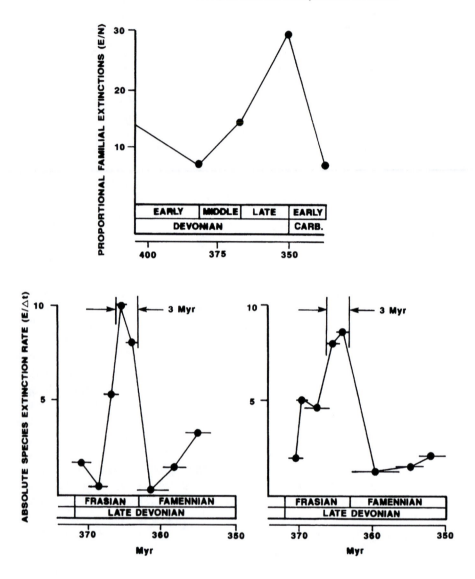

Fig. 14.2 An illustration of the problems of the analysis of a geological event, specifically the late Devonian (Frasnian–Fammenian) mass extinction, according to the resolution of the time-scale and stratigraphic unit chosen. The upper figures refer to familial diversity, respectively, at the stratigraphic level of epoch (left) and stage (right). The lower figures illustrate the extinction as resolved at finer stratigraphic levels. That on the left illustrates

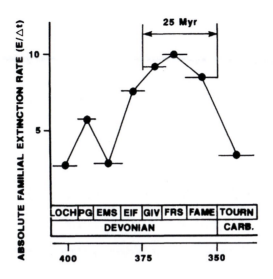

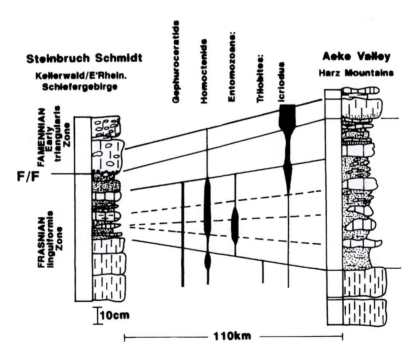

the diversity of brachiopods in New York State (left) and the southern Urals of Russia (right), while the right-hand figure depicts events as seen as the subzonal level with reference to two important sections in Germany. [Figures are redrawn from McGhee (1996, figs 3.1, 3.2, 3.4 and 3.5, respectively). Reprinted with permission from Columbia University Press.]

Valentine *et al.* 1996; Erwin *et al.* 1997). The abruptness, or otherwise, of the Cambrian 'explosion' is predicated on both the possibility of an extended Proterozoic history and the phyletic relevance, if any, of the Ediacaran faunas. If present attempts to identify plausible metazoans in the Ediacaran assemblages (e.g. Conway Morris 1993*a*, Waggoner 1996; Fedonkin and Waggoner 1997) win favour then the interval for this radiation might be extended to at least 30 Myr. Alternatively, even if the evidence for at least some Ediacaran fossils being unequivocal metazoans is accepted, this could still represent a radiation largely independent of the subsequent Cambrian event (Conway Morris 1993*b*). Some support for this idea may come from the record of Ediacaran trace fossils. Although a considerable diversity of Ediacaran ichnotaxa have been described, and some with peculiar morphology interpreted as representing 'bizarre' behaviours (Crimes 1994), a reinvestigation of the record suggests this diversity has been exaggerated so that the Ediacaran record of trace fossils in fact was very impoverished (Soren Jensen, personal communication). The revolution in benthic behaviour, and by implication neurological complexity, in the succeeding Cambrian would be consistent with the key steps in this radiation being achieved very rapidly, perhaps in less than 10 Myr. These uncertainties have an obvious bearing on the difficulties of identifying plausible triggers to explain the initiation of the Cambrian 'explosion', although the role of atmospheric oxygen or tectonic reconfiguration (Kirschvink *et al.* 1997) remain perennial favourites.

The Cambrian diversifications are of pivotal importance both for the establishment of a wide range of metazoan designs, codified in orthodox taxonomy as phyla, and a corresponding occupation of marine ecologies. Nevertheless, as has been long recognized diversification at the lower levels of the taxonomic hierarchy, such as the family, seems to have been more restricted. The figures for the number of genera and families will certainly require upward revision as our knowledge of Cambrian faunas, especially those preserved in Burgess Shale-type settings (e.g. Chen *et al.* 1996) and also by phosphatization (e.g. Müller and Hinz-Schallreuter 1993), continues to improve. Nevertheless, in comparison with the preceding Cambrian, the subsequent radiations in the Ordovician (Figure 14.1) remain far more dramatic in terms of increase of diversity, notably of cnidarians, brachiopods, cephalopods, echinoderms and ectoprocts. Not only that, but the major increase in diversity appears to have been achieved in only a few million years, at least for a number of important shelly groups that included the trilobites, brachiopods and certain molluscs (Miller and Foote 1996). Although new evidence supports the biological distinctiveness of these emerging faunas (see Sprinkle and Guensburg 1995) and the pattern of diversification can be modelled with some detail, as yet we still have no convincing explanation for what may have initiated this Ordovician radiation. Thereafter the diversity, at least of well-skeletonized marine families, appears to have stayed at a plateau, possibility representing an equilibrium, for about 200 Myr and weathering two significant mass extinctions (late Ordovician, late Devonian) (Figure 14.1). The origins of the third major rise in marine diversity can be traced to the beginning of the Jurassic, and notwithstanding the end-Cretaceous extinctions this diversity has continued to rise towards the Recent. Until very recently it seems that our planet housed the richest biotas it had ever seen.

14.4 Extinctions

Parallel to the depiction of the history of diversity there has been extensive documentation of the evidence for past extinctions, notably the five mass extinctions. Of these the episode at the end of the Cretaceous (K/T) has attracted unparalleled attention. A convenient distinction between mass and background extinctions receives some support from an analysis of the factors favouring species survival in certain groups, such as the molluscs (Jablonski 1989), according to the severity of the event. On the other hand, Raup's (1991, 1992) formulation of the so-called 'kill curve' suggests that its end-members—relatively benign with limited loss of species to planet-wide trauma, respectively—are connected by a continuum of increasing disaster. Another influential view that has received wide support is Gould's (1985) identification of tiers of influence for evolutionary process that transcend those that operate at the level of the species and presumably are open to neodarwinian explanations. In this formulation mass extinctions are said to play a key role in resetting ecologies and shaping the future development of post-catastrophe diversity (see also Courtillot and Gaudemer 1996). But are matters as simple as this?

The one instance that has received general agreement is the end-Permian event (Erwin 1993, 1995), marking the replacement of the Palaeozoic faunas by those of the Mesozoic. Given the severity of this episode (see Eshet *et al.* 1995), with extreme estimates of species loss reaching 96 per cent (Raup 1979), it is not surprising that after the protracted time for recovery (see below) the marine communities were markedly different. Yet there is still some danger of exaggeration. Thus, once important groups of Palaeozoic marine organisms such as the trilobites (Fortey and Owens 1990) and rugose corals (Scrutton 1988) were already minor components of even early Permian ecosystems. Other groups, such as the ammonoids and sea urchins, certainly suffered badly but subsequently rebounded to even greater levels of success. The possible pattern of extinction in the gastropods is particularly intriguing (Erwin 1996b; Erwin and Pan 1996). This is because the post-catastrophe faunas are markedly depauperate, yet a simple reading of the data may be misleading. One possibility is that absences are only apparent, and reflect a taphonomic lacuna imposed by a general failure of aragonitic preservation. Alternatively, when the Triassic gastropods do reappear they have some striking similarities to the preceding Permian forms. Is this because they are so-called 'Lazarus taxa', or are they convergent with strongly home-omorphic shells? Erwin and Pan (1996, p. 229) conclude by noting 'few phylogenetic studies have investigated clade relationships across the Permo-Triassic boundary, but those that have (for brachiopods, bryzoans and asteroids) all suggest that traditional systematics may exaggerate the magnitude of extinctions at higher levels and miss phylogenetic links at lower levels'. Even the celebrated transition between brachiopod-dominated Palaeozoic seas and the post-Permian abundance of bivalve molluscs, which had been cast into a fortuitous correlation of respective decline and rise by Gould and Calloway (1980), has been reanalysed by Sepkoski (1996; see also Rhodes and Thompson 1993). The Permian mass extinction certainly remains very significant for both groups, but Sepkoski (1996) convincingly reconstrues their histories into a new context of dynamic interaction and competition.

In terms of three of the other mass extinctions, the evidence for wholesale remodelling of ecosystems is rather weaker. This is especially true of the late Ordovician event (see Droser *et al.* 1997), which is also enigmatic in as much as its proposed correlation with a short but severe glaciation has no parallel with either of the other two episodes of Phanerozoic glaciation in the Permo-Carboniferous and Plio-Pleistocene respectively. The Devonian mass extinction certainly marks some significant changes in marine ecosystems, especially among the reefs (Scrutton 1988; McGhee 1996). Unfortunately, the terrestrial record of animals is rather inadequate, but that of plants does not point to any particular trauma (Raymond and Metz 1995; Willis and Bennett 1995). The Triassic event also has some noteworthy victims, including the conodonts, but here too the differences between pre- and post-catastrophe biotas perhaps are not that dramatic in terms of overall composition.

Such reservations concerning the severity and ultimate effects of mass extinctions would, however, appear to be dispelled when the K/T event is considered. The overwhelming evidence for an impact [iridium, shocked mineral grains (especially quartz), stishovite, microtektites with relict glass, extraterrestrial amino acids, microdiamonds; see Koerberl 1996 for a recent review] and severe environmental perturbations (soot and pyrotoxins, acid rain, the collapse of oceanic productivity, tsunamis) underline the catastrophic nature of this event. And yet important qualifications need to be made. First it is agreed that some significant extinctions occurred well before the time of impact, most notably in the inoceramid bivalve molluscs (MacLeod and Orr 1993) and possibly the rudists (Johnson and Kauffman 1996). In other cases the disappearance of a group at the K/T interval may be more apparent than real. In the case of the ammonites there are occurrences in Seymour Island, albeit as loose specimens in float, up to 13 m above the locally defined K/T boundary. These may be reworked from underlying Maastrichtian strata, but are possibly genuine survivors (Zinsmeister *et al.* 1989; see also Table 14.1). In addition, in a number of Cretaceous ammonites there is evidence that the shell had become internal to enveloping soft tissues (comparable with the living *Spirula*) (Doguzhaeva and Mutvei 1993), and it is possible that the ammonitic survivors of this mass extinction were effectively soft-bodied (Lewy 1996). In other cases the longer term differences across this boundary are muted or even negligible (MacLeod *et al.* 1997). This appears to be the case for such groups as the foraminifers, brachiopods, gastropods, echinoids, fish, and terrestrial plants. The emphasis on catastrophism has been reinforced, however, by the demise of the dinosaurs and the subsequent radiation of the mammals, including

Table 14.1 Time-scales for destruction and recovery of biotas across the Cretaceous-Tertiary boundary. All figures are approximate and reflect uncertainties of sedimentation rates, magnetostratigraphy and radiometric dating. Sources of information: 1, this assumes a crater diameter of ca. 180 km (Hildebrand et al. 1995); 2, see Toon et al. (1997); 3, see Alvarez et al. (1995); 4, see Covey et al. (1994); 5, see Pope et al. (1994); 6, see Tschudy and Tschudy (1986); no precise estimate of the duration of the 'Fern Spike' appears to be available; 7, see Coccioni and Galeotti (1994); these data refer to Spanish stratigraphic sections and elsewhere the benthic foraminifera seem to have been less effected (see, e.g. Kaiho 1992); 8, see Kajiwara and Kaiho (1992); this section is from Japan and may not be representative of other areas; 9, see Rigby et al. (1987) for a possible instance, but Lofgren et al. (1990) for a rebuttal; 10, see Zinsmeister et al. (1989) for possible post-catastrophe ammonites, and also Marshall (1995); 11, see Barrera and Keller (1994) for data based on foraminiferans; 12, see Alcala-Herrera et al. (1992) for data based on calcareous nannoplankton; 13, see D'Hondt et al. (1996), in addition many other workers on microfossils have quoted similar recovery time; 14, see Hansen et al. (1993)

Time	Effect
1 second	Annihilation around impact site (ca. 30 000 km^2) of Chicxulub[1]
1 minute	Earthquakes, Richter scale ten[2]
10 minutes	Spontaneous ignition of North American forests[2]
60 minutes	Impact ejecta crosses North America[3]
10 hours	Tsunamis swamp Tethyean coastal margins[2]
1 week?	First extinctions
9 months	Dust clouds begin to clear[4]
10 years	Very severe climatic disturbance (especially cooling) ends[5]
1000 years	Continental vegetation begins to recover: end of 'Fern Spike'[6]
1500 years	Initial recovery of some deeper-water benthic ecosystems[7]
7000 years	Recovery of some deeper-water benthic ecosystems[7]
70 000 years	Oceanic anoxia diminishes[8]
100 000 years?	Final extinction of dinosaurs[9]
300 000 years?	Final extinction of ammonites[10]
500 000 years	Major fluctuations in oceanic ecosystems begin to moderate[11,12]
1 000 000 years	Open oceanic ecosystems partially recovered[13]
2 000 000 years	Marine mollusc faunas largely recovered[14]
2 500 000 years	Global ecosystems normal[12]

of course the primates. Here too there may need to be some qualification of the evidence. To date the only strong evidence for dinosaurs living immediately before the impact, and possibly surviving for a short period afterwards (Rigby et al. 1987; see also Argast et al. 1987 and Lofgren et al. 1990), comes from North America. There is evidence to suggest that this region suffered extensive devastation not only because of its proximity to the impact site at Chicxulub, Mexico, but also because of the oblique nature of the impact and the movement of the ejecta curtain northwards (Schultz and D'Hondt 1996). While evidence for Maastrichtian dinosaurs is widespread, in at least some areas there is a possibility of extinction prior to the impact (e.g. Galbrun 1997).

There is another feature of the aftermath of mass extinctions that deserves further consideration, and that is the extraordinarily protracted lag-times for recovery to ecological stability and biotic diversity. This certainly seems to be the case for the end-Permian and Cretaceous events, whereas for at least the late Ordovician extinctions recovery appears to have been relatively rapid and this feature may further underline the apparent peculiarity of this event. The evidence for the post-Permian trauma has long been appreciated in terms of the conspicuously depauperate Lower Triassic faunas (e.g. Schubert and Bottjer 1995). More recently an even grimmer scenario has been presented by Bottjer and coworkers (1995, 1996), who record a notable abundance of stromatolites that grew in what are interpreted as open marine settings. While other such examples are known from elsewhere in the Phanerozoic, including the present day (Rasmussen *et al.* 1993; Reid *et al.* 1995), they remain exceptions. The relative abundance of the Triassic stromatolites recalls the state of affairs in the Precambrian when, as noted above, the abundance of stromatolites is generally interpreted as a direct reflection of the absence or at least ineffectiveness of metazoan grazers. As also mentioned previously the connection between metazoan activity and stromatolite diversity is not accepted by all workers, and Grotzinger (1990) has invoked an hypothesis that looks to amounts of carbon dioxide and degree of carbonate saturation in the oceans as the factors controlling stromatolite growth. This too may provide a link with the Permian debacle, because Knoll *et al.* (1996; see also Grotzinger and Knoll 1995) have argued that rapid ocean turnover and release of carbon dioxide led to severe hypercapnia in animals and consequent physiological stress. In any event for the marine environment of the Triassic to be pushed towards a situation reminiscent of the Precambrian underlines yet further the severity of the Permian extinctions, and the scope of the ecological vacancies awaiting reoccupation later in the Mesozoic. On present evidence it seems more likely that the protracted lag-time in the Triassic was due to extrinsic stress, rather than for any innate inability for communities to reaggregate and restore diversity.

The case for protracted ecological instability in the aftermath of the Cretaceous event is better documented, although overall the extent of the crisis seems to have fallen short of the equivalent early Triassic episode. Although the degree of stratigraphic resolution is still relatively crude it is already clear that there was a time-scale of both disaster and more importantly recoveries, with some environments and biotas appearing to rebound much more promptly than others (Table 14.1). This compilation is very preliminary, and apart from considerable uncertainty about the time intervals, especially at the shorter end of the scale, it also begs the question of what is meant by the term 'recovery'. Nevertheless, such a dissection of response to the catastrophe is overdue, as is an extensive investigation of regional variations (e.g. Barrera and Keller 1994; MacLeod 1995; Keller *et al.* 1997; but see Raup and Jablonski 1993).

The evidence for perturbations, before eventual recovery, comes from several sources. These include the slow re-emergence from unspecified refugia of the socalled 'Lazarus' taxa, such as certain generalized micromorphic brachiopods (e.g. Surlyk and Johansen 1984; see also Johansen 1989*a,b*). It is even more convincingly demonstrated in terms of the dramatic fluctuations in population size, morphological variability and species composition in the communities of planktonic foraminifera (e.g. Gerstel *et al.* 1986, 1987). Enhanced vital effects of autotrophic fractionation of carbon isotopes in the coccolithophorids may also be attributed to ecological trauma (Stott and Kennett 1989). Evidence for major perturbations in the oceanic environment are evident from fluctuations in the record of carbon isotopes (e.g. Stott and Kennett 1989; Zachos et al. 1989) and also barium (Zachos et al. 1989) which is a tracer for marine productivity. The extraordinary instability of the oceanic ecosystems has also been emphasized by a study of the magnetic susceptibility of sediments either side of the boundary (D'Hondt et al. 1996). This methodology is especially appropriate for deep-sea cores, both because of the ease of technical operation and more importantly its sensitivity to the ratio between the calcareous component of the sediment, derived from the infall of the plankton, and the clay fraction whose contained iron provides the magnetic signature. Those sediments deposited prior to the K/T impact show muted variations in magnetic susceptibility (Figure 14.3), but in the earliest Tertiary there is significant amplification. D'Hondt et al. (1996) plausibly suggest that this enhanced amplification results from a stressed ecosystem with a reduced ability to buffer climatically forced Milankovitch cyclicity, possibly compounded by the interference on solar insolation by an equatorial ring of orbiting debris (Schultz and Gault 1990). In addition a detailed examination of the relationship between the magnetic susceptibility and concentrations of carbonate reveals that although in general there is the expected inverse relationship, when the concentration of carbonate is low this relationship becomes positive. D'Hondt et al. (1996) suggest that one explanation for the latter observation is that in the immediately post-catastrophe ocean the palaeoproductivity of the siliceous plankton (diatoms and radiolaria) might have been negatively correlated with the calcareous plankton. This too could be consistent with ecological instability. This evidence agrees with the data from microfossils in suggesting major oceanic instability persisting for at least a million years (Table 14.1). Evidence from benthic molluscan faunas indicates an even more protracted period for recovery (Hansen et al. 1993; see also D'Hondt et al. 1998).

It seems, therefore, that the early Tertiary marine environment showed recovery across a range of time scales, and possibly also geographies. While it could be argued that different environments were more or less sensitive to both perturbation and subsequent recovery, the immense time before some sort of normalcy returned suggests that extrinsic forcing (e.g. atmospheric composition, especially enhanced CO_2, or the existence of orbiting ejecta derived from the original impact) were more important.

So much for the oceans, what of terrestrial ecosystems? Apart from the immediately post-impact 'fern-spike' (Table 14.1) in North America plant communities were evidently hard-hit (e.g. Johnson and Hickey 1990), but made a rapid recovery (e.g. Tschudy and Tschudy 1986). Discussion of pre- and post-impact floras, however, is not straightforward because in the early Tertiary there were also significant long-term climatic changes (e.g. Wolfe 1990). Although the radiation of the mammals has received wide attention, only recently has it become a focus for quantitative investigations which point to a very rapid radiation (J. Alroy, personal communication).

Magnetic Susceptibility

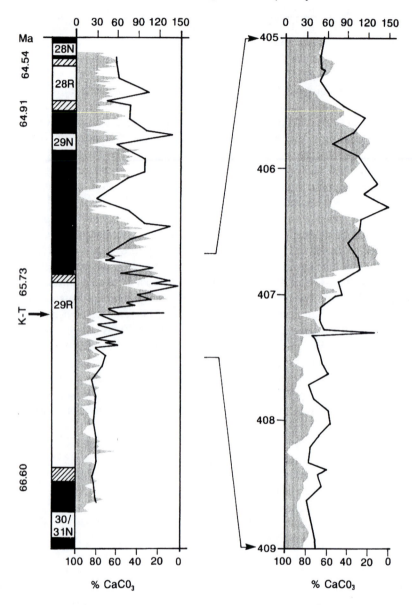

Fig. 14.3 A comparison of the sedimentary record at Deep Sea Drilling Project Site 528 (South Atlantic) in terms of magnetic susceptibility (solid curve) and carbonate content (solid line) across the Cretaceous/Tertiary (K/T) boundary. The magnetostratigraphy and radiometric ages (Ma) are depicted on the left-hand side and the right-hand column is an enlargement of the K/T section; figures refer to metres below the seafloor with the K/T boundary at 407.31 m. The magnetic susceptibility is largely controlled by iron concentrations, which occur mostly in detrital clays. The carbonate is largely derived from calcareous microplankton, including coccolithophorids and foraminifera. The figure emphasises the

As first construed the impact hypothesis for the K/T event emphasized global darkening due to the ejection of rock dust and, it was subsequently realized, soot into the upper atmosphere as the forcing mechanism for extinction by both the cessation of photosynthesis and climatic perturbations, especially cooling. This was no doubt highly deleterious, but present estimates suggest that severe darkening did not persist for more than a few months and that the postulated plummeting of surface temperatures was offset by the heat reservoir of the oceans so that climatic disturbances may have been severe rather than catastrophic (Covey *et al.* 1994). It is also possible that the trauma was enhanced by the production of acid rain, with a sulphur source either in the bolide itself or vaporized from the Chicxulub impact site that was rich in anhydrite ($CaSO_4$).

Paradoxically, the major accumulation of data pertaining to the K/T impact has led to the comparative neglect of the biological effects of other impacts. In part this is probably because a rather extensive search for extraterrestrial signatures at other stratigraphic intervals recording mass extinctions has been for the most part negative (Orth *et al.* 1984, 1986; McGhee *et al.* 1986; Goodfellow *et al.* 1992; Erwin 1995) or even when detected (Hodych and Dunning 1992; Bice *et al.* 1992; Claeys and Casier 1994) still seems to be inadequate as a forcing mechanism. Nevertheless, recent calculations of the energy released from major impacts (Figure 14.4) are sobering. In particular, the calculations by Toon *et al.* (1997) deserve detailed consideration. Some effects, such as major earthquakes and local blast, are perhaps more germane to the fragile infrastructures of our civilization and may be less significant in terms of the biosphere as a whole. Others, however, certainly are not. Calculations are, of course, constrained by the frequency of impact and the amount of energy released, which is a function of both size of the bolide and its impact velocity. Significant danger, however, is imposed by impact energies in excess of about 10^6 megatons, which if the infall rates are calculated correctly will occur on average about once every 10–50 Myr. In such cases, the prospects are hardly encouraging. As was almost certainly the case in the K/T event lofting of fine-grain particles, be they dust, soot or sulphate aerosols, leads to global darkening and the collapse of photosynthesis (Figure 14.5). Toon *et al.* (1997) estimate such disasters should occur approximately every 5–10 Myr. Only more recently has consideration been given to the effects of the generation of major tsunamis (Figure 14.6), which although not necessarily global would be locally devastating across the continental shelf and lowlands (see also Jansa 1993). At impact energies in excess of about 10^7 megatons the kinetic energy released by the infalling debris is transformed into radiant heat that can lead to spontaneous ignition of terrestrial floras in areas substantially in excess of the area of the present-day United States (Figure 14.7).

---

instability of the post-catastrophe oceanic system in comparison with the latest Cretaceous. [From Oscillatory marine response to the Cretaceous-Tertiary impact, *Geology* (vol. 24, pp. 611–614), S. D'Hondt, J. King and C. Gibson. Modified with permission of the publisher, the Geological Society of America, Boulder, Colorado, USA. Copyright 1996 Geological Society of America.]

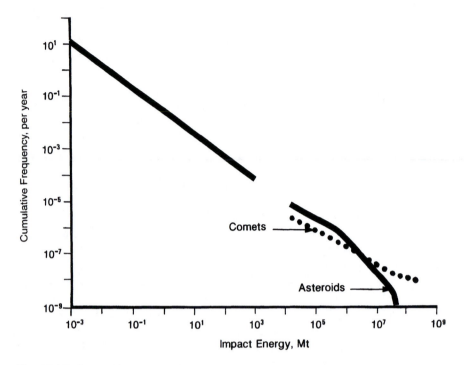

Fig. 14.4 Estimated frequency of impacts of extraterrestrial bodies on the Earth, based on several sources of data [Figure is redrawn from Toon *et al.* (1997, fig. 2). Reprinted from *Reviews of Geophysics* and the American Geophysical Union.]

Where then is the evidence for these traumatic episodes in the geological record? In an investigation of the Montagnais impact structure, about 45 km diameter and formed about 51 Myr ago, Jansa (1993) has noted the lack of evidence for extinction, so proposing that whatever local destruction is caused, only in the case of comets with a nuclear diameter in excess of about 4 km is a real threat imposed. As noted above the evidence for any mass extinction being driven by extraterrestrial impact is, apart from the K/T event, at best weak and for at least the late Ordovician (Orth *et al.* 1986) and end-Permian event (e.g. Clark *et al.* 1986) apparently entirely lacking. Nor do the puzzles end here. A potentially gigantic crater has been identified in the Kalahari desert of South Africa, possibly up to 350 km diameter, and dated at about 145 Myr (Koeberl *et al.* 1997). The near coincidence with the Jurassic Cretaceous boundary is intriguing, yet to date the evidence for severe extinctions across this interval is limited.

The proponents of co-ordinated stasis, a concept returned to in Section 14.5, also identify times of rapid turnover, and it might be possible to equate some of these with extraterrestrial impacts. It is also likely that some so-called 'event-horizons', the stratigraphic potential of which has received close attention by stratigraphers, are a consequence of highly energetic collisions. Nevertheless, the

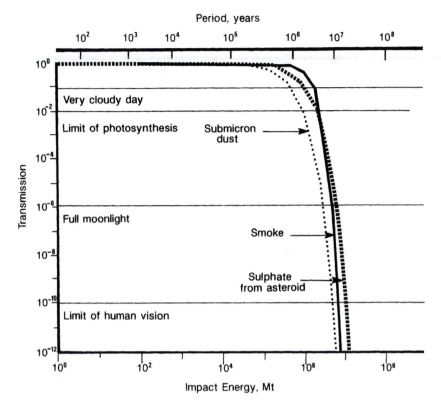

Fig. 14.5 Estimates of transmission of visible light to the Earth's surface through dust, smoke (soot) and sulphate aerosols following major impacts. Note that the three sources of particulate are independent of each other, and indicate that impacts yielding energies in excess of ca. 10^6 megatons will lead to darkness and the collapse of photosynthesis about once every 5–10 Ma. [Figure is redrawn from Toon *et al.* (1997, fig. 20a). Reprinted from *Reviews of Geophysics* and the American Geophysical Union.]

fossil record of extinctions appears to show little evidence for catastrophic extirpation, although one wonders if the vagaries of range completeness (or lack thereof) and taxonomic splitting are masking some events. It is also the case that sedimentologists and stratigraphers are seldom attuned to the possible evidence for impact driven processes, not least because so far as I am aware there are no coherent models for sedimentological consequences of the spontaneous ignition of several million square kilometres of land followed by its partial inundation beneath a 50 metre high tsunami.

While the influence of extraterrestrial impacts is surely more important than presently realized, certainly it may be premature to assume that mass extinctions *per se*, however caused, are the leading force in shaping the history of diversity (see also Jackson and McKinney 1990). Nor need they represent the only extrinsic factors that pummel the biosphere. The role of Milankovitch driven climatic

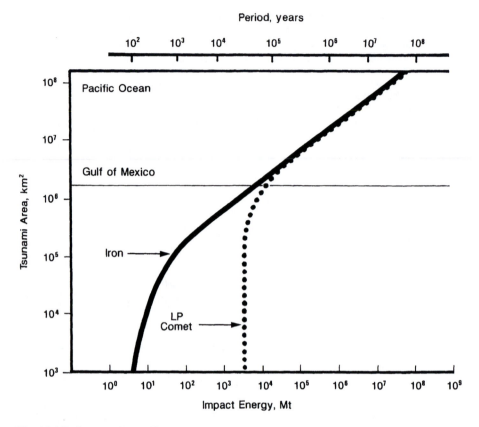

Fig. 14.6 Estimates of the effects of impacts by iron asteroids and long-period (LP) comets in terms of production of deep-water tsunamis with a 10 m amplitude in a 4 km deep ocean. [Figure is redrawn from Toon *et al.* (1997, fig. 7). Reprinted with permission from *Reviews of Geophysics* and the American Geophysical Union.]

cycles in stressing the environment and promoting extinctions has received the forceful advocacy of K. D. Bennett (1997). Here too we have the possibility of recurrent stress, and a mechanism that could, in principle, be equated to the turnover pulse hypothesis of Vrba (1995). And yet doubts remain (Conway Morris, in press). In part this is because of the difficulty in correlating Milanko-vitch driven climate change with any known extinction, combined with the fact that most species durations far exceed the observed cyclicities. With such low species mortalities versus the frequencies of oscillation this makes the hypothesis difficult to test. There is also the uncertainty of deciding whether the much-enhanced Milankovitich cyclicities of the last couple of million years and their accepted connection to wildly oscillating glacial and interglacials would apply with equal force in a past where the influence of such oscillations was apparently more muted. Not only that, but as Bennett (1997) himself is at pains to point out the possibly chaotic behaviour of the Solar System (e.g. Laskar 1989; Sussman

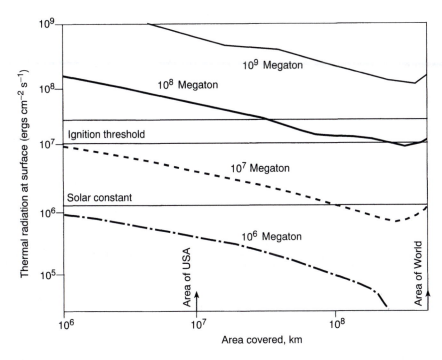

Fig. 14.7 The effect of ballistic ejecta re-entering the atmosphere in terms of the thermal radiative flux experienced on the Earth's surface. Debris infall is estimated to take 20 minutes, and that only half the energy generated by re-entry is radiated downward and half of that actually reaches the surface. The impactor is taken to be an icy object travelling at 50 km s^{-1}. The threshold for spontaneous ignition is bracketed by flash heating (upper line) and more protracted irradiation (lower line). Note that for impacts yielding an energy of 10^8 megatons fires will be ignited across much of the globe. [Figure is redrawn from Toon *et al.* (1997, fig. 15 (lower). Reprinted with permission from *Reviews of Geophysics* and the American Geophysical Union.]

and Wisdom 1992) may mean that the orbital forcing of the deep past is such that the present cyclicities provide a guide rather than a key to environmental stress and potential extinction.

14.5 The stability of communities

Ecologists have long emphasized the contrasting views of community structure as formulated by the individualistic and haphazard view of H. A. Gleason versus the recurrence and ecological interlocking identified by F. E. Clements (and subsequently C. Elton) (see Walter and Patterson 1994; Jablonski and Sepkoski 1996). In terms of the fossil record of the most recent past, especially with respect to terrestrial floras, there has been a strong emphasis on the Gleasonian view whereby driven by climatic shifts plant species reaggregate in novel combinations so that the present-day distributions and associations may have no precise counterpart in floras even a few thousand years old (e.g. Overpeck *et al.* 1992; see also Bennett 1997). Similar conclusions have been reached by Coope (1979, 1987) in his extensive analyses of Quaternary beetle assemblages, which also point to an exceptionally dynamic and unpredictable history of migration and reassociation. These influential studies refer to relatively short periods of geological time, and co-occur with the very marked shifts in the environment imposed by the recurrent swings between glacial and interglacial regimes. A very much longer perspective was taken by Buzas and Culver (1994), who documented the changing species compositions of communities of shallow marine benthic foraminifera of the North American Atlantic Coastal Plain during much of the Cainozoic. In this case embayments (known as the Salisbury and Albemarle embayments) were successively isolated and reconnected with the main species reservoir of the open shelf by regressions and transgressions, respectively. Buzas and Culver (1994) demonstrated that each reimmigration established a community of benthic foraminifera with a markedly distinct taxonomic profile, and in this sense represented only a subset, apparently drawn at random, of the main species pool.

This turbulent view of diversity as expressed in communities is almost directly antithetical to a widespread identification of ancient communities as stable, recurrent entities with such properties as persistence, stasis, ecological similarity, and self-regulation. This notion of community stability, that is only occasionally punctuated by turnover and extinction, encompasses quite a wide range of views. Nevertheless, broadly consistent approaches have been presented both by some palaeobotanists (e.g. DiMichele and Phillips 1996), as well as by extensive series of investigations in shallow marine communities (e.g. Aberhan 1993; Miller 1993; Brett *et al.* 1996; Schopf 1996; Tang and Bottjer 1996; see also DiMichele 1994) and also reef ecosystems (Pandolfi 1996). Not all palaeoecologists see such patterns, and in particular trenchant criticisms of the general concept of community stability as now enshrined in the concept of co-ordinated stasis has been offered by Alroy (1996). This analysis deserves particular attention, because it is based on a carefully prepared database of Tertiary North American mammals and also on account of the exactly formulated tests for the several hypotheses connected to the depiction of co-ordinated stasis. Significantly Alroy (1996) failed to identify either sudden turnovers of diversity or a correlation between the rates of extinction and origination. A critique in a related vein is offered by Bambach and Bennington (1996). Their review, of whether communities can in any sense be said to evolve, is more sympathetic to the ideas of co-ordinated stasis, but also notes that even within each interval of supposed stability there is still considerable turnover. Such ripostes may only be a partial solution to solving this divergence of interpretation. One wonders whether the climatic extremes of the last few million years are especially disruptive, notably for terrestrial communities (see also DiMichele 1994). There is also the obvious criticism of the problems of time-averaging (e.g. Bambach and Bennington 1996) and the difficulties of precise stratigraphic correlation in some older sequences. These factors might conspire to impose an apparent uniformity on community structure that originally was much more dynamic and labile.

14.6 Cryptic diversity

Despite the biases and imperfections of the fossil record the pattern of Phaner-ozoic diversity is generally accepted to be robust. One obvious pitfall, that of the relative proportion of soft-bodied to skeletalized taxa at various geological intervals, deserves specific mention because of the increasing level of documentation of fossil Lagerstätten such as the Burgess Shale-type faunas (Conway Morris 1998) or mid-Mesozoic Plattenkalk (Bernier and Gaillard 1994). Such an analysis, however, is not straightforward. The first problem is that there is a spectrum of taphonomic resistances of animals, meaning there is no simple distinction between skeletal and soft-bodied. Another problem is the wide variety of preservations encountered in fossil Lagerstätten, including carbonaceous films, diagenetic carbonate nodules, pyritization, phosphatization and polymerized tree resin (amber). In the last case, for example, Henwood (1993) ingeniously compared the taxonomic profile of insects in amber faunas from the mid Tertiary of the Dominican Republic with those obtained in similar circumstances by entomologists using standard trapping techniques (e.g. light trap, fogging) to show that notwithstanding the exquisite fossil preservation (e.g. Poinar 1992) amber faunas contain comparable sampling biases. At present all it seems possible to conclude is that fossil Lagerstätten are themselves incomplete repositories of diversity, may occur in 'unusual' environments (e.g. lagoons, deeper water), but still do not suggest that our existing ideas of ancient diversity are hopelessly warped by taphonomic factors.

This does not mean, however, that we can afford to be complacent because several potentially more severe problems exist. These concern the role of phylogenetic systematics, the apparent discrepancies between molecular and taxonomic data, and hitherto overlooked 'hot-spots' of diversity. The first point, that of the role of phylogenetic systematics and especially the cladistic methodology in the context of the study of diversity has not gone unnoticed. In brief, present-day diversity curves are for the most part plotted at the taxonomic level of family. Generic databases do exist, but are largely unpublished and seldom taxonomically up-to-date. The principal complaint of phylogenetic systematists is that many of the taxa are paraphyletic and even polyphyletic (see Smith 1994), and thereby give a skewed assessment of original diversity. The most powerful demonstration of this was by Patterson and Smith (1987) who reanalysed taxonomic data, specifically of fish and echinoderms of which they have respectively expertise, that had been used to identify a periodicity in mass extinctions (Raup and Sepkoski 1986). Using criteria of phylogenetic systematics the periodicity of extinction, in at least these two groups, vanished. Since then the debate as to whether Linnean taxa, especially at the level of family, can provide sufficient proxy for diversity has continued in a rather half-hearted and inconclusive manner (Sepkoski and Kendrick 1993). Another area of phylogenetic systematics that may deserve more attention in this context is the concept of ghost taxa (see Norell 1992). These are taxa identified as logically necessary in the construction of a cladogram, but unknown in actuality. Clearly where ghost taxa are abundant, again our perceptions of diversity may be skewed.

The other important area of cryptic diversity concerns the emerging discrepancies between established taxonomic data and information inferred from molecular biology. In brief, three categories of divergence stand out, of which the first two involve proposed times of origination. Consider angiosperms and mammals. The palaeontological record indicates the flowering plants originated in the early Cretaceous, while following the end-Cretaceous extinctions the latter underwent a major adaptive radiation in the Paleocene that resulted in appearance of most of the major mammal groups. The molecular data point to a very different story. In the case of the angiosperms dates of divergence varying from early Jurassic to Carboniferous are inferred (e.g. Wolfe *et al.* 1989; Brandl *et al.* 1992; Martin *et al.* 1993; Laroche and Bousquet 1995; Kolukisaoglu *et al.* 1995). The case of the mammals is somewhat less dramatic, but molecular data still suggest a significant divergence of mammal groups in the Mesozoic (Janke *et al.* 1994; Hedges *et al.* 1996; Springer *et al.* 1997), as well as discrepancies within the Cainozoic radiation (e.g. Frye and Hedges 1995). Similar conclusions may also apply to the divergence times of some groups of birds (Härlid *et al.* 1997).

The standard responses to these divergences in the estimated times of origination are well known, and they may be correct. The principal objection, of course, is that the basis for the molecular estimates assumes a clock-like behaviour for substitution. If molecular clocks periodically run fast, then it may be easier to reconcile these data with the fossil record. It is sometimes supposed that such speeding up of the molecular clock is coincident with adaptive radiations, although the consequences of accepting such an assumption appear to have been neglected. An alternative source of potential confusion, again widely acknowledged, is the lateral transfer of genetic material (Syvanen 1994). The third possibility is to decouple assumptions about molecular architecture and the taxonomic pigeon-holing that too often accompanies the recognition of major groups. One of the most valuable concepts of phylogenetic systematics is that of the stem-group, whereby the emergence of major groups is documented in the serial acquisition of characters. The implications of this approach for the angiosperms and mammals are interesting, albeit in somewhat different ways. In the former case there is a highly controversial record of pre-Cretaceous plants that have been interpreted as having angiospermous characteristics (see Crane *et al.* 1995). The case of the mammals is more disputed because the identification of such groups as the primates and ungulates in the Mesozoic depends on subtleties of interpretation, especially of the teeth. Archibald (1996), for example, identifies a late Cretaceous group (the zhelestids) of rat-sized ungulates, but even so these come from strata only 20 Myr before the Tertiary.

The significance of both these examples may be, however, that a substantial part of the organismal complexity had been achieved prior to the adaptive radiation, so that the triggers—still obscure in the case of the angiosperms, but evidently the ecological release in the post-catastrophe world of the Paleocene for the mammals—were those of opportunity rather than macroevolutionary reorganization. The implication of this for understanding adaptive radiations could be considerable because it implies that in at least some cases functional complexes were assembled over substantial periods of geological time. Such inherency is not meant to imply that there is an inevitability in an evolutionary outcome, but it may favour the issue in a particular direction.

In the above two examples at least there is a controversial fossil record that could match the estimated time of molecular divergence. There are, however, various other discrepancies whereby the molecular data indicates a protracted cryptic interval that predates the appearance of the group in the fossil record. Examples include the metazoans as a whole (Wray *et al.* 1996), and among the protistans examples may be taken from the foraminiferans (Merle *et al.* 1994; Darling *et al.* 1996, 1997; Pawlowski *et al.* 1994, 1996, 1997; Wade *et al.* 1996), the diatoms (Philippe *et al.* 1994), and using biomarkers rather than a molecular clock the dinoflagellates (Summons 1992; Moldowan *et al.* 1996).

In the case of the metazoans the assumption of a molecular clock in the case of both haemoglobin (Runnegar 1982*b*) and collagen (Runnegar 1985) had pointed to an origination at least 750 Myr ago, whereas the more recent analysis of Wray *et al.* (1996) indicates a substantially older origin that could have exceeded 1000 Myr.

A critical reanalysis of these data, however, suggests that such figures—arrived at by averaging—may be less reliable than a clock-by-clock comparison. This latter approach gives figures more in line with the estimates of Runnegar (1982b, 1985), but still points to an interval of pre-Ediacaran metazoan evolution (Conway Morris 1997; see also Ayala *et al.* 1998). The case for cryptic evolution in the foraminifera is more complex, because broadly there are two schools of thought. One (Darling *et al.* 1996, 1997; Wade *et al.* 1996) has emphasized an apparently very ancient origin for the planktonic foraminifera, hundreds of millions of years before their appearance in the Lower Jurassic. The extreme branch lengths of the planktonic foraminifera is emphasized by the other school (Pawlowski *et al.* 1994, 1996, 1997), who while acknowledging an ancient origin for the group as a whole place the planktonics in the crown group and suggest the very variable rates of substitution have distorted the phylogeny so that the apparently early origin of the planktonics is an artefact. The reasons for such extreme branch lengths in the planktonic foraminifera are obscure, although Pawlowski *et al.* (1997) proposed that ultraviolet radiation in the upper water column may be responsible. This needs to be tested against other planktonic groups, such as the coccolithophorids, and within the planktonic foraminifera themselves according to the positions they adopt in the water column. Philippe *et al.* (1994) have drawn attention to a major discrepancy in the estimated origination of diatoms from molecular data versus the fossil record, the latter indicating a Jurassic appearance of limited diversity followed by a Cretaceous effloresence. They conclude that a cryptic history of 'naked' diatoms, possibly extending as far back as the Precambrian, preceded their acquisition of siliceous skeletons. Philippe *et al.* (1994) also review (and reject) the alternative explanation of molecular rate changes. This is, of course, the standard riposte of those anxious to reconcile times of molecular divergence with the fossil record. This latter view raises some uncomfortable difficulties. First, if molecular clocks run fast during adaptive radiations it implies a linkage between phenotype and molecules that are generally thought to be 'invisible' to morphology (but see Omland 1997). Alternatively, clock dates could be telescoped if the initial stages ran fast before it slowed along a hyperbolic function. The idea of a molecular clock 'ageing' seems to find little favour, as yet.

The case for an appearance of dinoflagellates long before the first fossil records relies on molecular data, but in the form of chemical fossils extracted from sediments that date back at least as far as 700 Myr (Summons 1992). The dinoflagellates are especially appropriate in this respect because of their possessing characteristic biomarkers, and the potential for inferring a cryptic history of this group has been explored by Moldowan *et al.* (1996).

Not all the data from molecular biology, however, point to cryptic intervals of evolution pre-dating first appearances in the fossil record. In an extensive analysis of the molecular clock data of enzymes Doolittle *et al.* (1996; see also Feng *et al.* 1997) concluded that the times of divergence of many major groups of both prokaryotes and eukaryotes were substantially younger than indicated in the fossil record. This analysis has not been well received by palaeontologists, and in addition Martin (1996) has proposed that this analysis by Doolittle *et al.* (1996) is seriously confounded by the prevalence of lateral gene transfer. For the most part such transfer, while admitted for groups such as bacteria and plants, has been considered to be unimportant in animals. Such a view is coming under scrutiny with possible examples of hybridization in groups as disparate as cnidarians (Odorico and Miller 1997) and chordates (Spring 1997).

There is a somewhat different area of molecular biology which also hints at major gaps in our knowledge of the diversity of microbial communities, notably of prokaryotes. Research by Barns and co-workers (e.g. Barns *et al.* 1996*a*, *b*), using genetic probes in microbial systems such as hot-springs, is revealing an astonishing array of otherwise unknown prokaryotes. Many show affinities to the group of Archaebacteria known as Crenarchaeota, but these bacteria also expand the taxonomic range and suggest the divisions between the Kingdoms are less absolute than earlier thought. Although there is reason to believe that these prokaryotic groups (and especially the thermophiles) are phylogenetically very ancient, at first sight the prospects of recognition in the fossil record are not encouraging. Nevertheless, fossil hydrothermal systems are known (Walter 1996; Knoll and Walter 1996), and there is also the potential for diagnostic biomarkers to be recovered (Summons 1992; Summons *et al.* 1996).

14.7 Conclusions

The pattern of diversity as seen in the fossil record appears to be robust, but it is almost entirely based on a traditional Linnean taxonomy (see Foote 1996). The impact of phylogenetic systematics, with their insistence on the recognition of monophyly may lead to significant changes in the tabulation of diversity. Revisions may also stem from the recognition of cryptic diversities that may be inferred from the disconcerting gaps that appear to separate times of origination as seen in the fossil record as against the inferences made on the basis of molecular biology. Reconciliation of these seemingly contradictory data may go some way to resolving the problems in understanding the speed and extent of adaptive radiations if it transpires that many key features are acquired prior to the 'releasing event' such as a mass extinction or climatic shift.

Nevertheless, whatever shifts in the pattern of diversity emerge, it is surely the case that the known curve (Figure 14.1) is not simply constructed on the basis of artefacts. In particular, times of diversification are real. At present it is still very difficult to identify the key factors involved, and it would be unwise to identify an over-riding and unique principle. First, there is appeal to the many extrinsic factors. Here one might include: marine regressions, biogeographic changes of continental distribution and oceanic structure that are ultimately controlled by plate tectonics, changes in atmospheric composition and/or ocean chemistry, massive volcanism, and changes in substrate type as engendered by taphonomic feedback that includes skeletal production and microbial activity. Of those factors that are usually labelled as intrinsic there are also many examples. They tend to be rather specific, relating to particular groups, but they are many. Two examples, taken more or less at random, are the development of the mammalian hypocone (Hunter and Jernvall 1994), and the ability of Ordovician echinoderms to colonize new substrates (Sprinkle and Guensburg 1995).

Is the history of diversity then merely the product of a concatenation of a vast number of independent variables, whose chance associations and timings result in what for us is a familiar world but one which in actuality is almost entirely contingent? Alternatively, is there a deeper mathematical structure to diversity? As it happens this is a metaphysical question (Conway Morris 1998), and reveals the endless but potentially fruitful tension between the belief in a bedrock of physical realities and the operation of an historical process with what to many of us comprises a clear directionality. The historical dimension has formed the bulk of the discussion here, but can we also identify a deeper architecture that might provide a theoretical underpinning to our enterprise? One possible avenue is the work by Solé et al. (1997; but see Kirchner and Weil (1998)). These workers identified self-similarity in different time series of ancient diversity, mostly at the taxonomic level of family but also genera. The fractal structure, it is suggested, could be consistent with self-organized criticality. The data source they used (Benton 1993) is open to serious, albeit inevitable, criticisms of incompleteness and bias (Conway Morris 1994b). Thus, this analysis is perhaps more suggestive than compelling, especially as the observed power spectra consistently only approach the expected scaling exponent of $\beta = 1$. Nevertheless, the conclusion of Solé et al. (1997, p. 766) that 'the internal biotic organization [is] the basic component for the response of the biosphere to external perturbations' is consistent with a widely held view of the world-biota being interactive, coherent and possessing its own dynamic that while not immune to abiological factors was and is not in a state of recurrent lability on account of a ceaseless pummelling by the external environment. In this sense ecological factors are important determinants of diversity and during rare intervals, e.g. early Triassic or Paleocene, exert exceptional levels of stress on biological communities. Certainly on any familiar ecological time scale such intervals of one or more million years are protracted, but otherwise long-term disruption that might be predicted from such factors as cometary impacts, Milankovitch forcing, or severe climatic fluctuations (e.g. El Niño) in point of fact appear to be muted.

Let us accept, then, the four billion year history of evolution follows a script that is largely based on biological interactions. In this context, the role of interspecific competition as a force for moulding diversity might seem pre-eminent, yet in actuality it has encountered intense scepticism (e.g. Benton 1996), in part because of the problem of trying to define credible tests. In a 3-month trial Schluter (1994) identified competition as the driving force in morpho-logical divergence of stickleback fish, but many palaeontologists would question whether such results can be extrapolated into a geological time-scale, notwith-standing Schluter's explicit link to adaptive radiations. On a much larger time-scale Sepkoski's (1996) reformulation of diversity plots points towards interac-tions between brachiopods and bivalve molluscs. This study suggests that even on a relatively coarse scale diversity data may reveal competitive patterns. Sepkoski's (1996) analysis also echoes the earlier conclusions of Raup (1981) who in charting a secular decline in trilobite diversity noted that it could be explained by competitive displacement. Nevertheless, one of the key objections to the notion of competition is the persistence of survivors representing the supposedly inferior group. How seriously should we take this point? When we consider the range of groups for which extirpation via an obvious mass extinction is unlikely, e.g. graptolites, thelodonts, creodonts, or have at best relict status as the trilobites did in the Permian or monoplacophorans do in Recent seas, then competitive displacement remains a reasonable possibility. This does not provide a complete explanation. A well-known problem concerns the encrusting cyclostome and cheilostome bryozoans, in which notwithstanding the latter's conspicuous success in overgrowth has still failed to dislodge the weaker cyclostomes in competition for substrate space (Lidgard et al. 1993). Nevertheless, as McKinney (1995) points out subsequent to the rise of the cheilostomes, the cyclostomes are forced into a temporal refuge by restrictions in colony size and early sexual maturation. Thus, there are other ways of staying in the race. Not only that but in the real world there are other groups, notably algae, sponges and tunicates, that are even more effective at overgrowth. Competition need not spell the doom of a group, but the net result is a more complex and possibly faster-paced world (McKinney 1993).

In this way the history of diversity can be seen to be progressive in as much as the diversity of species, the richness of communities and the complexity of ecologies in the Recent appear to be unrivalled in comparison with the geological past. To point out that bacteria are still with us really misses the point. Not only is their 'expertise' biochemical, but they are displaced into the refugia of hot springs, sediments, and animal guts. As we have seen if the animal biosphere enters extreme crisis, as seems to have been the case in the early Triassic, then the microbial mats and stromatolites return. But if the history of diversity is now clearer, how we actually got to where we are is still obscure. And with our present predicament in one sense this hardly matters. The appearance of our species with unique capabilities means that, so to speak, all bets are now off. The fact we are a product of evolution and represent a tiny twig on the immense tree of life is not germane to our destruction of the richest ecosystem the world has ever seen.

Summary

On a perfect planet, such as might be acceptable to a physicist, one might predict that from its origin the diversity of life would grow exponentially until the carrying capacity, however defined, was reached. The fossil record of the Earth, however, tells a very different story. One of the most striking aspects of this record is the apparent evolutionary *longueur*, marked by the Precambrian record of prokaryotes and primitive eukaryotes, although our estimates of microbial diversity may be seriously incomplete. Subsequently there were various dramatic increases in diversity, including the Cambrian 'explosion' and the radiation of Palaeozoic-style faunas in the Ordovician. The causes of these events are far from resolved. It has also long been appreciated that the history of diversity has been punctuated by major extinctions. The subtleties and nuances of extinction as well as the survival of particular clades have to date, however, received rather too little attention, and there is still a tendency towards blanket assertions rather than a dissection of these extraordinary events. In addition, some but perhaps not all mass extinctions are characterized by long lag-times of recovery, which may reflect the slowing waning of extrinsic forcing factors or alternatively the incoherence associated with biological reassembly of stable ecosystems. The intervening periods between the identified mass extinctions may be less stable and benign then popularly thought and in particular the frequency of extraterrestrial impacts leads to predictions of recurrent disturbance on timescales significantly shorter than the intervals separating the largest extinction events. Even at times of quietude it is far from clear whether biological communities enjoy stability and interlocked stasis or are dynamically reconstituted at regular intervals. And finally can we yet rely on the present depictions of the rises and falls in the levels of ancient diversity? Existing data is almost entirely based on Linnean taxa, and the application of phylogenetic systematics to this problem is still in its infancy. Not only that, but even more intriguingly the pronounced divergence in estimates of origination times of groups as diverse as angiosperms, diatoms, and mammals in terms of the fossil record as against molecular data point to the possibilities of protracted intervals of geological time with a cryptic diversity. If this is correct, and there are alternative explanations, then some of the mystery of adaptive radiations may be dispelled, in as much as the assembly of key features in the stem groups could be placed in a gradualistic framework of local adaptive response punctuated by intervals of opportunity.

Acknowledgements

I thank the organizers of this Discussion meeting, Anne Magurran and Bob May, for their invitation to participate. Sandra Last is warmly thanked for typing numerous versions of this paper, as is Sharon Capon for drafting the figures. My research is supported by the Natural Environment Research Council and St John's College, Cambridge. Cambridge Earth Sciences Publication 5055.

References

Aberhan, M. 1993 Faunal replacement in the Early Jurassic of northern Chile: implications for the evolution in Mesozoic benthic shelf ecosystems. *Palaeogeogr. Palaeoclimatol. Palaeoecol.* **103**, 155–177.

Alcala-Herrera, J. A., Grossman, E. L. and Gartner, S. 1992 Nannofossil diversity and equitability and fine-fraction $\delta^{13}C$ across the Cretaceous/Tertiary boundary at Walvis Ridge Leg 74, South Atlantic. *Mar. Micropaleont.* **20**, 77–88.

Alroy, J. 1996 Constant extinction, constrained diversification, and uncoordinated stasis in North American mammals. *Palaeogeogr. Palaeoclimatol. Palaeoecol.* **127**, 285–311.

Alvarez, W., Claeys, P. and Kieffer, S. W. 1995 Emplacement of Cretaceous-Tertiary boundary shocked quartz from Chicxulub crater. *Science* **269**, 930–935.

Archibald, J. D. 1996 Fossil evidence for a late Cretaceous origin of 'hoofed' mammals. *Science* **272**, 1150–1153.

Argast, S., Farlow, J. O., Gabet, R. M. and Brinkman, D. L. 1987 Transport-induced abrasion of fossil reptilian teeth: Implications for the existence of Tertiary dinosaurs in the Hell Creek Formation, Montana. *Geology* **15**, 927–930.

Awramik, S. M. 1971 Precambrian columnar stromatolite diversity: reflection of metazoan appearance. *Science* **174**, 825–827.

Ayala, F. J., Rzhetsky, A. and Ayala, F. J. 1998 Origin of the Metazoan phyla: Molecular clocks conform paleontological estimates. *Proc. Natl. Acad. Sci. USA* **95**, 606–611.

Bambach, R. K. and Bennington, J. B. 1996 Do communities evolve? A major question in evolutionary paleoecology. In *Evolutionary paleobiology in honor of James W. Valentine* (ed. D. Jablonski, D. H. Erwin and J. H. Lipps), pp. 123–160. Chicago: University of Chicago Press.

Barns, S. M., Delwiche, C. F., Palmer, J. D., Dawson, S. C., Hershberger, K. L. and Pace, N. R. 1996*a* Phylogenetic perspective on microbial life in hydrothermal ecosystems, past and present. In *Evolution of hydrothermal ecosystems on Earth (and Mars)*, pp. 24–39. Ciba Foundation Symp. **202**. Chichester: Wiley.

Barns, S. M., Delwiche, C. F., Palmer, J. D. and Pace, N. R. 1996*b* Perspectives on archaeal diversity, thermophily and monophyly from environmental rRNA sequences. *Proc. Natl Acad. Sci USA* **93**, 9188–9193.

Barrera, E. and Keller, G. 1994 Productivity across the Cretaceous/Tertiary boundary in high latitudes. *Geol. Soc. Am. Bull.* **106**, 1254–1266.

Bennett, K. D. 1997 *Evolution and Ecology: The Pace of Life.* Cambridge: Cambridge University Press.

Bennett, P. and Owens, I. P. F. 1997 Variation in extinction risk among birds: chance or evolutionary predisposition? *Proc. R. Soc. London Ser. B* **264**, 401–408.

Benton, M. J. 1985 Mass extinctions among non-marine tetrapods. *Nature* **316**, 811–814.

Benton, M. J. 1990 The causes of the diversification of life. In *Major evolutionary radiations* (ed. P. D. Taylor and G. P. Larwood), pp. 409–430. Syst. Ass. Spec. Vol. **42**. Oxford: Clarendon Press.

Benton, M. J. (ed.) 1993 *The fossil record 2*. London: Chapman & Hall.

Benton, M. J. 1995 Diversification and extinction in the history of life. *Science* **268**, 52–58.

Benton, M. J. 1996 On the nonprevalence of competitive replacement in the evolution of tetrapods. In *Evolutionary Paleobiology: In honor of James W. Valentine* (ed. D. Jablonski, D. H. Erwin and J. H. Lipps), pp. 185–210. Chicago: University of Chicago.

Bernier, P. and Gaillard, C. (ed.) 1994 Les calcaires lithographiques, sédimentologie, paléontologie, taphonomie. *Geobios. Mem. Spec.* **16**.

Bice, D. M., Newton, C. R., McCauley, S., Reiners, P. W. and McRoberts, C. A. 1992 Shocked quartz at the Triassic-Jurassic boundary in Italy. *Science* **255**, 443–446.

Bottjer, D. J., Campbell, K. A., Schubert, J. K. and Droser, M. L. 1995 Palaeoecological models, non-uniformitarianism, and tracking the changing ecology of the past. In *Marine environmental analysis from fossils* (ed. D. W. J. Bosence and P. A. Allison), pp. 7–26. Geol. Soc. Spec. Publ. **83**.

Bottjer, D. J., Schubert, J. K. and Droser, M. L. 1996 Comparative evolutionary palaeoecology: assessing the changing ecology of the past. In *Biotic recovery from mass extinction events* (ed. M. B. Hart), pp. 1–13. Geol. Soc. Spec. Publ. **102**.

Brandl, R., Mann, W. and Sprinzl, M. 1992 Estimation of the monocot-dicot age through tRNA sequences from the chloroplast. *Proc. R. Soc. London Ser. B* **249**, 13–17.

Brasier, M. D. 1992 Background to the Cambrian explosion [Introduction to thematic set: Precambrian-Cambrian boundary (10 papers)]. *J. Geol. Soc., London* **149**, 585–587.

Brett, C. E., Ivany, G. C. and Schopf, K. M. 1996 Coordinated stasis: An overview. *Palaeogeogr. Palaeoclimatol. Palaeoecol.* **127**, 1–20.

Budd, A. F. and Johnson, K. G. 1991 Size-related evolutionary patterns among species and subgenera in the *Strombina* group (Gastropoda: Columbellidae). *J. Paleontol.* **65**, 417–434.

Butterfield, N. J. 1994 Burgess Shale-type fossils from a Lower Cambrian shallow-shelf sequence in northwest Canada. *Nature* **369**, 477–479.

Butterfield, N. J. 1997 Plankton ecology and the Proterozoic-Phanerozoic transition. *Paleobiology* **23**, 247–262.

Buzas, M. A. and Culver, S. J. 1994 Species pool and dynamics of marine paleocommunities. *Science* **264**, 1439–1441.

Chen, J. Y., Zhou, G. Q., Zhu, M. Y. and Yeh, K. Y. ca. 1996 *The Chengjiang biota. A unique window of the Cambrian explosion*. Taiwan: National Museum of Natural Science.

Claeys, P. and Casier, J. G. 1994 Mickotektite-like impact glass associated with the Frasnian-Fammenian boundary mass extinction. *Earth Planet. Sci. Lett.* **122**, 303–315.

Clark, D. L., Wang C-Y., Orth, C. J. and Gilmore, J. S. 1986 Conodont survival and low iridium abundances across the Permian-Triassic boundary in South China. *Science* **233**, 984–986.

Clarke, A. 1993 Temperature and extinction in the sea: a physiologist's view. *Paleobiology* **19**, 499–518.

Coccioni, R. and Galeotti, S. 1994 K-T boundary extinction: geologically instantaneous or gradual event? Evidence from deep-sea benthic foraminifera. *Geology* **22**, 779–782.

Conway Morris, S. 1993*a* Ediacaran-like fossils in Cambrian Burgess Shale-type faunas of North America. *Palaeontology* **36**, 593–635.

Conway Morris, S. 1993*b* The fossil record and the early evolution of the Metazoa. *Nature* **361**, 219–225.

Conway Morris, S. 1994*a* Why molecular biology needs palaeontology. *Development, Suppl.* **1994**, 1–13.

Conway Morris, S. 1994*b* [Review of Benton 1993]. *Geol. Mag.* **131**, 706–707.

Conway Morris, S. 1997 Molecular clocks: Defusing the Cambrian 'explosion'? *Curr. Biol.* **7**, R71–R74.

Conway Morris, S. 1998 *The Crucible of Creation: The Burgess Shale and the Rise of Animals.* Oxford: Oxford University Press.

Conway Morris, S. (in the press). Orbital cycles: Distant drummer or frenzied timpanist? [A review of Bennett 1997]. *J. Biogeogr.*

Coope, G. R. 1979 Late Cenozoic fossil Coleoptera: Evolution, biogeography and ecology. *Annu. Rev. Ecol. Syst.* **10**, 247–267.

Coope, G. R. 1987 The response of late Quaternary insect communities to sudden climatic changes. In *Organization of communities: past and present* (ed. J. H. R. Gee and P. S. Giller), pp. 421–438. 27th Symp. British Ecol. Soc. Aberystwyth, 1986. Oxford: Blackwell Scientific.

Coope, G. R. 1994 The response of insect faunas to glacial-interglacial climatic fluctuations. *Phil. Trans. R. Soc. London Ser. B* **344**, 19–26.

Courtillot, V. and Gaudemer, Y. 1996 Effects of mass extinctions on biodiversity. *Nature* **381**, 146–148.

Covey, C., Thompson, S. L., Weissman, P. R. and MacCracken, M. C. 1994 Global climatic effects of atmospheric dust from an asteroid or comet impact on Earth. *Global Planet. Change* **9**, 263–273.

Crane, P. R., Friis, E. M. and Pederson, K. R. 1995 The origin and early diversification of angiosperms. *Nature* **374**, 27–33.

Crepet, W. J. 1979 Insect pollination: a paleontological perspective. *BioScience* **29**, 102–108.

Crimes, T. P. 1994 The period of early evolutionary failure and the dawn of evolutionary success: the record of biotic changes across the Precambrian-Cambrian boundary. In *The palaeobiology of trace fossils* (ed. S. K. Donovan), pp. 105–133. Chichester: Wiley.

Cronin, T. M. 1985 Speciation and stasis in marine Ostracoda: climatic modulation of evolution. *Science* **227**, 60–63.

Darling, K. F., Kroon, D., Wade, C. M. and Leigh Brown, A. J. 1996 Molecular phylogeny of the planktic foraminifera. *J. Foram Res.* **26**, 324–330.

Darling, K. F., Wade, C. M., Kroon, D. and Leigh Brown, A. J. 1997 Planktic foraminiferal molecular evolution and their polyphyletic origins from benthic taxa. *Mar. Micropaleontol.* **30**, 251–266.

D'Hondt, S., Dunaghay, P., Zachos, J. C., Luttenberg, D. and Lindinger, M. 1998 Organic carbon flukes and ecological recovery from the Cretaceous–Tertiary mass extinction. *Science* **282**, 276–279.

D'Hondt, S., King, J. and Gibson, C. 1996 Oscillatory marine response to the Cretaceous-Tertiary impact. *Geology* **24**, 611–614.

DiMichele, W. A. 1994 Ecological patterns in time and space. *Paleobiology* **20**, 89–92.

DiMichele, W. A. and Phillips, T. L. 1996 Climate change, plant extinctions and vegetational recovery during the Middle Pennsylvanian transition: the case of tropical peat-forming environments in North America. In *Biotic recovery from mass extinction events* (ed. M. B. Hart), pp. 201–221. Geol. Soc. Lond. Special Publication **102**.

Doguzhaeva, L. and Mutvei, H. 1993 Structural features in Cretaceous ammonoids indicative of semi-internal or internal shells. In *The Ammonoidea: environment, ecology and evolutionary change* (ed. M. R. House), pp. 99–114. System. Ass. Spec. Vol. **47**. Oxford: Clarendon Press.

Doolittle, R. F., Feng, D-F., Tsang, S., Cho, G. and Little, E. 1996 Determining divergence times of the major kingdoms of living organisms with a protein clock. *Science* **271**, 470–477.

Droser, M. L. 1991 Ichnofabric of the Paleozoic *Skolithos* ichnofacies and the nature and distribution of *Skolithos* piperock. *Palaios* **6**, 316–325.

Droser, M. L. and Bottjer, D. J. 1993 Trends and patterns of Phanerozoic ichnofabrics. *Annu. Rev. Earth Planet. Sci.* **21**, 205–225.

Droser, M. L., Bottjer, D. J. and Sheehan, P. M. 1997 Evaluating the ecological architecture of major events in the Phanerozoic history of marine invertebrate life. *Geology* **25**, 167–170.

Erwin, D. H. 1993 The great Paleozoic crisis: life and death in the Permian. New York: Columbia University Press.

Erwin, D. H. 1995 Permian global bio-events. In *Global events and event stratigraphy* (ed. O. H. Walliser), pp. 251–264. Berlin: Springer-Verlag.

Erwin, D. H. 1996a The geologic history of diversity. In *Biodiversity in managed landscapes* (ed. R. C. Szaro and D. W. Johnston), pp. 3–16. New York: Oxford University Press.

Erwin, D. H. 1996b Understanding biotic recoveries: extinction, survival, and preservation during the end-Permian mass extinction. In *Evolutionary paleobiology: in honor of James W. Valentine* (ed. D. Jablonski, D. H. Erwin and J. H. Lipps), pp. 398–414. Chicago: University of Chicago Press.

Erwin, D. H. and Pan, H-Z. 1996 Recoveries and radiations: gastropods after the Permo-Triassic mass extinction. In *Biotic recovery from mass extinction events* (ed. M. B. Hart), pp. 223–224. Geol. Soc. Lond. Spec. Publ. **102**.

Erwin, D., Valentine, J. and Jablonski, D. 1997 The origin of animal body plans. *Am. Sci.* **85**, 126–137.

Eshet, Y., Rampino, M. R. and Visscher, H. 1995 Fungal event and palynological record of ecological crisis and recovery across the Permian-Triassic boundary. *Geology* **23**, 967–970.

Fedonkin, M. A. and Waggoner, B. M. 1997 The late Precambrian fossil *Kimberella* is a mollusc-like bilaterian organism. *Nature* **388**, 868–871.

Feduccia, A. 1980 *The age of birds.* Cambridge, Massachusetts: Harvard University Press.

Feng, D.-F., Cho, G. and Doolittle, R. F. 1997 Determining divergence times with a protein clock. Update and reevaluation. *Proc. Natl. Acad. Sci. USA* **94**, 13028–13033.

Foote, M. 1996 Perspective: Evolutionary patterns in the fossil record. *Evolution* **50**, 1–11.

Foote, M. and Raup, D. M. 1996 Fossil preservation and the stratigraphic ranges of taxa. *Paleobiology* **22**, 121–140.

Fortey, R. A. and Owens, R. M. 1990 Trilobites. In *Evolutionary trends* (ed. K. J. McNamara), pp. 121–142. London: Belhaven.

Fortey, R. A., Briggs, D. E. G. and Wills, M. A. 1996 The Cambrian evolutionary 'explosion': decoupling cladogenesis from morphological disparity. *Biol. J. Linn. Soc.* **57**, 13–32.

Frye, M. J. and Hedges, S. B. 1995 Monophyly of the order Rodentia inferred from mitochondrial DNA sequences of the genes for 12S rRNA, 16S rRNA, and tRNA-valine. *Mol. Biol. Evol.* **12**, 168–176.

Galbrun, B. 1997 Did the European dinosaurs disappear before the K-T event? Magnetostratigraphic evidence. *Earth Planet. Sci. Lett.* **148**, 569–579.

Garrett, P. 1970 Phanerozoic stromatolites: noncompetitive ecologic restriction by grazing and burrowing animals. *Science* **169**, 171–173.

Gehling, J. G. 1991 The case for Ediacaran fossil roots to the metazoan tree. *Geol. Soc. India Mem.* **20**, 181–224.

Gehling, J. G. and Rigby, J. K. 1996 Long-expected sponges from the Neoproterozoic Ediacara fauna of South Australia. *J. Paleontol.* **70**, 185–195.

Gerstel, J., Thunell, R., Zachos, J. C. and Arthur, M. A. 1986 The Cretaceous/Tertiary boundary event in the North Pacific: planktonic foraminiferal results from Deep Sea Drilling Project site 577, Shatksy Rise. *Paleoceanography* **1**, 97–117.

Gerstel, J., Thunell, R. and Ehrlich, R. 1987 Danian faunal succession: Planktonic foraminiferal response to a changing marine environment. *Geology* **15**, 665–668.

Goodfellow, W. D., Nowlan, G. S., McCracken, A. D., Lenz, A. C. and Grégoire, D. C. 1992 Geochemical anomalies near the Ordovician-Silurian boundary, northern Yukon territory, Canada. *Hist. Biol.* **6**, 1–23.

Gould, S. J. 1985 The paradox of the first tier: an agenda for paleobiology. *Paleobiology* **11**, 2–12.

Gould, S. J. and Calloway, C. B. 1980 Clams and brachiopods—ships that pass in the night. *Paleobiology* **6**, 383–396.

Grotzinger, J. P. 1990 Geochemical model for Proterozoic stromatolite decline. *Am. J. Sci.* **290A**, 80–103.

Grotzinger, J. P. and Knoll, A. H. 1995 Anomalous carbonate precipitates: is the Precambrian the key to the Permian? *Palaios* **10**, 578–596.

Han, T. M. and Runnegar, B. 1992 Megascopic eukaryotic algae from the 2.1 billion-year-old Negaunee Iron-Formation, Michigan. *Science* **257**, 232–235.

Hansen, T. A., Farrell, B. R. and Upshaw, B 1993 The first 2 million years after the Cretaceous-Tertiary boundary in east Texas: rate and paleoecology of the molluscan recovery. *Paleobiology* **19**, 251–265.

Härlid, A., Janke, A. and Arnason, U. 1997 The mtDNA sequence of the ostrich and the divergence between paleognathous and neognathous birds. *Mol. Biol. Evol.* **14**, 754–761.

Harper, E. M., Palmer, T. J. and Alphey, J. R. 1997 Evolutionary response by bivalves to changing Phanerozoic sea-water chemistry. *Geol. Mag.* **134**, 403–407.

Hedges, S. B., Parker, P. H., Sibley, C. G. and Kumar, S. 1996 Continental breakup and the ordinal diversification of birds and mammals. *Nature* **381**, 226–229.

Henwood, A. A. 1993 Ecology and taphonomy of Dominican Republic amber and its inclusions. *Lethaia* **26**, 237–245.

Hildebrand, A. R., Pilkington, M., Connors, M., Ortiz-Aleman, C. and Chavez, R. E. 1995 Size and structure of the Chicxulub crater revealed by horizontal gravity gradients and cenotes. *Nature* **376**, 415–417.

Hodych, J. P. and Dunning, G. P. 1992 Did the Manicouagan impact trigger end-of-Triassic mass extinction? *Geology* **20**, 51–54.

Hughes, N. C. and Labandeira, C. C. 1995 The stability of species in taxonomy. *Paleobiology* **21**, 401–403.

Hunter, J. P. and Jernwall, J. 1995 The hypocone as a key innovation in mammalian evolution. *Proc. Natl Acad. Sci. USA* **92**, 10718–10722.

Jablonski, D. 1989 The biology of mass extinctions: a palaeontological view. *Phil. Trans. R. Soc. London Ser. B* **325**, 357–368.

Jablonski, D. and Sepkoski, J. J. 1996 Paleobiology, community ecology, and scales of ecological pattern. *Ecology* **77**, 1367–1378.

Jackson, J. B. C. 1994 Constancy and change of life in the sea. *Phil. Trans. R. Soc. London Ser. B* **344**, 55–60.

Jackson, J. B. C. and McKinney, F. K. 1990 Ecological processes and progressive macroevolution of marine clonal benthos. In *Causes of evolution: a paleontological perspective* (ed. R. M. Ross and W. D. Allmon), pp. 173–209. Chicago: University of Chicago Press.

Janis, C. M. 1989 A climatic explanation for patterns of evolutionary diversity in ungulate mammals. *Palaeontology* **32**, 463–481.

Janke, A., Feldmaier-Fuchs, G., Thomas, W. K., von Haeseler, A. and Pääbo, S. 1994 The marsupial mitochondrial genome and the evolution of placental mammals. *Genetics* **137**, 243–256.

Jansa, L. F. 1993 Cometary impacts into ocean: their recognition and the threshold constraint for biological extinctions. *Paleogeogr. Palaeoclimatol. Palaeoecol.* **104**, 271–286.

Jensen, S. 1997 Trace fossils from the Lower Cambrian Mickwitzia sandstone, south-central Sweden. *Fossils Strata* **42**, 1–111.

Johansen, M. B. 1989*a* Adaptive radiation, survival and extinction of brachiopods in the northwest European Upper Cretaceous-Lower Paleocene Chalk. *Paleogeogr. Palaeoclimatol. Palaeoecol.* **74**, 147–204.

Johansen, M. B. 1989*b* Background extinction and mass extinction of the brachiopods from the Chalk of northwest Europe. *Palaios* **4**, 243–250.

Johnson, C. C. and Kauffman, E. G. 1996 Maastrichtian extinction patterns of Caribbean province rudistids. In *The Cretaceous-Tertiary mass extinction: Biotic and environmental changes* (ed. N. Macleod and G. Keller), pp. 231–274. New York: Norton.

Johnson, K. R. and Hickey, L. J. 1990 Megafloral change across the Cretaceous/Tertiary boundary in the northern Great Plains and Rocky Mountains, U.S.A. *Geol. Soc. Am. Spec. Pap.* **247**, 433–444.

Kaiho, K. 1992 A low extinction rate of intermediate-water benthic foraminifera at the Cretaceous/Tertiary boundary. *Mar. Micropaleontol.* **18**, 229–259.

Kajiwara, Y. and Kaiho, K. 1992 Oceanic anoxia at the Cretaceous/Tertiary boundary supported by the sulphur isotope record. *Paleogeogr. Palaeoclimatol. Palaeoecol.* **99**, 151–162.

Kammer, T. W., Baumiller, T. K. and Ausich, W. I. 1997 Species longevity as a function of niche breadth: Evidence from fossil crinoids. *Geology* **25**, 219–222.

Keller, G., Adatte, T., Hollis, C., Ordóñez, M., Zambrano, I., Jiménez, N., Stinnesbeck, W., Aleman, A. and Hale-Erlich, W. 1997 The Cretaceous/Tertiary boundary event in Ecuador: reduced biotic effects due to eastern boundary current setting. *Mar. Micropaleontol.* **31**, 97–133.

Kenrick, P. and Crane, P. R. 1997 The origin and early evolution of plants on land. *Nature* **389**, 33–39.

Kidwell, S. M. and Brenchley, P. J. 1996 Evolution of the fossil record: Thickness trends in marine skeletal accumulations and their implications. In *Evolutionary paleobiology in honor of James W. Valentine* (ed. D. Jablonski, D. H. Erwin and J. H. Lipps), pp. 290–336. Chicago: University of Chicago Press.

Kirchner, J. W. and Weil, A. 1998 No fractals in fossil extinction statistics. *Nature* **395**, 337–338.

Kirschvink, J. L., Ripperdan, R. L. and Evans, D. A. 1997 Evidence for a large-scale reorganization of Early Cambrian continental masses by inertial interchange true polar wander. *Science* **277**, 541–545.

Koeberl, C., Armstrong, R. A. and Reimold, W. U. 1997 Morokweng, South Africa: A large impact structure of Jurassic-Cretaceous age. *Geology* **25**, 731–734.

Knoll, A. H. 1994 Proterozoic and early Cambrian protists: evidence for accelerating evolutionary tempo. *Proc. Natl Acad. Sci. USA* **91**, 6743–6750.

Knoll, A. H. and Walter, M. R. 1996 The limits of palaeontological knowledge: finding the gold among the dross. In *Evolution of hydrothermal ecosystems on Earth (and Mars?)*, pp. 198–213. Ciba Foundation Symposium **202**. Chichester: Wiley.

Knoll, A. H., Bambach, R. K., Canfield, D. E. and Grotzinger, J. P. 1996 Comparative Earth history and Late Permian mass extinction. *Science* **273**, 452–457.

Koerberl, C. 1996 Chicxulub—The K-T boundary impact crater: A review of the evidence, and an introduction to impact crater studies. *Abhandl. Geol. Bund.* **53**, 23–50.

Kolukisaoglu, H.Ü., Marx, S., Wiegmann, C., Hanelt, S. and Schneider-Poetsch, H. A. W. 1995 Divergence of the phytochrome gene family predates angiosperm evolution and suggests that *Selginella* and *Equisetum* arose prior to *Psilotum*. *J. Mol. Evol.* **41**, 329–337.

Labandeira, C. C. and Beall, B. S. 1990 Arthropod terrestriality. In *Arthropod Paleobiology* (13th Annual Short Course of the Paleontological Society; convened by D. G. Mikulic). Short Courses Paleontol. **3**, 214–256.

Labandeira, C. C. and Sepkoski, J. J. 1993 Insect diversity in the fossil record. *Science* **261**, 310–315.

Laroche, J., Li, P. and Bousquet, J. 1995 Mitochondrial DNA and monocot-dicot divergence time. *Mol. Biol. Evol.* **12**, 1151–1156.

Laskar, J. 1989 A numerical experiment on the chaotic behaviour of the Solar System. *Nature* **338**, 237–238.

Levinton, J. S. and Ginzburg, L. 1984 Repeatability of taxon longevity in successive foraminifera radiations and a theory of random appearance and extinction. *Proc. Natl Acad. Sci. USA* **81**, 5478–5481.

Lewy, Z. 1996 Octopods: Nude ammonoids that survived the Cretaceous-Tertiary boundary mass extinction. *Geology* **24**, 627–630.

Lidgard, S., McKinney, F. K. and Taylor, P. D. 1993 Competition, clade replacement, and a history of cyclostome and cheilostome bryozoan diversity. *Paleobiology* **19**, 352–371.

Lieberman, B. S. 1995 Phylogenetic trends and speciation: analyzing macroevolutionary processes and levels of selection. In *New approaches to speciation in the fossil record* (ed. D. H. Erwin and R. L. Anstey), pp. 316–337. New York: Columbia University Press.

Lipps, J. H. and Signor, P. W. (ed.) 1992 Origin and early evolution of the Metazoa. *Topics in Geobiology* **10**, xiii + 570 pp. New York: Plenum.

Lofgren, D. L., Holton, C. L. and Runkel, A. C. 1990 Reworking of Cretaceous dinosaurs into Paleocene channel deposits, Upper Hell Creek Formation, Montana. *Geology* **18**, 874–877.

Logan, G. A., Hayes, J. M., Hieshima, G. B. and Summons, R. E. 1995 Terminal Proterozoic reorganization of biogeochemical cycles. *Nature* **376**, 53–56.

McGhee, G. R. 1996 *The late Devonian mass extinction: the Frasnian/Fammenian crisis.* New York: Columbia University Press.

McGhee, G. R., Orth, C. J., Quintana, L. R., Gilmore, J. S. and Olsen, E. J. 1986 Late Devonian 'Kellwasser Event' mass-extinction horizon in Germany: No geochemical evidence for a large body impact. *Geology* **14**, 776–779.

McKinney, F. K. 1993 A faster-paced world?: contrasts in biovolume and life-process rates in cyclostome (Class Stenolaemata) and cheilostome (Class Gymnolaemata) bryozoans. *Paleobiology* **19**, 335–351.

McKinney, F. K. 1995 One hundred million years of competitive interactions between bryozoan clades: asymmetrical but not escalating. *Biol. J. Linn. Soc.* **56**, 465–481.

MacLeod, K. G. and Orr, W. N. 1993 The taphonomy of Maastrichtian inoceramids in the Basque region of France and Spain and the pattern of their decline and disappearance. *Paleobiology* **19**, 235–250.

MacLeod, N. 1995 Biogeography of Cretaceous/Tertiary (K/T) planktic foraminifera. *Hist. Biol.* **10**, 49–101.

MacLeod, N. *et al.* 1997 The Cretaceous-Tertiary biotic transition. *J. Geol. Soc. London* **154**, 265–292.

McShea, D. W. 1996 Metazoan complexity and evolution: is there a trend? *Evolution* **50**, 477–492.

Marshall, C. R. 1995 Distinguishing between sudden and gradual extinctions in the fossil record: Predicting the position of the Cretaceous-Tertiary iridium anomaly using the ammonite fossil record on Seymour Island, Antarctica. *Geology* **23**, 731–734.

Marshall, C. R. 1997 Confidence intervals on stratigraphic ranges with nonrandom distributions of fossil horizons. *Paleobiology* **23**, 165–173.

Martin, W. F. 1996 Is something wrong with the tree of life? *BioEssays* **18**, 523–527.

Martin, W., Lydiate, D., Brinkmann, H., Forkmann, G. Saedler, H. and Cerff, R. 1993 Molecular phylogenies in angiosperm evolution. *Mol. Biol. Evol.* **10**, 140–162.

Merle, C., Moullade, M., Lima, O. and Perasso, R. 1994 Essai de caractérisation phylogénétique de Foraminifères planctoniques à partir de séquences partielles d'ADNr28S. *C. R. Acad. Sci. Paris* **319 (II)**, 149–153.

Miller, A. I. and Foote, M. 1996 Calibrating the Ordovician radiation of marine life: implications for Phanerozoic diversity trends. *Paleobiology* **22**, 304–309.

Miller, W. 1993 Models of recurrent fossil assemblages. *Lethaia* **26**, 182–183.

Moldowan, J. M., Dahl, J., Jacobsen, S. R. Huizanga, B. J., Fago, F. J., Shetty, R., Watt, D. S. and Peters, K. E. 1996 Chemostratigraphic reconstruction of biofacies: Molecular evidence linking cyst-forming dinoflagellates with pre-Triassic ancestors. *Geology* **24**, 159–162.

Müller, K. J. and Hinz-Schallreuter, I. 1993 Palaeoscolecid worms from the Middle Cambrian of Australia. *Palaeontology* **36**, 549–592.

Newell, N. D. 1967 Revolutions in the history of life. *Geol. Soc. Am. Spec. Pap.* 89, 63–91.

Niklas, K. J., Tiffney, B. H. and Knoll, A. H. 1983 Patterns in vascular land plant diversification. *Nature* **303**, 614–616.

Norell, M. A. 1992 Taxic origin and temporal diversity: the effect of phylogeny. In *Extinction and phylogeny* (ed. M. J. Novacek and Q. D. Wheeler), pp. 89–118. New York: Columbia University Press.

Odorico, D. M. and Miller, D. J. 1997 Variation in the ribosomal internal transcribed spacers and 5.8S rDNA among five species of *Acropora* (Cnidaria, Scleractinia): Patterns of variation consistent with reticulate evolution. *Mol. Biol. Evol.* **14**, 465–473.

Oji, T. 1996 Is predation intensity reduced with increasing depth? Evidence from the west Atlantic stalked crinoid *Endoxocrinus parrae* (Gervais) and implications for the Mesozoic marine revolution. *Paleobiology* **22**, 339–351.

Omland, K. E. 1997 Correlated rates of molecular and morphological evolution. *Evolution* **51**, 1381–1393.

Orth, C. J., Knight, J. D., Quintana, L. R., Gilmore, J. S. and Palmer, A. R. 1984 A search for iridium abundance anomalies at two late Cambrian biomere boundaries in western Utah. *Science* **223**, 163–165.

Orth, C. J., Gilmore, J. S., Quintana, L. R. and Sheehan, P. M. 1986 Terminal Ordovician extinctions: Geochemical analysis of the Ordovician/Silurian boundary, Anticosti Island, Quebec. *Geology* **14**, 433–436.

Overpeck, J. T., Webb, R. S. and Webb, T. 1992 Mapping eastern North American vegetation change of the past 18 ka: No-analogs and the future. *Geology* **20**, 1071–1074.

Pandolfi, J. M. 1996 Limited membership in Pleistocene reef coral assemblages from the Huon Peninsula, Papua New Guinea: constancy during global change. *Paleobiology* **22**, 152–176.

Patterson, C. and Smith, A. B. 1987 Is the periodicity of extinctions a taxonomic artefact? *Nature* **330**, 248–251.

Patterson, R. T. and Fowler, A. D. 1996 Evidence of self organization in planktic foraminiferal evolution: Implications for interconnectedness of palaeoecosystems. *Geology* **24**, 215–218.

Pawlowski, J., Bolivar, I., Guiard-Maffia, J. and Gouy, M. 1994 Phylogenetic position of Foraminifera inferred from LSU rRNA gene sequences. *Mol. Biol. Evol.* **11**, 929–938.

Pawlowski, J., Bolivar, I., Fahrni, J. F., Cavalier-Smith, R. and Gouy, M. 1996 Early origin of Foraminifera suggested by SSU ribosomal-RNA gene-sequences. *Mol. Biol. Evol.* **13**, 445–450.

Pawlowski, J., Bolivar, I., Fahrni, J. F., de Vargus, C., Gouy, M. and Zaninetti, L. 1997 Extreme differences in rates of molecular evolution of Foraminifera revealed by comparison of ribosomal DNA sequences and the fossil record. *Mol. Biol. Evol.* **14**, 498–505.

Philippe, H., Sörhannus, U., Baroin, A., Perasso, R., Gasse, F. and Adoutte, A. 1994 Comparison of molecular and paleontological data in diatoms suggests a major gap in the fossil record. *J. Evol. Biol.* **7**, 247–265.

Phillips, J. 1860 *Life on the Earth: its origin and succession.* Cambridge: Macmillan.

Poinar, G. O. 1992 *Life in amber.* Stanford: Stanford University Press.

Pope, K. O., Baines, K. H., Ocampo, A. C. and Ivanov, B. A. 1994 Impact winter and the Cretaceous/Tertiary extinctions: Results of a Chicxulub asteroid impact model. *Earth Planet. Sci. Lett.* **128**, 719–725.

Rasmussen, K. A., Macintyre, I. G. and Prufert, L. 1993 Modern stromatolite reefs fringing a brackish coastline, Chetumal Bay, Belize. *Geology* **21**, 199–202.

Raup, D. M. 1979 Size of the Permo-Triassic bottleneck and its evolutionary implications. *Science* **206**, 217–218.

Raup, D. M. 1981 Extinction: bad genes or bad luck? *Acta Geol. Hispanica* **16**, 25–33.

Raup, D. M. 1991 *Extinction: bad genes or bad luck?* New York: Norton.

Raup, D. M. 1992 Large-body impact and extinction in the Phanerozoic. *Paleobiology* **18**, 80–88.

Raup, D. M. and Jablonski, D. 1993 Geography of end-Cretaceous marine bivalve extinctions. *Science* **260**, 971–973.

Raup, D. M. and Sepkoski, J. J. 1986 Periodic extinction of families and genera. *Science* **231**, 853–836.

Raymond, A. and Metz, C. 1995 Laurussian land-plant diversity during the Silurian and Devonian: mass extinction, sampling bias, or both? *Paleobiology* **21**, 70–95.

Reid, R. P., Macintyre, I. G., Browne, K. M., Steneck, R. S. and Miller, T. 1995 Modern marine stromatolites in the Exuma Cays, Bahamas: Uncommonly common. *Facies* **33**, 1–18.

Rhodes, M. C. and Thompson, R. J. 1993 Comparative physiology of suspension-feeding in living brachiopods and bivalves: evolutionary implications. *Paleobiology* **19**, 322–334.

Rigby, J. K. Jr., Newman, K. R., Smit, J., van den Kaars, S., Sloan, R. E. and Rigby, J. K. 1987 Dinosaurs from the Paleocene part of the Hell Creek Formation, McCone County, Montana. *Palaios* **2**, 296–302.

Rigby, S. 1997 A comparison of the colonization of the planktic realm and land. *Lethaia* **30**, 11–17.

Rosenweig, M. L. and McCord, R. D. 1991 Incumbent replacement: evidence for long-term evolutionary progress. *Paleobiology* **17**, 202–213.

Runnegar, B. 1982a The Cambrian explosion: animals or fossils? *J. Geol. Soc. Austr.* **29**, 395–411.

Runnegar, B. 1982b A molecular-clock date for the origin of the animal phyla. *Lethaia* **15**, 199–205.

Runnegar, B. 1985 Collagen gene construction and evolution. *J. Mol. Evol.* **22**, 141–149.

Schluter, D. 1994 Experimental evidence that competition promotes divergence in adaptive radiation. *Science* **266**, 798–801.

Schopf, J. W. (ed.) 1983 *Earth's earliest biosphere: its origin and evolution.* Princeton: Princeton University Press.

Schopf, K. M. 1996 Coordinated stasis: biofacies revisited and the conceptual modeling of whole-fauna dynamics. *Palaeogeogr. Palaeoclimatol. Palaeoecol.* **127**, 157–175.

Schubert, J. K. and Bottjer, D. J. 1995 Aftermath of the Permian-Triassic mass extinction event: Paleoecology of Lower Triassic carbonates in the western USA. *Palaeogeogr. Palaeoclimatol. Palaeoecol.* **116**, 1–39.

Schultz, P. H and D'Hondt, S. 1996 Cretaceous-Tertiary (Chicxulub) impact angle and its consequences. *Geology* **24**, 963–967.

Schultz, P. H. and Gault, D. E. 1990 Prolonged global catastrophes from oblique impacts. *Geol. Soc. Am. Spec. Pap.* **247**, 239–261.

Scrutton, C. T. 1988 Patterns of extinction and survival in Palaeozoic corals. In *Extinction and survival in the fossil record* (ed. G. P. Larwood), pp. 65–88. Syst. Ass. Spec. Vol. 34.

Sepkoski, J. J. 1978 A kinetic model of Phanerozoic taxonomic diversity. I. Analysis of marine orders. *Paleobiology* **4**, 223–251.

Sepkoski, J. J. 1979 A kinetic model of Phanerozoic taxonomic diversity. II. Early Phanerozoic families and multiple equilibria. *Paleobiology* **5**, 222–251.

Sepkoski, J. J. 1981 A factor analytic description of the Phanerozoic marine fossil record. *Paleobiology* **7**, 36–53.

Sepkoski, J. J. 1988 Alpha, beta, or gamma: where does all the diversity go? *Paleobiology* **14**, 221–234.

Sepkoski, J. J. 1991 A model of onshore-offshore change in faunal diversity. *Paleobiology* **17**, 58–77.

Sepkoski, J. J. 1992 A compendium of fossil marine animal families. 2nd edition. *Milwaukee Public Mus. Contrib. Biol. Geol.* **83**, 1–156.

Sepkoski, J. J. 1993 Ten years in the library: new data confirm paleontological patterns. *Paleobiology* **19**, 43–51.

Sepkoski, J. J. 1996 Competition in macroevolution: the double wedge revisited. In *Evolutionary paleobiology: in honor of James W. Valentine* (ed. D. Jablonski, D. H. Erwin and J. H. Lipps), pp. 211–255. Chicago: University of Chicago Press.

Sepkoski, J. J. 1997 Biodiversity: Past, present, and future. *J. Paleontol.* **71**, 533–539.

Sepkoski, J. J. and Kendrick, D. C. 1993 Numerical experiments with model monophyletic and paraphyletic taxa. *Paleobiology* **19**, 168–184.

Sepkoski, J. J., Bambach, R. K., Raup, D. M. and Valentine, J. W. 1981 Phanerozoic marine diversity and the fossil record. *Nature* **293**, 435–437.

Signor, P. W. 1990 The geologic history of diversity. *Annu. Rev. Ecol. Syst.* **21**, 509–539.

Smith, A. B. 1994 *Systematics and the fossil record: documenting evolutionary patterns.* Oxford: Blackwell Scientific.

Solé, R. V. and Bascompte, J. 1996 Are critical phenomena relevant to large-scale evolution? *Proc. R. Soc. London Ser. B* **263**, 161–168.

Solé, R. V., Bascompte, J. and Manrubia, S. C. 1996 Extinction: bad genes or weak chaos? *Proc. R. Soc. London Ser. B* **263**, 1407–1413.

Solé, R. V., Manrubia, S. C., Benton, M. and Bak, P. 1997 Self-similarity of extinction statistics in the fossil record. *Nature* **388**, 764–767.

Spring, J. 1997. Vertebrate evolution by interspecific hybridisation—are we polyploid? *FEBS Lett.* **400**, 2–8.

Springer, M. J., Cleven, G. C., Madsen, O., de Jong, W. W., Waddell, V. G., Amrine, H. M. and Stanhope, M. J. 1997 Endemic African mammals shake the phylogenetic tree. *Nature* **388**, 61–64.

Sprinkle, J. and Guensburg, T. E. 1995 Origin of echinoderms in the Paleozoic evolutionary fauna: the role of substrates. *Palaios* **10**, 437–453.

Stott, L. D. and Kennett, J. P. 1989 New constraints on early Tertiary palaeoproductivity from carbon isotopes in foraminifera. *Nature* **342**, 526–529.

Summons, R. E. 1992. Abundance and composition of extractable organic matter. In *The

Proterozoic biosphere. A multidisciplinary study, (ed. Schopf, J. W. and Klein, C.) pp. 101–115. Cambridge: Cambridge University Press.

Summons, R. E., Jahnke, L. L. and Simoneit, B. R. T. 1996 Lipid biomarkers for bacterial ecosystems: Studies of cultured organisms, hydrothermal environments and ancient sediments. In *Evolution of hydrothermal ecosystems on Earth (and Mars?)*, pp. 174–194. Ciba Foundation Symp. **202**. Chichester: Wiley.

Surlyk, F. and Johansen, M. B. 1984 End-Cretaceous brachiopod extinctions in the Chalk of Denmark. *Science* **223**, 1174–1177.

Sussman, G. J. and Wisdom, J. 1992 Chaotic evolution of the Solar System. *Science* **257**, 56–62.

Syvanen, M. 1994 Horizontal gene transfer: Evidence and possible consequences. *Annu. Rev. Genet.* **28**, 237–261.

Tang, C. M. and Bottjer, D. J. 1996 Long-term faunal stasis without evolutionary coordination: Jurassic benthic marine paleocommunities, Western Interior, United States. *Geology* **24**, 815–818.

Tappan, H. 1969 Microplankton, ecological succession and evolution. *Proc. N. Am. Paleontol. Conv. Chicago, 1969, Proc.* **H**, 1058–1103.

Toon, O. B., Zahnle, K., Morrison, D., Turco, R. P. and Covey, C. 1997 Environmental perturbations caused by the impacts of asteroids and comets. *Rev. Geophys.* **35**, 41–78.

Tschudy, R. H. and Tschudy, B. D. 1986 Extinction and survival of plant life following the Cretaceous/Tertiary boundary event, Western Interior, North America. *Geology* **14**, 667–670.

Valentine, J. W. 1969 Patterns of taxonomic and ecological structure of the shelf benthos during Phanerozoic time. *Palaeontology* **12**, 684–709.

Valentine, J. W. 1973 Phanerozoic taxonomic diversity: a test of alternate models. *Science* **180**, 1078–1079.

Valentine, J. W. 1990 The fossil record: a sampler of life's diversity. *Phil. Trans. R. Soc. London Ser.* B **330**, 261–268.

Valentine, J. W., Collins, A. G. and Meyer, C. P. 1994 Morphological complexity increase in metazoans. *Paleobiology* **20**, 131–142.

Valentine, J. W., Erwin, D. H. and Jablonski, D. 1996 Developmental evolution of metazoan bodyplans: The fossil evidence. *Dev. Biol.* **173**, 373–381.

Vrba, E. S. 1995 On the connections between paleoclimate and evolution. In *Paleoclimate and evolution with emphasis on human origins* (ed. E. S. Vrba, G. H. Denton, T. C. Partridge and L. K. Burckle), pp. 24–45. New Haven: Yale University Press.

Wade, C. M., Darling, K. F., Kroon, D. and Brown, A. J. L. 1996 Early evolutionary origin of the planktic Foraminifera inferred from small rDNA sequence comparisons. *J. Mol. Evol.* **43**, 672–677.

Waggoner, B. M. 1996 Phylogenetic hypothesis of the relationships of arthropods to Precambrian and Cambrian problematic taxa. *Syst. Biol.* **45**, 190–222.

Walter, G. H. and Patterson, H. E. H. 1994. The implications of paleontological evidence for theories of ecological communities and species richness. *Austr. J. Ecol.* **19**, 241–250.

Walter, M. R. 1996 Ancient hydrothermal ecosystems on Earth: a new palaeobiological frontier. In *Evolution of hydrothermal ecosystems on Earth (and Mars?)*, pp. 112–130. Ciba Foundation Symp. **202**. Chichester: Wiley.

Walter, M. R. and Heys, G. R. 1985 Links between the rise of the Metazoa and the decline of stromatolites. *Precambrian Res.* **29**, 149–174.

Willis, K. J. and Bennett, K. D. 1995 Mass extinction, punctuated equilibrium and the fossil plant record. *Trends Ecol. Evol.* **10**, 308–309.

Wolfe, J. A. 1990 Palaeobotanical evidence for a marked temperature increase following the Cretaceous/Tertiary boundary. *Nature* **343**, 153–156.

Wolfe, K. H., Gouy, M., Yang, Y-W., Sharp, P. M. and Li, W-H. 1989 Date of the monocot-dicot divergence estimated from chloroplast DNA sequence data. *Proc. Natl Acad. Sci. USA* **86**, 6201–6205.

Wray, G. A., Levinton, J. S. and Shapiro, L. H. 1996 Molecular evidence for deep Precambrian divergences among metazoan phyla. *Science* **274**, 568–573.

Zachos, J. C., Arthur, M. A. and Dean, W. E. 1989 Geochemical evidence for suppression of pelagic marine productivity at the Cretaceous/Tertiary boundary. *Nature* **337**, 61–64.

Zinsmeister, W. J., Feldmann, R. M., Woodburne, M. O. and Elliot, D. H. 1989. Latest Cretaceous/earliest Tertiary transition on Seymour Island, Antarctica. *J. Paleontol.* **63**, 731–738.

Index